RENEWABLE ENERGY AND WILDLIFE CONSERVATION

Wildlife Management and Conservation

Paul R. Krausman, *Series Editor*

RENEWABLE ENERGY AND WILDLIFE CONSERVATION

EDITED BY
Christopher E. Moorman,
Steven M. Grodsky,
and Susan P. Rupp

Published in Association with *THE WILDLIFE SOCIETY*

 JOHNS HOPKINS UNIVERSITY PRESS | BALTIMORE

Johns Hopkins University Press
2715 North Charles Street
Baltimore, Maryland 21218-4363
www.press.jhu.edu

Library of Congress Cataloging-in-Publication Data

Names: Moorman, Christopher E., editor.
Title: Renewable energy and wildlife conservation / edited by
    Christopher E. Moorman, Steven M. Grodsky, and
    Susan P. Rupp.
Description: Baltimore : Johns Hopkins University Press,
    2019. | Series: Wildlife management and conservation |
    Includes bibliographical references and index.
Identifiers: LCCN 2019001473 | ISBN 9781421432724
    (hardcover : alk. paper) | ISBN 1421432722 (hardcover :
    alk. paper) | ISBN 9781421432731 (electronic) | ISBN
    1421432730 (electronic)
Subjects: LCSH: Renewable energy sources—
    Environmental aspects. | Wildlife conservation.
Classification: LCC TJ808 .R41735 2019 | DDC
    333.95/416—dc23
LC record available at https://lccn.loc.gov/2019001473

A catalog record for this book is available from the British
Library.

*Special discounts are available for bulk purchases of this
book. For more information, please contact Special Sales
at 410-516-6936 or specialsales@press.jhu.edu.*

Contents

Contributors

Edward B. Arnett is Chief Scientist, Theodore Roosevelt Conservation Partnership, Washington, DC, USA

Brian B. Boroski is Vice President, H. T. Harvey & Associates, Ecological Consultants, Fresno, CA, USA

Regan Dohm is Research Associate, Department of Forest and Wildlife Ecology, University of Wisconsin Madison, Madison, WI, USA

David Drake is Professor and Extension Wildlife Specialist, Department of Forest and Wildlife Ecology, University of Wisconsin Madison, Madison, WI, USA

Sarah R. Fritts is Assistant Professor, Department of Biology, Texas State University, San Marcos, TX, USA

Rachel Greene is Research Associate II, Department of Wildlife, Fisheries, and Aquaculture, Forest and Wildlife Research Center, Mississippi State University, Mississippi State, MS, USA

Steven M. Grodsky is Postdoctoral Scholar, Department of Land, Air & Water Resources, University of California Davis, Davis, CA, USA

Amanda M. Hale is Associate Professor, Department of Biology, Texas Christian University, Fort Worth, TX, USA

Cris D. Hein is Senior Project Leader-Wind Energy & Wildlife, National Renewable Energy Laboratory, National Wind Technology Center, Golden, CO, USA

Rebecca R. Hernandez is Assistant Professor, Department of Land, Air & Water Resources, University of California Davis, Davis, CA, USA

Jessica A. Homyack is Western Environmental Research Manager, Weyerhaeuser Company, Longview, WA, USA

Henriette I. Jager is Senior Scientist, Environmental Sciences Division, Oak Ridge National Laboratory, Oak Ridge, TN, USA, and Joint Faculty, University of Tennessee Ecology and Evolutionary Biology Department and Bredesen Center for Energy Science and Engineering, Knoxville, TN, USA

Nicole M. Korfanta is Director, Ruckelshaus Institute, Haub School of Environment and Natural Resources, University of Wyoming, Laramie, WY, USA

James A. Martin is Associate Professor, Warnell School of Forestry and Natural Resources, University of Georgia, Athens, GA, USA

Christopher E. Moorman is Professor and Coordinator, Fisheries, Wildlife, and Conservation Biology Program, North Carolina State University, Raleigh, NC, USA

Clint Otto is Research Ecologist, U.S. Geological Survey, Northern Prairie Wildlife Research Center, Jamestown, ND, USA

Christine A. Ribic is Unit Leader, U.S. Geological Survey, Wisconsin Cooperative Wildlife Research Unit, University of Wisconsin Madison, Madison, WI, USA

Susan P. Rupp is CEO, Enviroscapes Ecological Consulting, LLC, Gravette, AR, USA

Jake Verschuyl is Director of Forestry Research, Western United States and British Columbia, National Council for Air and Stream Improvement, Anacortes, WA, USA

Lindsay M. Wickman is Research Associate, Department of Marine Science, University of Otago, Dunedin, New Zealand

T. Bently Wigley is Senior Research Fellow (Retired), National Council for Air & Stream Improvement, Clemson, SC, USA

Victoria H. Zero is Research Biologist and Habitat Conservation Plan Specialist, Western EcoSystems Technology, Inc., Laramie, WY, USA

RENEWABLE ENERGY AND WILDLIFE CONSERVATION

INTRODUCTION

CHRISTOPHER E.
MOORMAN, STEVEN M.
GRODSKY, AND
SUSAN P. RUPP

Renewable Energy and Wildlife Conservation

Renewables as the Original Energy

Renewable energy is any theoretically inexhaustible source of energy (e.g., biomass, hydroelectric, solar, tidal, wind) not derived from fossil or nuclear fuels. Renewable sources of energy—often referred to synonymously as alternative energy—are largely rooted in the sun (the notable exception being geothermal), either directly from solar energy or indirectly from a variety of other sources (Panwar et al. 2011). Plants capture the sun's energy, which, in turn, can be converted to bioenergy (i.e., heat, power, and fuel derived from plant biomass). The sun's heat evaporates water, which then falls as precipitation and flows into streams and rivers that can be harnessed as hydropower. And, the differential heating caused by the sun creates circulation of air, which can be turned into wind power. These renewable sources of energy can be replenished naturally on short time scales and reduce greenhouse gas (GHG) emissions (carbon dioxide, CO_2; methane, CH_4; nitrous oxide, N_2O; fluorinated gases) associated with fossil fuel consumption and other anthropogenic activities (e.g., agriculture, industry, waste).

Though interest in renewables has increased in recent years, these forms of energy have been around for generations. Humans have burned wood for heat for eons, and even now more than two billion people around the globe depend on wood energy for cooking and heating, especially in nonindustrialized countries (Food and Agriculture Organization of the United Nations, http://www.fao.org/forestry/energy /en/). Moreover, our human ancestors harnessed energy from yoked animals, sails, waterwheels, and windmills. In addition to the previously mentioned sources of energy (e.g., animals, water, and wind), humans also used camelina (*Camelina sativa*), an oilseed crop now used as a feedstock for jet fuel, to provide lamp oil in the Bronze Age. The first authentic internal combustion engine, which ran on a mixture of ethanol, turpentine, and camphor, was developed by Samuel Morey in 1826 to power a small boat up the Connecticut River (McGuire 2012). Even the Model T introduced by Henry Ford in 1908 was a flexible-fuel vehicle that could run on ethanol (McGuire 2012). However, emphasis on renewables declined in Europe and the United States in the eighteenth and nineteenth centuries as the Industrial Revolution flourished. Coal, and later oil and gas, became the primary energy sources for the industrialized nations of the world.

The Demand for Fossil Fuels and the Environmental Costs

Today, demand for energy is unprecedented. The global human population is projected to exceed nine billion by 2050 (Figure I.1), and demand for energy to produce food, generate electricity, fuel automobiles, and support daily human activities is burgeoning. Furthermore, the globalizing economy has driven expansion in domestic and international transportation infrastructure and rapid advances in telecommunication technologies, all of which require substantial energy inputs. In fact, global energy consumption is expected to nearly double between 2012 and 2040, as China, Indonesia, Brazil, and other less-developed countries grow economically (US Energy Information Administration 2017). Even in developed nations, where population growth is less rapid, energy demand per capita far outpaces that of developing nations. In these developed countries, the number of people per household is decreasing, the number of houses and the average size of homes is increasing, and the sprawling nature of development requires more widespread land conversion to support automobile-centric infrastructure (Peterson et al. 2013). With fewer residents per household, heating, cooling, lighting, and water consumption in the home in developed countries correlates with more energy usage per person.

Currently, more than 80% of the world's energy demand is met with fossil fuels, which, when burned, emit GHGs that contribute to global warming (US Energy Information Administration 2017). Oil, coal, and natural gas are the leading sources of global energy, and natural gas is expected to surpass coal in the coming decades. Although the availability of fossil fuels is finite, current projections have them remaining prominent energy sources for many decades into the future (US Energy Information Administration 2017). Hence, the GHG emissions associated with fossil fuel use are likely to remain high, even as alternative energy sources replace fossil fuels.

Fossil fuel consumption and associated GHG emissions are rapidly changing the Earth's climate. The Intergovernmental Panel on Climate Change (IPCC 2014) stated that it is "extremely likely" that warming temperatures around the globe since the mid-twentieth century were caused by anthropogenic increases in GHGs. The planet's surface has warmed an average of 0.12°C (0.22°F) per decade since 1950, with similar trends for the ocean surface (0.11°C warming per decade since 1970). Although

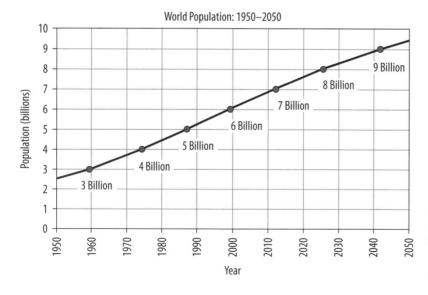

Figure I.1   The global human population is expected to exceed 9 billion by 2050. *US Census Bureau, International Database, June 2011 update.*

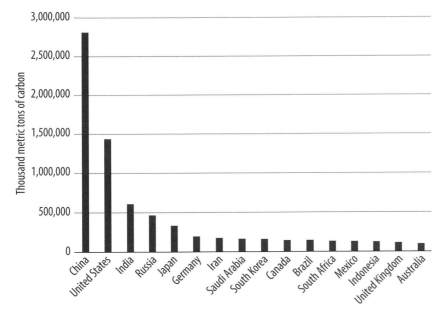

Figure I.2 Ranking of the leading countries by 2014 for total CO_2 emissions from fossil fuel burning, cement production, and gas flaring. Emissions are expressed in thousand metric tons of carbon (not CO_2). *Source: T. A. Boden, G. Marland, and R. J. Andres. 2017. Global, Regional, and National Fossil-Fuel CO_2 Emissions. Oak Ridge, TN: Carbon Dioxide Information Analysis Center, Oak Ridge National Laboratory, US Department of Energy.*

the planet has been warmer in the past, these modern increases in global temperature are unique in that they are driven by human activities. Simultaneous with global warming, glaciers are shrinking around the globe, and mean annual snow cover extent has decreased. Since 1900, mean sea level rose by nearly 0.2 m, with the rate of rise increasing over time. Concurrent with these climatic changes, atmospheric concentrations of carbon dioxide, methane, and nitrous oxide increased, and current atmospheric concentrations of these GHGs are greater than they have been in at least the last 800,000 years (IPCC 2014).

Global surface temperatures are projected to increase between 0.3°C (0.54°F) and 0.7°C (1.26°F) by 2035, and temperatures are projected to increase between 0.3°C and 4.8°C (8.64°F) during the twenty-first century, indicative of continuing influence of anthropogenic-driven climate change (IPCC 2014). The wide range of variability for twenty-first-century projections depends largely on whether GHG emissions are lessened by changes in energy policy and efficiencies and whether fossil fuels are replaced with "cleaner" energy technologies. Additional environmental changes projected to occur with rising

global temperatures include increases in ocean acidification, reductions in arctic sea ice, decreased area of permafrost, and rise in sea level at approximately 70% of the coastlines around the world (IPCC 2014).

Greenhouse gas emissions associated with fossil fuel use are greatest in areas with the greatest human population densities and where economic growth is most substantial (Figure I.2). China and the United States contribute approximately 26% and 14% of the global emission of GHGs, respectively (Boden et al. 2017). Per capita GHG emissions in Canada, the United States, and Russia are more than double the global average per person, highlighting the connection between emissions and the industrialized economies of the world. However, the human populations at greatest risk from climate change are often those that contribute least to GHG emissions. Environmental costs of climate change, including lower crop yields, declining stores of drinking water, reductions in fisheries productivity, rise in human health problems, and increased risk of natural disasters (e.g., drought, heat stress, hurricanes, wildfire) can have devastating effects on poorer, less industrialized nations of the world (Table I.1; Mirza 2003).

Table I.1.   Selected impacts of climate-related extreme events in developing regions (Mirza 2003)

Region	Expected regional impact of extreme events
Africa	Increases in droughts, floods, and other extreme events will add to stress on water resources, food security, human health, and infrastructure, and would constrain development in Africa (*high confidence*)
	Sea-level rise would affect coastal settlements, flooding, and coastal erosion, especially along the southeastern African coast (*high confidence*)
	Desertification, exacerbated by reductions in average annual rainfall, runoff, and soil moisture (*medium confidence*)
	Major rivers highly sensitive to climate variation: average runoff and water availability would decrease in Mediterranean and southern countries in Africa, affecting agriculture and hydropower systems (*medium confidence*)
Asia	Extreme events increase in temperate Asia, including floods, droughts, forest fires, and tropical cyclones (*high confidence*)
	Thermal and water stress, flood and drought, sea-level rise, and tropical cyclones would diminish food security in countries of arid, tropical, and temperate Asia; agriculture would expand and increase in productivity in northern areas (*medium confidence*)
	Sea-level rise and increase in intensity of tropical cyclones would displace tens of millions of people in low-lying coastal areas of temperate and tropical Asia; increased intensity of rainfall would increase flood risks in temperate and tropical Asia (*high confidence*)
	Climate change increases energy demand, decreases tourism, and influences transportation in some regions of Asia (*medium confidence*)
Latin America	Loss and retreat of glaciers would adversely affect runoff and water supply in areas where glacier melt is an important water source (*high confidence*)
	Floods and drought would increase in frequency, higher sediment loads would degrade water quality in some areas (*high confidence*)
	Increases in the intensity of tropical cyclones would alter the risks to life, property, and ecosystems from heavy rain, flooding, storm surges, and wind damages (*high confidence*)
	Coastal human settlements, productive activities, infrastructure, and mangrove ecosystems would be negatively affected by sea-level rise (*medium confidence*)
Small Island States	Projected sea-level rise of 5 mm per year for the next 100 years would cause enhanced coastal erosion, loss of land and property, dislocation of people, increased risk from storm surges, reduced resilience of coastal ecosystems, saltwater intrusion into freshwater resources, and high resource costs for adaptation (*high confidence*)
	Islands are highly vulnerable to impacts of climate change on water supplies; agricultural productivity, including exports of cash crops; coastal ecosystems; and tourism as an important source of foreign exchange for many islands (*high confidence*)

Source: IPCC (2001).

Note: The IPCC uses the following words to indicate judgmental estimates of confidence: *very high* (95% or higher), *high* (67%–95%), *medium* (33%–67%), *low* (5%–33%), and *very low* (5% or less).

The environmental changes associated with rising global temperatures, including less snow cover, higher extreme temperatures, drought, and weather extremes, can have direct and indirect effects on wildlife populations and their associated habitats (Mawdsley et al. 2009). These effects are projected to vary among wildlife taxa, and effects can be variable among species within taxonomic groups (i.e., winners and losers—some species will benefit, whereas others will be affected negatively). In a comprehensive report, the National Audubon Society (2015) predicted that climate change could affect the ranges of 588 species of birds in North America. Of those 588 species, 314 were predicted to lose more than 50% of their current ranges by 2080. Earlier warming in the spring may lead to peak availability of caterpillars and other food sources before most migrant birds arrive and begin reproduction; this mismatch could lead to reproductive failure and long-term population declines for birds and for other insectivorous wildlife taxa (Visser and

Both 2005). Amphibians and reptiles are ectothermic, making many species vulnerable to altered temperatures (i.e., thermal stress) associated with climate change (Araujo et al. 2006). Additionally, many reptiles and amphibians have reproductive strategies that are tied to narrow temperature and moisture conditions (e.g., ephemeral pools that hold water in the late winter and spring and dry out during other portions of the year), so changes in average environmental conditions (i.e., average temperature extremes or hydroperiod) or more variable climatic conditions from one year to the next could lead to boom and bust periods of reproductive output in herpetofauna populations. Altered hydroperiods or more extended periods of drought could lead to phenological mismatches in reproductive timing and larval development of amphibians also (Blaustein et al. 2001). Mammals may be forced to shift ranges to maintain appropriate thermal conditions, and declining snow cover could lead to coat-color mismatch and increased predation risk for species like snowshoe hare (*Lepus americanus*; Mills et al. 2013). Additionally, increases in drought severity and duration could lead to reductions in forage quality and quantity for herbivores (Lashley and Harper 2012).

A number of strategies have been proposed to minimize or mitigate the projected effects of climate change on wildlife (see Mawdsley et al. 2009). Examples of potential mitigation approaches include purchasing or protecting land, improving landscape permeability through targeted habitat management (e.g., corridors), captive breeding, and assisting migrations or translocations. However, strategies to slow and reverse GHG concentrations in the atmosphere may prove the most effective at slowing climate change and therefore reducing long-term impacts on biodiversity.

The Return of Renewables

By 2040, fossil fuel contribution to energy consumption is projected to decline by more than 10% as cleaner renewable energy replaces fossil fuels (Casler et al. 2009, US Energy Information Administration 2017; Figure I.3). The demand for renewable energy in the United States has grown with efforts to reduce GHG emissions, expand domestic energy production with less dependence on foreign oil, and develop policy that requires a lower percentage of energy production from fossil fuels (Allison et al. 2014, Ellabban et al. 2014). In turn, production of biomass for biofuels that replace fossil fuels could pose lower risks to biodiversity because bioenergy resources are largely produced in areas already heavily affected by

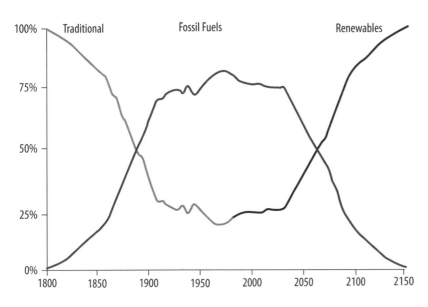

Figure I.3 Trends in human use of traditional fuels, fossil fuels, and renewable fuels from 1800 to present and predicted to the mid-twenty-second century. *Modified from Casler et al. 2009.*

human activities (Dale et al. 2015), and because bio-energy production may help lessen conversion of agricultural lands to urban development (Graves et al. 2016). In 2010, renewables accounted for 14% of global energy production, but production from renewable sources is expected to increase ~270% between 2010 and 2035 (Panwar et al. 2011, Ellabban et al. 2014). Bioenergy is expected to account for ~50% of renewable energy consumption by 2040, followed by large hydropower, geothermal, wind, solar, and marine tidal energy (Panwar et al. 2011). However, current trends indicate that solar energy and, to a lesser extent, wind energy, will be the prominent renewable energy technologies by the end of the century. These diverse sectors are more or less available regionally, so renewable energy portfolios can vary locally. For example, solar and wind energy are less available in many forested regions, whereas the potential for wood-based bioenergy there may be greater.

Emerging domestic and international policies require renewable power generation or provide incentives for renewable transportation fuels. Examples of adopted legislative policy in the United States include state renewable portfolio standards (RPSs) and the national Renewable Fuel Standard included in the Energy Independence and Security Act of 2007 passed by Congress. State-level RPS policies set target amounts of energy production that must be met by renewables by a target date, with obligations often increasing over time. The RPS language varies greatly among states, but overall the policies promote renewable energy development (Yin and Powers 2010). At the international level, signatories of the Kyoto Protocol (of which the United States is not one) agreed to reduce global GHG emissions, in part, by increasing the contribution of renewable energy. Reviews of the policy's effectiveness have been mixed, but researchers largely agree that the Kyoto Protocol has not met target reductions in global GHG emissions (Aichele and Felbermayr 2012, Grunewald and Martinez-Zarzoso 2016).

The relative contribution of renewable energy sectors to reduction in GHG emissions varies, and related calculations for each sector are exceptionally complex. Reports on the carbon neutrality of wood-based bioenergy, for example, are mixed. McKechnie et al. (2011) conducted a life-cycle assessment and reported that tree harvest for wood pellets and ethanol production reduces forest carbon stores, and the GHG emissions associated with forest harvesting initially exceed fossil fuel–related emissions; but, electricity generation from wood would reduce overall emissions relative to coal over longer time periods. Conversely, Sedjo and Tian (2012) suggested that carbon emissions from wood bioenergy are offset by increased forest land cover connected to increased market values and by the carbon captured by regrowth in these forests. The variable and often contradictory results of these studies are based largely in the variable frameworks of the models. Factors that may influence model results include the time intervals analyzed, the reference systems used, the infrastructure required to capture and transport fuels, and the source of feedstocks (Schlamadinger et al. 1997). Similar to the bioenergy sector, wind and solar power require consideration for complexities of the production process and the relative role of fossil fuels in their carbon footprint. In short, even renewable energy production is not free from environmental cost.

Fifty Shades of Green

Renewable energy is often termed "green energy," but its production is not free from impacts on wildlife and other forms of biodiversity (Fargione et al. 2010, Northrup and Wittemeyer 2013). Effects of renewable energy production on wildlife vary among wildlife taxa and among energy sectors and can be positive or negative (Northrup and Wittemeyer 2013). The expanding demand for renewable energy and the associated advances in technology generate environmental effects that often are novel and understudied. At the same time, sociopolitical forces direct rapid changes in energy policy that may call for sustainability measures, which may or may not include wildlife conservation. Some sustainability standards discuss general impacts to biodiversity (i.e., a measure of species richness and evenness across the landscape) without recognizing specific species or taxa that contribute to biodiversity, whether species are native or not, or species' potential for invasiveness. Further, public discourse about the pros and cons of various renewable energy technologies often hinges on debates about effects on wildlife. It is critical that these conversations be directed by current and science-based information related to the ecological effects of renewable energy production. Additionally, mitigation measures to prevent or minimize negative effects on wildlife behavior, demographics (i.e., reproduction and survival), and high-quality habitat are critically important to conservation efforts (Northrup and Wittemeyer 2013).

Renewables have low energy density relative to fossil fuels and therefore often require more infrastructure; in other words, the diffuseness of renewables means that more land area must be allocated to energy production than with fossil fuels. Hence, the conversion of wildlife habitat to presumably lower-quality habitat, or non-habitat, in the form of energy facilities and infrastructure generally is the most substantial impact on wildlife (Fargione et al. 2010). In fact, over 200,000 km² of new land are projected to be developed for energy production in the United States by 2035, and the majority of the impact would come from biofuels production (McDonald et al. 2009). These landscape-level changes can fragment or even eliminate habitat, but the relative magnitude of effects on wildlife depends on the context of the surrounding landscape, the extent of land conversion, and the condition of the land before conversion. For example, conversion of cropland to short-rotation woody crops could improve conditions for woodland birds, whereas conversion of mature forest to short-rotation woody crops could degrade habitat value for the same group of animals. Moreover, bioenergy crops are typically character-ized by less-diverse plant communities than naturally occurring vegetation. Additionally, the structure of bioenergy crops may be lower quality for some wildlife species than the structure of naturally occurring vegetation communities. For example, switchgrass (*Panicum virgatum*) fields grown for bioenergy often grow dense and tall, with a lower forb component than native prairie. Yet, forbs constitute an important component of the plant community, providing food and cover for wildlife. Increasingly, nonnative plant species are introduced as bioenergy crops, and these species may escape cultivation, cross-pollinate with closely related native species, invade nearby natural areas, and degrade habitat for native wildlife. Other indirect effects of growing bioenergy crops may include high water use and extensive use of fertilizers and pesticides. Finally, dams erected to generate hydropower flood surrounding upland areas and alter the hydrology of the watershed dramatically. The resulting reservoirs eliminate habitat for terrestrial species, serve as barriers for migrating fish, and isolate individuals from critical spawning grounds; but the same reservoirs create new habitat for many aquatic species, including waterfowl and lacustrine fish.

Renewable energy production also can lead to a variety of other impacts on wildlife, including behavioral changes and direct mortality. Wildlife collisions with energy infrastructure are widely documented, especially bird and bat collisions with wind turbines (Grodsky et al. 2011, 2012, 2013; Figure I.4). Birds, especially raptors and other large-bodied birds, often collide with or are electrocuted by the powerlines that transport energy throughout the grid (Harness and Wilson 2001). Effects of solar energy facilities on wildlife are relatively understudied, but bird mortality at solar facilities also has been documented (McCrary et al. 1986, Walston et al. 2016). Wildlife may exhibit avoidance behaviors around wind and solar energy facilities, especially in areas with the greatest human traffic and associated impacts (e.g., noise and light pollution). Additionally, road networks required to support these energy facilities

Figure I.4   A dead silver-haired bat (*Lasionycteris noctivagans*) under a US wind turbine in southeastern Wisconsin. *Photo credit: Tom Underwood.*

may facilitate additional traffic by humans (e.g., hunters, hikers) into more remote areas, further exacerbating the effects on local wildlife populations (e.g., introduction of non-native plants).

Goals and Organization

Our goal for this book was to synthesize the extensive and rapidly growing base of scientific literature on renewable energy and wildlife into a single, comprehensive resource for wildlife ecologists, university students, policy makers, industry representatives, and environmental nongovernmental organizations. In this book, we (1) describe the forms of renewable energy, the processes used to generate each form, and the scale and types of effects of each on wildlife; (2) suggest known or proposed means to mitigate potential negative effects on wildlife; (3) identify areas where information is lacking and additional research is most needed; and (4) provide Deep Dive sections that cover specific topics in greater detail and highlight the three aforementioned objectives.

The book comprises four parts. Parts I through III are structured around major renewable energy sectors: bioenergy (Chapters 1 through 4), wind energy (Chapters 5, 6, and 7), and solar energy and waterpower (Chapters 8 and 9). In each part, chapters describe the processes used to generate energy, review the documented effects on wildlife, provide potential mitigation strategies to lessen effects on wildlife, and identify research needs related to wildlife conservation. Individual chapters within Part I (bioenergy) focus on the types of feedstocks and their effects on wildlife, whereas chapters within Part II (wind energy) focus on the taxa potentially influenced by wind energy development. Extensive research has been conducted on the effects of wind energy on bats and birds aloft, so we dedicate a chapter to each. We include a third chapter in Part II that deals with terrestrial wildlife, including prairie grouse, ungulates, and invertebrates. In Part III, solar energy and waterpower are covered with a single chapter each. Each chapter in Parts I through III includes a Deep Dive section that highlights in detail potential wildlife impacts, research efforts, or mitigation strategies that can be employed to help conserve wildlife. The book concludes with Part IV, on the future of renewable energy and wildlife conservation. Chapter 10 focuses on policy directives now and into the future, as these actions could result in dramatic direct or indirect effects on wildlife conservation. Chapter 11 ties together the consistent themes of the book and describes emerging opportunities related to renewable energy and wildlife conservation with recommendations for future research. We thank the chapter authors, each of whom is nationally or internationally recognized as an expert in their respective renewable energy sector and area of policy focus.

LITERATURE CITED

Aichele, R., and G. Felbermayr. 2012. Kyoto and the carbon footprint of nations. Journal of Environmental Economics and Management 63:336–354.

Allison, T. D., T. L. Root, and P. C. Frumhoff. 2014. Thinking globally and siting locally: Renewable energy and biodiversity in a rapidly warming world. Climate Change 126:1–6.

Araujo, M. B., W. Thuiller, and R. G. Pearson. 2006. Climate warming and the decline of amphibians and reptiles in Europe. Journal of Biogeography 33:1712–1728.

Blaustein, A. R., L. K. Belden, D. H. Olson, D. M. Green, T. L. Root, and J. M. Kiesecker. 2001. Amphibian breeding and climate change. Conservation Biology 15:1804–1809.

Boden, T. A., G. Marland, and R. J. Andres. 2017. Global, regional, and national fossil-fuel CO_2 emissions. Carbon Dioxide Information Analysis Center. Oak Ridge, TN: Oak Ridge National Laboratory, US Department of Energy.

Casler, M. D., E. Heaton, K. J. Shinners, H. G. Jung, P. J. Weimer, M. A. Liebig, R. B. Mitchell, and M. F. Digman. 2009. Grasses and legumes for cellulosic bioenergy. Chapter 12 in Grassland: Quietness and Strength for a New American Agriculture, edited by W. F. Wedin and S. L. Fales, 205–219. Madison, WI: American Society of Agronomy.

Dale, V. H., E. S. Parish, and K. L. Kline. 2015. Risks to global biodiversity from fossil-fuel production exceed those from biofuel production. Biofuels, Bioproducts, and Biorefining 9:177–189.

Ellabban, O., H. Abu-Rub, and F. Blaabjerg. 2014. Renewable energy sources: Current status, future prospects and their enabling technology. Renewable and Sustainable Energy Reviews 39:748–764.

Fargione, J. E., R. J. Plevin, and J. D. Hill. 2010. The ecological impact of biofuels. Annual Review of Ecology, Evolution, and Systematics 41:351–377.

Graves, R. A., S. M. Pearson, and M. G. Turner. 2016. Landscape patterns of bioenergy in a changing climate: Implications for crop allocation and land-use competition. Ecological Applications 26:515–529.

Grodsky, S. M., M. J. Behr, A. Gendler, D. Drake, B. D. Dieterle, R. J. Rudd, and N. L. Walrath. 2011. Investigating the causes of death for wind turbine-associated bat fatalities. Journal of Mammalogy 92:917–925.

Grodsky, S. M., C. S. Jennelle, D. Drake, and T. Virzi. 2012. Bat mortality at a wind-energy facility in southeastern Wisconsin. Wildlife Society Bulletin 36:773–783.

Grodsky, S. M., C. S. Jennelle, and D. Drake. 2013. Bird mortality at a wind-energy facility near a Wetland of International Importance. Condor: Ornithological Applications 115:700–711.

Grunewald, N., and I. Martinez-Zarzoso. 2016. Did the Kyoto Protocol fail? An evaluation of the effect of the Kyoto Protocol on CO_2 emissions. Environment and Development Economics 21:1–22.

Harness, R. E., and K. R. Wilson. 2001. Electric-utility structures associated with raptor electrocutions in rural areas. Wildlife Society Bulletin 29:612–623.

IPCC (Intergovernmental Panel on Climate Change). 2001. Climate change 2001—impacts, adaptation, and vulnerability: Summary for policymakers and technical summary of the Working Group II report. Geneva: IPCC.

IPCC (Intergovernmental Panel on Climate Change). 2014. Climate change 2014: Synthesis report. Contribution of Working Groups I, II and III to the fifth assessment report of the Intergovernmental Panel on Climate Change. Core writing team, R. K. Pachauri and L. A. Meyer. Geneva: IPCC.

Lashley, M. A., and C. A. Harper. 2012. The effects of extreme drought on native forage nutritional quality and white-tailed deer diet selection. Southeastern Naturalist 11:699–710.

Mawdsley, J. R., R. O'Malley, and D. S. Ojima. 2009. A review of climate-change adaptation strategies for wildlife management and biodiversity conservation. Conservation Biology 23:1080–1089.

McCrary, M. D., R. L. McKernan, R. W. Schreiber, W. D. Wagner, and T. C. Sciarrotta. 1986. Avian mortality at a solar energy plant. Journal of Field Ornithology 57:135–141.

McDonald, R. I., J. Fargione, J. Kiesecker, W. M. Miller, and J. Powell. 2009. Energy sprawl or energy efficiency: Climate policy impacts on natural habitat for the United States of America. PLOS One 4:e6802.

McGuire, B. D. 2012. Assessment of bioenergy provisions in the 2008 Farm Bill. Association of Fish and Wildlife Agencies, funding provided through the Sport Fish and Wildlife Restoration Program of the US Fish and Wildlife Service, pursuant to the Stevens Amendment to P.L.100-463. https://www.fishwildlife.org/application/files/3515/2846/3862/08_22_12_bioenergy_report_web_final_1.pdf.

McKechnie, J., S. Colombo, J. Chen, W. Mabee, and H. L. MacLean. 2011. Forest bioenergy or forest carbon? Assessing trade-offs in greenhouses gas mitigation with wood-based fuels. Environmental Science and Technology 45:789–795.

Mills, L. S., M. Zimova, J. Oyler, S. Running, J. Abatzoglou, and P. Lukacs. 2013. Camouflage mismatch in seasonal coat color due to decreased snow duration. Proceedings of the National Academy of Sciences 110:7360–7365.

Mirza, M. M. Q. 2003. Climate change and extreme weather events: Can developing countries adapt? Climate Policy 3:233–248.

National Audubon Society. 2015. Audubon's birds and climate change report: A primer for practitioners. Version 1.2. Contributors Gary Langham, Justin Schuetz, Candan Soykan, Chad Wilsey, Tom Auer, Geoff LeBaron, Connie Sanchez, Trish Distler. New York: National Audubon Society.

Northrup, J. M., and G. Wittemyer. 2013. Characterizing the impacts of emerging energy development on wildlife, with an eye towards mitigation. Ecology Letters 16:112–125.

Panwar, N. L., S. C. Kaushik, and S. Kothari. 2011. Role of renewable energy sources in environmental protection: A review. Renewable and Sustainable Energy Reviews 14:1513–1524.

Peterson, M. N., T. R. Peterson, and J. Liu. 2013. The housing bomb. Baltimore, MD: Johns Hopkins University Press.

Schlamadinger, B., M. Apps, F. Bohlin, L. Gustavsson, G. Jungmeier, G. Marland, K. Pingoud, and I. Savolainen. 1997. Towards a standard methodology for greenhouse gas balances of bioenergy systems in comparison with fossil energy systems. Biomass and Bioenergy 13:359–375.

Sedjo, R., and X. Tian. 2012. Does wood bioenergy increase carbon stocks in forests? Journal of Forestry 110:304–311.

US Energy Information Administration. 2017. Energy information outlook, 2017. September 14. https://www.eia.gov/outlooks/ieo/pdf/0484(2017).pdf.

Visser, M. E., and C. Both. 2005. Shifts in phenology due to global climate change: The need for a yardstick. Proceedings of the Royal Society B 272:2561–2569.

Walston, L. J. Jr., K. E. Rollins, K. E. LaGory, K. P. Smith, and S. A. Meyers. 2016. A preliminary assessment of avian mortality at utility-scale solar energy facilities in the United States. Renewable Energy 92:405–414.

Yin, H., and N. Powers. 2010. Do state renewable portfolio standards promote in-state renewable generation? Energy Policy 38:1140–1149.

PART I BIOENERGY AND WILDLIFE CONSERVATION

1 — Short-Rotation Woody Crops and Wildlife Conservation

RACHEL GREENE,
JAMES A. MARTIN, AND
T. BENTLY WIGLEY

Introduction

Increasing energy demand alongside concerns about energy security and climate change have generated interest and funding for renewable energy development in many developed countries and emerging economies. In the United States, the largest single source of renewable energy is biomass, accounting for 3.7 quadrillion of 9.2 quadrillion British thermal units (Btu) in 2016 (US Energy Information Administration 2017). Biomass includes agricultural and forest resources grown explicitly for energy production (e.g., herbaceous and woody energy crops, forest thinning for pellet production), residues and wastes from agriculture and forestry (e.g., grain hulls, logging slash, and mill waste), municipal solid waste, and algae. Residues and wastes from agriculture, forestry, and municipalities currently are available and may represent a large percentage of biomass-based energy in the near future. The US Department of Energy (US DOE) concluded that at least 1 billion dry tons of biomass resources could be produced annually without negatively affecting food production or environmental resources (US DOE 2016). The 1 billion dry ton benchmark is equivalent to approximately 30% of US petroleum consumption in 2005, prior to the economic recession of the late 2000s (US DOE 2016). Woody biomass is a promising energy feedstock (i.e., raw material) that can simultaneously control soil erosion, recycle nutrients, and sequester carbon. The volume of woody crops planted solely for energy, although limited at present, could expand rapidly with market development because of their short crop rotations and play a vital role in the long-term production of biomass-based energy (US DOE 2016). Here, we discuss the cultivation and potential impacts to wildlife of these short-rotation woody crops (SRWCs) (forest-based biomass, annual energy crops, and second-generation feedstocks from energy crops are treated in Chapter 2, Homyack and Verschuyl; Chapter 3, Otto; and Chapter 4, Rupp and Ribic, respectively).

Short-Rotation Woody Crop Production and Management

Short-rotation woody crops are produced through a silvicultural system based on short cutting cycles (e.g., 1–15 years). Genetically improved stock and intensive agricultural techniques, such as fertilization and weed control, are often used in conjunction with coppice regeneration (i.e., new shoots of vegetation from stumps or roots following cutting; Drew et al. 1987, Dickmann 2006, Rupp et al. 2012). Many

Figure 1.1   Example of short-rotation loblolly pine (*Pinus taeda*) stand (*left*) and *Eucalyptus* stand (*right*) in Georgia, US. *Photos courtesy of Kevin Fouts.*

hardwood tree species and a few conifers have been evaluated for use as SRWCs, including clones of poplars and aspens (*Populus* spp.) and willows (*Salix* spp.) in temperate zones of North America and Europe; sweetgum (*Liquidambar styraciflua*) and loblolly pine (*Pinus taeda*) in the southeastern United States; and *Eucalyptus* species in subtropical and tropical climates (Figure 1.1). Poplars and willows have generated special interest because of their rapid growth rates in northern temperate climates, easy propagation, and wealth of genetic diversity that lends itself to various stock improvement methodologies (Dickmann 2006, Zalesny et al. 2011).

Site Selection

Site quality is an important determinant of SRWC productivity that is influenced by soil properties, climate, and topography. In general, high-quality agriculture sites are the ideal setting for SRWC plantations. Sites with poorly or excessively drained soil, highly acidic or alkaline pH, steep slopes, or degradation caused by erosion or compaction should be avoided for SRWC production (Dickmann 2006). Transportation costs associated with SRWC production should be considered, because hauling distances (i.e., proximity to a processing facility) dictate SRWC economic feasibility. High-quality forested sites may be suitable for SRWCs if the current forest is harvested and stump sprouts are suppressed. However, conversion of forestland to SRWC plantations may

have negative effects on plant and wildlife communities, groundwater recharge, and ecosystem services, especially if best management practices (BMPs) are not implemented. All US states and Canadian provinces with commercial forests have forestry BMPs to protect the environment from sedimentation, nutrient deposition, and pesticide use, typically via riparian buffers or streamside management zones. In the United States, forestry BMPs are implemented at a high national rate (>90%) and a large body of research confirms that BMPs, when properly implemented, protect water quality, retain riparian forest structure, and should benefit aquatic and riparian species (e.g., crayfish, mussels) sensitive to changes in water quality and forest structure (Broadmeadow and Nisbet 2004, Cristan et al. 2016, 2018, Warrington et al. 2017).

Woody Crop Silviculture

Woody crop silviculture includes consideration for planting density, rotation length, and inputs during and after the establishment phase, such as fertilization, herbicide, and pest control. Planting density is influenced by designated use of harvested raw material, species and clone, and site quality, among other factors. In the 1960s, trees used for SRWCs were planted at high densities (5,000 to 20,000 stems/ha) with tight inter- and intra-row spacing to produce high yields on short rotations (1 to 5 years; Dickmann 2006). Mean annual increment (i.e., measure of annual growth) peaked early, especially in coppice rotations, and weed suppression efforts were usually limited to the first and second year before the overstory canopy closed. However, this production strategy incurs high establishment costs due to the large number of hardwood cuttings or seedlings planted, and the harvested raw material has limited uses because of its small diameter and low wood-to-bark ratio (Dickmann 2006). Harvested material from these high-density plantations is acceptable for bioenergy feedstock production, where the maximum conversion of solar energy is the primary objective and flexibility of raw material is unimportant. When high wood-to-bark ratio or product flexibility is essential, lower densities (1,000 to 2,500 stems/ha) and longer rotations (8 to 12 years) are used (Dickmann 2006).

Because of the enormous number of species and clones used for SRWC production, variability in site quality, and the continuing development of a bioenergy market, industry standard spacing and planting densities have not been established. Results from research plantings suggest that spacing and planting density decisions should be specific to management objectives, site, tree species, and clone (DeBell et al. 1996, Eufrade et al. 2016, Paris et al. 2011). Like planting density, potential benefits and rates of fertilization differ by species, clone, and site quality. Although fertilization may not be necessary for every rotation, it may be required after successive SRWC harvests, but rates and blend of nutrients will differ among SRWC systems (Dickmann 2006). Irrigation is not vital to sustain biomass yields or coppicing capacity (e.g., Dillen et al. 2013), but may be beneficial during drought years, on degraded land, and for particularly "thirsty" species (Coleman et al. 2004). Competing vegetation can be managed through cultivation, herbicide, or a combination thereof, but intensity and type of vegetation management will vary with site, species, and establishment method.

Pests and diseases can affect some SRWC species. Perennial biomass crops generally need fewer pesticide applications than annual agricultural crops (Janowiak and Webster 2010) and, because of high pesticide costs, SRWC plantations have a high economic threshold to pest damage (Sage 1998). Polyclonal cultures and checkerboarding (i.e., establishing small, 10 to 20 ha islands of SRWC in a matrix of other land uses) can be employed as a means of minimizing the spread and impact of pests and disease (Dickmann 2006).

Short-Rotation Coppice Plantations

Short-rotation coppice plantations often are hailed as a promising bioenergy option because of their net energy efficiency and ability to mitigate greenhouse gas emissions (Volk et al. 2004, Styles and Jones 2007, Lasch et al. 2010, Berhongaray et al. 2015, Pereira and Costa 2017). In the United States, short-rotation coppice crops are dominated by willow and poplar hybrids planted at high densities on former arable land. In short-rotation coppice plantations, aboveground biomass is harvested during winter on a cutting cycle of 3 or more years (3 years for willow, 3 to 7 years for poplar; Karp and Shield 2008). Rather than removing woody crops during harvest and replanting with genetically improved seedlings or clones, coppicing allows the belowground rootstock ("stool") to produce new shoots each following spring (Weih 2009) until productivity declines or rootstocks no longer are viable (~15 to 30 years; Aylott et al. 2008). Thus, several cutting cycles can be derived from a single rootstock establishment, and aboveground shoot age may differ substantially from rootstock and plantation age.

Harvesting

Harvesting is the single largest cost in the production of SRWCs, sometimes equaling about 45% of total production costs, and it requires the second largest input of fossil fuel energy in coppicing systems after fertilization (El Kasmioui and Ceulemans 2013, Eisenbies et al. 2014). Short-rotation coppice crops have been harvested using conventional forestry machinery, given its proven performance, availability, and reliability. Harvester prototypes dedicated to SRWC cultivation have been field-tested to improve efficiency, although few prototypes have advanced toward commercial use (Stuart et al. 1983, Ulich 1983, Ehlert and Pecenka 2013). SRWCs may be harvested by several methods, including the most common cut-and-store and cut-and-chip systems (Vanbeveren et al. 2015). The cut-and-store system is a 2-step process in which the entire shoot is harvested, and chipping typically occurs near the stand edge. In contrast, the cut-and-chip system is a 1-step operation that converts standing biomass to woody chips in the stand. Field sites with poor ground conditions (e.g., steep slopes, poorly drained) and lower field stocking have reduced harvest efficiencies (Eisenbies et al. 2014, Vanbeveren et al. 2017). Inter-row spacings, also known as headland widths, need to provide sufficient room for harvester and collection vehicles to turn around at the ends of crop rows. When appropriate headland widths are not used, residual woody biomass can accumulate at the edges of fields and ends of rows, reducing efficiency and profitability (Eisenbies et al. 2014).

Economic Feasibility

The most significant impediment to implementation of SRWCs for bioenergy is economic viability. Current returns on investment are poor in the United States because of high establishment and maintenance costs, low productivity on marginal land, low stumpage values, competition with other fuel sources, and an uncertain energy policy environment (Ghezehei et al. 2015). Although SRWC production is expected to become more lucrative as global markets for bioenergy develop, other concerns may discourage widespread use of SRWC plantations. For example, bioenergy can compete with traditional forestry and food production systems on a shrinking land base (Tyndall et al. 2011), leading some SRWC investors to turn to marginal lands with lower productivity (Ghezehei et al. 2015). Production costs are

strongly influenced by previous land use, and yields of all potential woody crops are sensitive to site history and soil category (Downing and Graham 1996). However, genetic modifications continue to facilitate yield improvements, and harvesting efficiency could reduce price differentials between biomass and its energy competitors (Volk et al. 2006). Policy changes also influence feasibility; federal and state programs (e.g., renewable portfolio standards) have attempted to foster biomass markets in the United States (Volk et al. 2006).

Potential Impacts to Plant and Wildlife Communities

As a foundation for wildlife foraging, roosting, and nesting sites, plant communities are central to wildlife habitat management. There is apparent consensus among plant and wildlife studies in SRWC plantations that woody crops resemble shrubland structural conditions rather than those of grasslands or forests (Christian 1997, Christian et al. 1998, Sage 1998, Riffell et al. 2011). Light availability, plantation age, tree canopy coverage, and crop species planted are the major factors influencing habitat quality in SRWC plantations and are closely related to one another. Tree species and plantation age will strongly affect the amount of tree canopy and subsequently the amount of light reaching the forest floor. Forest canopy coverage often is negatively associated with disturbance-adapted species throughout much of the southeastern United States. For example, in loblolly pine (*Pinus taeda*) plantations, wildlife community measures such as the number of species, number of individuals, and average conservation priority of the community are generally greater in stands with more open canopies, especially early in a rotation (Greene et al. 2016). However, wildlife species that require mature trees (e.g., cavity-nesting birds) are negatively associated with younger forest stands (Gottlieb et al. 2017), although these features typically are retained in unharvested buffers along streams and in other portions of the landscape.

Other determinants of habitat quality include the surrounding landscape, seed bank, site history, and soil condition. Plant communities in SRWC systems generally consist of commonly occurring species, with a shift in species composition from annual and perennial pioneer and ruderal species to perennial woodland species as woody crops age (Baum et al. 2012). Ground cover is generally low during the plantation establishment phase because of intensive vegetation management. Vegetation understory coverage increases, however, during the subsequent 4 or more years, followed by a decline associated with canopy closure from longer rotations in traditional plantations (Lane et al. 2011). Understory cover and plant diversity typically increase again post-thinning in traditional plantations; however, SRWC stands are unlikely to be thinned. Thus, planting conditions and, thereby, habitat modifications are repeated following each harvest. Also, conditions at planting dictate the structure and composition of plant communities later in the rotation. In loblolly pine plantations, the intensity of site preparation changes the plant communities in early rotation, especially with use of chemicals. For example, Lane et al. (2011) showed that chemical site preparation had effects on the herbaceous plant community up to 4 years post-planting, and effects persisted even longer for woody cover. However, Jones et al. (2012) reported that plant communities in stands established with practices that varied in intensity converged as crown closure approached, and structural differences were largely insignificant by year 5.

Knowledge about wildlife responses to SRWCs is incomplete and biased toward avian research (Rupp et al. 2012). Short-rotation woody crop plantations provide habitat conditions more favorable for shrubland- or forest-associated birds and mammals than do agricultural croplands but are of poorer quality for some species groups (e.g., cavity-nesting birds) than is more mature forest. Bird and small mammal species composition on SRWC plantations is a unique blend of grassland and forest species (Christian et al. 1998). Landscape composition plays a central role in determining occupancy for highly mobile species (e.g., birds), whereas within-plantation structure and composition is more critical for less mobile species, indicating that SRWC forest managers may need to purposefully integrate more plant species and structural heterogeneity across the landscape (i.e., a shifting mosaic) to retain diverse and abundant wildlife communities (Rupp et al. 2012, Root and Betts 2016).

Woody crop cultivation for biomass could contribute to overall wildlife diversity in forest-dominated landscapes and provide habitat for shrubland bird species (Christian et al. 1998, Riffell et al. 2011, Tarr et al. 2017). In a meta-analysis, Riffell et al. (2011) indicated that an abundance of shrub-associated birds are likely to be present on SRWC plantations, but mature forest-associated and cavity-nesting birds tend to be less abundant. Although abundance of individual bird species is expected to vary, diversity and abundance of avian guilds is predicted to be lower on SRWC plantations compared with surrounding woodlands (Riffell et al. 2011). Midsize mammals (e.g., squirrels, rabbits, hares) have shown limited use of SRWC patches, and white-tailed deer (*Odocoileus virginianus*) use of SRWC areas has been documented (Christian 1997). In additional, the headlands, glades, and woodland edges commonly associated with SRWC plantations may provide habitat for butterflies (Sage 1998) because of the abundant forbs and grasses prevalent following frequent disturbances.

Woody crop cultivation performs well on high-quality agricultural sites. Food production and economic concerns, however, may facilitate conversion of forested systems, rather than traditional agricultural land, to energy crop production. Because SRWCs are typically even-aged stands dominated by a single tree species, there is concern about potential impacts when replacing multispecies, structurally diversified forests with non-native energy crops (Rupp et al. 2012). Extensive conversion of mature forest of high conservation value to SRWC production would likely decrease wildlife diversity and negatively impact populations of rare forest-dwelling species (Riffell et al. 2011, Tarr et al. 2017). Additional effects of SRWC production could include degradation of downstream water quality, with impacts to aquatic and riparian species sensitive to water quality. Longer SRWC rotations can, however, facilitate shifts in plant and wildlife species composition from ruderal to shrubland communities (Baum et al. 2012) and potentially increase vertical structure and heterogeneity, thereby increasing the number of nesting and foraging substrates (Rupp et al. 2012).

Conclusion

Short-rotation woody crops are projected to play a vital role in the US biomass energy portfolio. Production costs, stumpage values, competition with other fuels, and an uncertain political environment put the current financial viability of SRWC at risk until development of biomass markets are realized. Short-rotation woody crop plantations provide compositional and structural conditions that are a hybrid of grasslands and woodlands that could benefit a myriad of shrubland species, particularly birds. Longer rotation lengths and fewer disturbances may further facilitate transition from ruderal to shrub communities. Although high-quality agricultural land is well suited to supporting growth of SRWCs, high agricultural crop prices, agricultural site availability, and policies aimed at alleviating food production concerns could prevent SRWC plantations from being established on agriculture land. Instead, SRWC plantations may be established on currently forested land, with possible negative impacts to wildlife associated with mature forest (e.g., cavity-nesting birds). Although SRWC plantations may augment diversity and wildlife abundance in forested landscapes on a small scale, replacing forests of high conservation value and large-scale conversion of mature forests to SRWCs likely would decrease overall wildlife diversity. Although the concept of SRWC plantations has roots in the 1960s and 1970s, the environmental implications of different options for energy crop culture (e.g., planting density, rotation length, and crop species) are still not fully understood. Landscape-scale experiments that investigate the environmental and economic trade-offs among varying amounts of SRWC plantations and other land-use types would be particularly helpful (see Deep Dive). Integration of precision agricultural production technologies to improve productivity of SRWC plantations that also include environmental considerations, such as protection of sensitive plant communities, could also advance wildlife conservation.

Deep Dive: Thinning Is for the Birds, Short-Rotation Woody Crops Outperform Corn

Researchers with the University of Florida and the Joseph W. Jones Ecological Research Center worked across an array of bioenergy feedstocks from 2013 to 2015 as part of a study funded by the United States Department of Agriculture National Institute of Food and Agriculture (Gottlieb 2017). The study's aim was to understand plant and bird community response to 8 land-use types: cornfields, clearcut pine forests with harvest residues removed, clearcut pine forests with harvest residues not removed, young pine plantations (8 to 10 years old), unthinned pine plantations and thinned pine plantations (12 to 16 years old), mature pine plantations (20–32 years old), and reference longleaf pine (*Pinus palustris*) forest (>40 years old). The young plantations, unthinned plantations, and the 2 types of clearcuts mimicked a few key phases of short-rotation pine stands in the region. The authors used statistical contrasts to tease apart these management scenarios and make inferences about the effects of bioenergy production on wildlife. The study areas spanned the southeastern coastal plain of the United States in 3 states: Alabama, Georgia, and Florida. Birds and plants were surveyed at 85 sites. Researchers observed >6,600 individual detections across 81 bird species and were able to estimate occupancy of 31 bird species.

Compared with all other land-use types, cornfields had the lowest average avian occupancy rate. The relative effect of young, unthinned pine plantations to >12-year-old conventional pine plantations was negligible (i.e., similar bird occupancy). However, inclusion of a residue-removal scenario reduced average avian occupancy rates. Generally speaking, land-use types associated with short-rotation scenarios reduced occupancy rates for cavity-nesting birds and species that

forage by gleaning tree bark. Moreover, scenarios with residue removal almost universally reduced occupancy of all nesting and foraging bird guilds. Notably, 4 species of conservation concern—Bachman's sparrow (*Peucaea aestivalis*), brown-headed nuthatch (*Sitta pusilla*), eastern towhee (*Pipilo erythrophthalmus*), and red-headed woodpecker (*Melanerpes erythrocephalus*)—showed a negative response to the harvest of residues or short-rotation management.

The authors emphasized 2 key conclusions from their results. First, corn production has the greatest overall negative ecological effect compared with pine biomass, suggesting that if corn fields were converted to pine biomass systems, bird communities could be improved. Second, thinning was generally a more bird-friendly way to acquire pine biomass than were short rotations and harvesting pine residue.

LITERATURE CITED

Aylott, M. J., E. Casella, I. Tubby, N. R. Street, P. Smith, and G. Taylor. 2008. Yield and spatial supply of bioenergy poplar and willow short-rotation coppice in the UK. New Phytologist 178:358–370.

Baum, S., M. Weih, and A. Bolte. 2012. Stand age characteristics and soil properties affect species composition of vascular plants in short rotation coppice plantations. BioRisk 7:51–71.

Berhongaray, G., M. S. Verlinden, L. S. Broeckx, and R. Ceulemans. 2015. Changes in belowground biomass after coppice in two *Populus* genotypes. Forest Ecology and Management 337:1–10.

Broadmeadow, S., and T. R. Nisbet. 2004. The effects of riparian forest management on the freshwater environment: A literature review of best management practice. Hydrology and Earth System Sciences Discussions, European Geosciences Union 8:286–305.

Christian, D. P. 1997. Wintertime use of hybrid poplar plantations by deer and medium-sized mammals in the midwestern U.S. Biomass and Bioenergy 12:35–40.

Christian, D. P., W. Hoffman, J. M. Hanowski, G. J. Neimi, and J. Beyea. 1998. Bird and mammal diversity of woody

biomass plantations in North America. Biomass and Bioenergy 14:395–402.

Coleman, M. D., D. R. Coyle, J. Blake, K. Britton, M. Buford, R. G. Campbell, J. Cox, et al. 2004. Production of short-rotation woody crops grown with a range of nutrient and water availability: Establishment report and first-year responses. General Technical Report SRS-72. Asheville, NC: USDA Forest Service, Southern Research Station.

Cristan, R., W. M. Aust, M. C. Bolding, S. M. Barrett, J. F. Munsell, and E. Schilling. 2016. Effectiveness of forestry best management practices in the United States: Literature review. Forest Ecology and Management 360:133–151.

Cristan, R., W. M. Aust, M. C. Bolding, S. M. Barrett, and J. F. Munsell. 2018. National status of state developed and implemented forestry best management practices for protecting water quality in the United States. Forest Ecology and Management 418:73–84.

DeBell, D. S., G. W. Clendenen, C. A. Harrington, and J. C. Zasada. 1996. Tree growth and stand development in short-rotation *Populus* plantings: 7-year results for two clones at three spacings. Biomass and Bioenergy 11:253–269.

Dickmann, D. I. 2006. Silviculture and biology of short-rotation woody crops in temperate regions: Then and now. Biomass and Bioenergy 30:696–705.

Dillen, S. Y., S. N. Djomo, N. Al Afas, S. Vanbeveren, and R. Ceulemans. 2013. Biomass yield and energy balance of a short-rotation poplar coppice with multiple clones on degraded land during 16 years. Biomass and Bioenergy 56:157–165.

Downing, M., and R. L. Graham. 1996. The potential supply and cost of biomass from energy crops in the Tennessee Valley Authority region. Biomass and Bioenergy 11:282–303.

Drew, A. P., L. Zsuffa, and C. P. Mitchell. 1987. Terminology related to woody plant biomass and its production. Biomass 12:79–82.

Ehlert, D., and R. Pecenka. 2013. Harvesters for short rotation coppice: Current status and new solutions. International Journal of Engineering 24:170–182.

Eisenbies, M. H., T. A. Volk, J. Posselius, C. Foster, S. Karapetyan, and S. Shi. 2014. Evaluation of single-pass, cut and chip harvest system on commercial-scale, short-rotation shrub willow biomass crops. Willow Harvest Publications, Paper 5. http://digitalcommons.esf.edu/hvstpub/5.

El Kasmioui, O., and R. Ceulemans. 2013. Financial analysis of the cultivation of short rotation woody crops for bioenergy in Belgium: Barriers and opportunities. Bioenergy Research 6:336–350.

Eufrade, H. J. Jr., R. X. de Melo, M. M. Pereira Sartori, S. P. Sebastião Guerra, and A. Wagner Ballarin. 2016. Sustainable use of eucalypt biomass grown on short rotation coppice for bioenergy. Biomass and Bioenergy 90:15–21.

Ghezehei, S. B., S. D. Shifflett, D. W. Hazel, and E. G. Nichols. 2015. SRWC bioenergy productivity and economic feasibility on marginal lands. Journal of Environmental Management 160:57–66.

Gottlieb, I. G. W., R. J. Fletcher Jr., Nunez-Regueiro, M. M., H. Ober, L. Smith, B. J. Brosi. 2017. Alternative biomass strategies for bioenergy: Implications for bird communities across the southeastern United States. Global Change Biology Bioenergy 9:1606–1617.

Greene, R. E., R. B. Iglay, K. O. Evans, D. A. Miller, T. B. Wigley, and S. K. Riffell. 2016. A meta-analysis of biodiversity responses to management of southeastern pine forests: Opportunities for open pine conservation. Forest Ecology and Management 360:30–39.

Janowiak, M. K., and C. R. Webster. 2010. Promoting ecological sustainability in woody biomass harvesting. Journal of Forestry 108:16–23.

Jones, P. D., S. Demarais, and A. W. Ezell. 2012. Successional trajectory of loblolly pine (Pinus taeda) planta-tions established using intensive management in Southern Mississippi, USA. Forest Ecology and Management 265:116–123.

Karp, A., and I. Shield. 2008. Bioenergy from plants and the sustainable yield challenge. New Phytologist 179:15–32.

Lane, V. R., K. V. Miller, S. B. Castleberry, D. A. Miller, T. B. Wigley, G. M. Marsh, and R. L. Mihalco. 2011. Plant community responses to a gradient of site preparation intensities in pine plantations in the Coastal Plain of North Carolina. Forest Ecology and Management 262:370–378.

Lasch, P., C. Kollas, J. Rock, and F. Suckow. 2010. Potentials and impacts of short-rotation coppice plantation with aspen in Eastern Germany under conditions of climate change. Regional Environmental Change 10:83–94.

Paris, P., L. Mareschi, M. Sabatti, A. Pisanelli, A. Ecosse, F. Nardin, and G. Sarascia-Mugnozza. 2011. Comparing hybrid Populus clones for SRF across northern Italy after two biennial rotations: Survival, growth and yield. Biomass and Bioenergy 35:1524–1532.

Pereira, S., and M. Costa. 2017. Short rotation coppice for bioenergy: From biomass characterization to establishment—A review. Renewable and Sustainable Energy Reviews 74:1170–1180.

Riffell, S., J. Verschuyl, D. Miller, and T. B. Wigley. 2011. A meta-analysis of bird and mammal response to short-rotation woody crops. Global Change Biology Bioenergy 3:313–321.

Root, H. T., and M. G. Betts. 2016. Managing moist temperate forests for bioenergy and biodiversity. Journal of Forestry 114:66–74.

Rupp, S. P., L. Bies, A. Glaser, C. Kowaleski, T. McCoy, T. Rentz, S. Riffell, J. Sibbing, J. Verschuyl, and T. Wigley. 2012. Effects of bioenergy productions on wildlife and wildlife habitat. Wildlife Society Technical Review 12-03. Bethesda, MD: Wildlife Society.

Sage, R. B. 1998. Short rotation coppice for energy: Towards ecological guidelines. Biomass and Bioenergy 15:39–47.

Stuart, W. B., D. S. Marley, and J. B. Teel. 1983. A prototype short rotation harvester. Abstract from the 7th International FPRS Industrial Wood Energy Forum. September 19–21, Nashville, TN.

Styles, D., and M. B. Jones. 2007. Energy crops in Ireland: Quantifying the potential life-cycle greenhouse gas reductions of energy-crop electricity. Biomass and Bioenergy 31:759–772.

Tarr, N. M., M. J. Rubino, J. K. Costanza, A. J. McKerrow, J. A. Collazo, and R. C. Abt. 2017. Projected gains and losses of wildlife habitat from bioenergy-induced landscape change. Global Change Biology Bioenergy 9:909–923.

Tyndall, J. C., L. A. Schulte, R. B. Hall, and K. R. Grubh. 2011. Woody biomass in the U.S. Cornbelt? Constraints and opportunities in the supply. Biomass and Bioenergy 35:1561–1571.

Ulich, W. L. 1983. Development of a biomass combine. Lubbock: Texas Technical University, Department of Agricultural Engineering and Technology.

US DOE (US Department of Energy). 2016. 2016 Billion-ton report: Advancing domestic resources for a thriving bioeconomy, Volume 1: Economic availability of feedstocks. Lead authors, M. H. Langholtz, B. J. Stokes, and L. M. Eaton, ORNL/TM-2016/160. Oak Ridge, TN: Oak Ridge National Laboratory. doi:10.2172/1271651. http://energy.gov/eere/bioenergy/2016-billion-ton -report.

US Energy Independence and Security Act of 2007 (Pub. L. 110-140).

US Energy Information Administration. 2017. Short-term energy outlook. https://www.eia.gov/forecasts/steo /report/renew_co2.cfm.

Vanbeveren, S. P. P., J. Schweier, G. Berhongaray, and R. Ceulemans. 2015. Operational short rotation woody crop plantations: Manual or mechanised harvesting? Biomass and Bioenergy 72:8–18.

Vanbeveren, S. P. P., R. Spinelli, M. Eisenbies, J. Schweier, B. Mola-Yudego, N. Magagnotti, M. Acuna, I. Dimitriou, and R. Ceulemans. 2017. Mechanised harvesting of short-rotation coppices. Renewable and Sustainable Energy Reviews 76:90–104.

Volk, T. A., T. Verwijst, P. J. Tharakan, L. P. Abrahamson, and E. H. White. 2004. Growing fuel: A sustainability assessment of willow biomass crops. Frontiers in Ecology and the Environment 2: 411–418.

Volk, T. A., L. P. Abrahamson, C. A. Nowak, L. B. Smart, P. J. Tharakan, and E. H. White. 2006. The development of short-rotation willow in the northeastern United States for bioenergy and bioproducts, agroforestry and phytoremediation. Biomass and Bioenergy 30:715–727.

Warrington, B. M., M. W. Aust, S. M. Barrett, W. M. Ford, A. C. Dolloff, E. B. Schilling, T. B. Wigley, and C. M. Bolding. 2017. Forestry best management practices relationships with aquatic and riparian fauna: A review. Forests 8:1–16.

Weih, M. 2009. Perennial energy crops: Growth and management. In Soils, Plant Growth and Crop Production—Vol. III, edited by W. H. Verhaye. Encyclopedia of Life Support Systems (EOLSS), developed under the Auspices of the UNESCO. Oxford: Eolss.

Zalesny, R. S., M. W. Cunningham, R. B. Hall, J. Mirck, D. L. Rockwood, J. A. Stanturf, and T. A. Volk. 2011. Woody biomass from short rotation energy crops. In Sustainable Production of Fuels, Chemicals, and Fibers from Forest Biomass, edited by J. Zhu, X. Zhang, and X. J. Pan. ACS Symposium Series. Washington, DC: American Chemical Society.

2

JESSICA A. HOMYACK AND
JAKE VERSCHUYL

Effects of Harvesting Forest-Based Biomass on Terrestrial Wildlife

Introduction

In North America, wood product manufacturing facilities have converted residual materials produced at lumber and pulp mills (e.g., black liquor, sawdust, chips) to electricity for decades. However, scaling up woody biomass (i.e., slash, tops, limbs, and small-diameter trees unmarketable as higher-value timber products) production to supply feedstocks (i.e., raw materials) for conversion to municipal electricity or liquid transportation biofuels has a shorter and rapidly evolving narrative. Modern advances in technologies have greatly improved our ability to quickly regenerate forests, efficiently harvest and transport wood commodities, and convert forest biomass to power (Talbert and Marshall 2005, Fox et al. 2007, Demarias et al. 2016). Forest-based biomass feedstocks have a low economic return for large volumes of materials, which has limited their viability in renewable energy markets until recent changes in social and political climates increased interest in diminishing fossil fuel usage. Both United States and European Union policies and regulations have mandated increases in the proportion of energy from renewable sources, and forests are poised to be significant contributors to meeting these demands (Rupp et al. 2012, Donner et al. 2017).

Excluding manufacturing by-products, forest-based biomass is used primarily for wood pellet production, in cogeneration power plants, and, to a lesser extent, to produce liquid transportation biofuels. Expanding markets to meet renewable energy mandates of the European Union have led to an increase in wood pellet production in the southeastern United States, growing from negligible amounts in the early 2000s to 5.4 million green metric tons in 2015 (Dale et al. 2017). The southeastern United States wood basket provides the greatest export of wood pellets globally and transported 98% of pellets produced in the region to Europe in 2015 (Dale et al. 2017). Further, cogeneration plants where forest-based feedstocks are burned in the same facilities as coal or other fuels to generate electricity and heat (i.e., co-firing) are becoming more common as municipalities and businesses strive to decrease their carbon footprint and source local materials. From 2010 to 2015, the amount of biomass-based electricity produced in the United States grew from 56 to 64 terawatt hours, taking advantage of abundant forest resources in regions where other sources of renewable energy (e.g., wind, solar) were less favorable (US Energy Information Administration 2016). The increase in biomass-based electricity occurred in large part from new facilities that source biomass feedstocks in the southeastern United States and California and from conversion of energy facilities from coal to biomass in Virginia. Although many challenges exist for scaling up the supply chain and for technologies converting forest feedstocks to liquid transportation fuels, one of the US Department of Agriculture's Regional Bioenergy System Coordinated Agricultural Projects, the Northwest Advanced Renewables Alliance (NARA), made international headlines in November 2016: amid great fanfare, the first commercial airline flight powered by alternative jet fuel produced from forest residuals made a cross-country journey from Seattle to Washington, DC (NARA 2017). Currently, corn provides the majority of plant-based biofuel feedstocks in the United States (see Chapter 3, Otto); however, advancing technologies and societal pressures to minimize land-use change from food to biofuels feedstock production may shift the feedstock portfolio toward forest-based biomass (Phalan et al. 2011). Thus, projected increases and continued use of forest-based biomass for these energy pathways depend on the economic viability of feedstock production in forests, efficient transportation to processing facilities, and conversion and connection to the grid or pipeline systems for liquid fuels.

Unsurprisingly, the major geographic sources of forest-based biomass occur in US regions with the greatest available wood volume, but regional differences in sources and predicted future forest biomass harvest potentials differ geographically (Donner et al. 2017). In the United States, the Southeast has the greatest forest biomass potential, primarily from pine plantations on private land, followed by the North Central region, New England, Pacific Northwest, and Interior West. Based on economic modeling, logging residuals, which include tops, limbs, stumps, and other woody materials not merchandized as intact logs during commercial harvesting operations, offer the greatest biomass potential. Regionally, logging residuals are predominant in the Pacific Northwest and southeastern United States as

by-products of either planted softwood or naturally regenerated softwood stands (Donner et al. 2017). Whole-tree biomass harvests, which are the most common source of harvest in the Interior West, include operations in which merchantable or unmerchantable smaller-diameter trees are harvested. These biomass harvests often source woody material from thinning operations during the middle of a forest-stand rotation that reduce stand density to improve growth and vigor of remaining crop trees (deStefano et al. 2016). Thus, forest-based biomass typically is produced as a secondary product on managed forests, thereby avoiding competition with lands used for food production. Although geographic regions differ in sources and abundance of forest biomass, many of the environmental concerns associated with harvesting forest-based biomass are similar across regions.

Environmental Effects of Forest Biomass Production

Sustainability concerns for forest-based biomass harvests have focused on potential loss of soil productivity and degradation of wildlife habitat at multiple spatiotemporal scales (Van Hook et al. 1982, Riffell et al. 2011). By removing downed wood or trees that may eventually contribute coarse and fine woody debris, forest-based biomass harvests could reduce or remove structures that wildlife use for foraging, nesting, estivating, and escape cover. For example, lower-quality trees (i.e., hardwood species in a pine plantation; trees with a broken top or other defect) that are harvested as renewable energy feedstock could otherwise provide structures to the 85 species of North American birds that use snags for foraging, nesting, roosting, and communicating (Scott et al. 1977, Loman et al. 2013, Grodsky et al. 2016a). Similarly, limbs, stumps, and other wood gleaned (i.e., collected by equipment during harvests) for biomass could provide rest sites for small mammals and furbearers or sites for egg deposition for amphibians and reptiles, if downed woody material remained in forest stands (Otto et al. 2013, Fritts et al. 2015a, 2015b). Finally, changes in abundance or structure of decaying wood could affect numerous species of invertebrates through several pathways (Grodsky et al. 2018a, 2018b), including a reduction in food sources for saproxylic beetles that feed on decaying wood as larvae and a more complex reconfiguration of the invertebrate food web (e.g., reduction in prey for invertebrate predators; Work et al. 2014).

Thinning and harvesting woody residues can simplify vertical and horizontal structure compared with managed stands without a biomass harvest by reducing crop tree mortality and removing mid- and understory vegetation (Homyack et al. 2004, 2011b; Figure 2.1). Structural simplification can occur when residues are harvested during a clearcut harvest, during which all overstory trees are harvested in a sin-

gle entry, or over a longer time period, when thinning and abbreviated stand rotations reduce recruitment of deadwood into managed forests (Spies et al. 1988, Homyack et al. 2004, 2011b). In these scenarios, biomass harvesting can represent an "intensification" of traditional forest management when activities such as harvest of forest residues include additional equipment passes to collect and transport woody materials. Many studies have identified strong relationships between vegetation structure and diversity and abundance of terrestrial wildlife. Therefore, changes to habitat complexity from biomass harvesting are predicted to translate into changes in abundance and diversity of terrestrial wildlife.

In addition to providing habitat structure, decaying wood ameliorates thermal extremes, particularly

Figure 2.1 Forest thinnings used for bioenergy production, such as those from this commercial thinning of ponderosa pine (*Pinus ponderosa*) forest in Oregon, are often a by-product of management prescriptions for other goals. *Photo credit: Jordan Benner, Oregon Forest Resources Institute.*

the increased temperatures and lowered humidity that occur after forest harvests remove canopy trees (Chen et al. 1999, Brooks and Kyker-Snowman 2008, Kluber et al. 2009). Ground-dwelling animals, particularly amphibians, are sensitive to changes in microclimate caused by increased solar radiation on the forest floor. For ectotherms such as salamanders, metabolic energy use increases and digestive efficiency can decrease when temperatures are above their optimal range, with potential negative effects on survival, growth, and reproduction (Rothermel and Luhring 2005, Homyack et al. 2011a). Woodland salamanders from the family Plethodontidae are lungless and require cool, moist conditions to respire cutaneously. Without large, well-decayed woody material available in forest stands, ground-dwelling wildlife like woodland salamanders may lack refuges from extreme temperatures following biomass harvests. However, recruitment and decay dynamics of woody debris in managed forests, particularly after biomass harvests, are not well documented. Snag and woody debris availability and decay dynamics vary across geographic regions, climates, and tree species (Moorman et al. 1999, Fraver et al. 2002). Therefore, determination of thresholds of woody debris necessary to support wildlife populations needs to account for exogenous factors and life-history characteristics of wildlife species of interest (Otto et al. 2013). In short, it's complex.

Environmental concerns regarding extraction of forest-based biomass at larger spatial scales often focus on perceived targeting of old-growth forest stands or bottomland hardwood forest for woody biomass harvests. Some fear that increasing demand for biomass will create incentives to clearcut and replant these lands into faster-growing, intensively managed stands, causing additional reduction in their presence on the landscape and consequential negative effects on wildlife diversity. However, economic analyses indicate that conversion of native forestland to short-rotation plantations solely to produce biomass feedstocks is unlikely because of low revenues from bioenergy feedstocks and high availability of al-

ternate biomass sources as secondary products (Dale et al. 2017). Even in regions where forest-based biomass production is substantial (e.g., southeastern United States), conversion of forestland to urban or suburban uses is a greater threat to maintaining forested landscapes than is forest-based biomass production (Wear and Greis 2013). Further, third-party forest sustainability certification programs, regulations, forestry best management practices, and conservation programs all contribute to promoting maintenance and enhancement of biodiversity and ecosystem services on both private and public forestlands.

Biomass Harvesting Guidelines (BHGs) have been developed in several states and by conservation organizations to provide additional protections to forest resources beyond existing best management practices or regulations (e.g., Forest Guild Southeast Working Group 2012). These voluntary guidelines typically assume a positive and linear response between wildlife population size or species diversity and amount of woody debris retained in a forest stand following a biomass harvest (Fritts et al. 2016). Most BHGs describe retention targets as either a specified volume of woody material or a percentage of preharvest woody material to be retained. Some BHGs also suggest placement of residual materials in piles versus dispersed across the harvest unit. However, there is limited understanding of effects of manipulating sources of downed woody debris on many taxa, particularly at operational spatial and temporal scales (Riffell et al. 2011, Otto et al. 2013). Riffell et al. (2011) reported that cavity-nesting birds were most likely to be negatively affected by removal of woody materials across forest-stand types and geographic regions, and Verschuyl et al. (2011) reported that intensity of forest thinning influenced whether observed effects on wildlife were neutral or positive. These two meta-analyses also highlighted a geographical bias toward research of woody debris and wildlife populations in the southeastern United States and a lack of knowledge of interactive effects among species following woody debris manipulations (Riffell et al. 2011, Verschuyl et al. 2011). Finally, most published research on the topic is not in the context of biomass harvests for bioenergy, but rather are comparative studies on thinning for other objectives or sites with varying amounts of downed wood (Table 2.1). Although useful for understanding relationships of species or communities with similar habitat characteristics, these studies may miss operational realities of spatial scale, timing of harvests, and economics that limit their relevance to current discussions of bioenergy and sustainability.

Table 2.1. Summary of manipulative studies published from 2010 to 2017 included in meta-analyses of effects of forest-based biomass harvesting on biodiversity

Taxa	Total Effect Sizes	Number of Studies	Study
Mammals	80	6	Bagne and Finch 2010[a]; Fritts et al. 2015[b,g]; Fritts et al. 2017[b,g]; Homyack et al. 2014[b,g]; Manning et al. 2012[d]; Sullivan and Sullivan 2012[c]
Birds	219	5	Cahall et al. 2013[d]; Gaines et al. 2010[d]; Grodsky et al. 2016a[b,g]; Grodsky et al. 2016b[b,g]; Kendrick et al. 2015[d]
Invertebrates	66	3	Castro and Wise 2010[b]; Parrish and Summerville 2015[b,g]; Work et al. 2014[b,g]
Amphibians and reptiles	99	4	Davis et al. 2010[b,c,e,f]; Fritts et al. 2016[b,g]; Hocking et al. 2013[d]; Homyack et al. 2013[b,g]

[a] Fuels treatment thinning where understory and non-merchantable timber are removed.

[b] Removal of downed woody material.

[c] Addition of downed woody material.

[d] Commercial thinning to extract merchantable timber.

[e] Snag removal.

[f] Snag addition.

[g] Treatment completed in the context of bioenergy production.

Deep Dive: What's in a Guideline? An Experimental Look at Biomass Harvesting Guidelines

A broad collaboration of universities, forest industry representatives, and other partners completed an experimental manipulation of retained harvest residues across eight clearcut harvest units in North Carolina ($n=4$) and Georgia ($n=4$), 2010 to 2015 (Figures 2.2 and 2.3). Each study site had 6 biomass retention practices applied, ranging from no retention (i.e., operators removed all merchantable biomass as operationally feasible) to no removal of biomass (i.e., no biomass harvest executed). The intent of the research was to inform

Figure 2.2 Implementing successful land management experiments requires collaboration and communication across organizations. Shown here, partners from private industry, universities, and research organizations visit a biomass harvesting research site in Beaufort County, North Carolina. *Photo credit: Darren Miller.*

Figure 2.3   Downed woody material is sparse between windrows, as shown following clearcut and woody biomass harvest in southeastern Georgia, US. *Photo credit: Steven M. Grodsky.*

agencies and nongovernmental organizations that have developed BHGs that aim to conserve biodiversity, soil productivity, or other aspects of sustainability following harvest of forest residuals. Previously, most BHGs were developed without empirical support from replicated experiments and relied on the assumption that wildlife responds positively and linearly to the amount of downed woody materials retained after clearcut harvesting. Further, whether retention targets could be met in operational settings had not been evaluated.

Across a 5-year period, researchers examined responses of vegetation structure, amphibians, reptiles, shrews, rodents, breeding and wintering birds, and invertebrates

(Figure 2.4). The research occurred on operationally sized clearcut harvest units on properties owned by multiple landowners across 2 states and investigated whether residual downed woody material was reduced by biomass harvests. Forest stands with an operational biomass harvest had up to 80% less downed woody material than treatments without a biomass harvest, but biomass harvest treatments still maintained more downed woody material than what was typically recommended in BHGs (Fritts et al. 2014). Abundance and diversity were similar across all levels of biomass retention for most vertebrate taxa. For these vertebrates, negative effects were minimal or short-term. Instead, clearcut harvesting and the subsequent rapid

Figure 2.4   Habitat generalists, such as these southern toads (*Anaxyrus terrestris*), shown in amplexus, often have neutral responses to removal of woody material for bioenergy production. *Photo credit: Christopher E. Moorman.*

regrowth of vegetation that occurred in southeastern US forests had a greater influence on wildlife. Wildlife abundances and communities changed through time as the near-ground vegetation structure increased and vegetation characteristics shifted from being dominated by bare ground and herbaceous plants to woody vegetation and the rapid growth of planted crop trees by 4 years post-treatment. On the other hand, removal of harvest residues negatively affected key invertebrate taxa of early successional vegeta-

tion communities, including ground beetles, crickets, fungus gnats, and wood roaches, and facilitated colonization of the invasive red imported fire ant (*Solenopsis invicta*).

What do these results mean for implementing BHGs? They suggest that current rates of biomass harvest, even without additional retention targets, are sufficient to support vertebrate wildlife communities in managed forests. However, these results are from the hot, humid climate of the southeastern United States, where woody material decays rapidly and where historically frequent, low-intensity fires may have limited the amount of woody debris present across the landscape. Communities were dominated by habitat generalist species, but several species of conservation concern, such as state-level priority species, were observed in treatments. Thus, many terrestrial wildlife species of the region may be less reliant on downed woody material than in other regions, such as the Pacific Northwest, where individual pieces of downed wood can persist on the landscape for decades to centuries. For invertebrates, additional research on the effects of woody biomass harvesting on individual species, species-process interactions, and biological invasions is warranted.

Synthesizing Our Knowledge with a Meta-analysis

In recent years, and coinciding with rapidly growing markets for woody biomass, the base of empirical research documenting potential effects of biomass harvesting on wildlife populations and diversity has increased substantially. Further, harvest of low-value woody material during final harvests or forest thinning operations continue to be viewed as primary sources for expanding renewable energy feedstock production in the United States (Donner et al. 2017).

Here, we incorporate contemporary literature into the prior meta-analyses of Riffell et al. (2011) and Verschuyl et al. (2011) that examined effects of woody debris manipulation and thinning on abundance and diversity of wildlife populations in North America. Our goal is to summarize current knowledge to develop a more informed understanding of wildlife responses to woody biomass harvest, provide suggestions for mediating negative effects, and identify knowledge gaps to guide future research efforts.

Deeper Dive: Into the Meta-analysis . . .

We supplemented the meta-analyses by Riffell et al. (2011) and Verschuyl et al. (2011) by including results from studies published from 2010 to 2017 that reported effects on wildlife populations or diversity from snag or downed woody debris alteration or forest thinning. We included results from both manipulative experiments and observational studies that compared forest stands or plots with different management histories. We considered snags as standing dead trees ≥ 1.8 m in height and ≥ 10.2 cm in diameter at breast height (dbh) and coarse woody debris (CWD) as logs, stumps, and piles of limbs and other woody material on the forest floor. Most studies considered CWD as > 10 cm in diameter and > 60 cm in length, but we use the term downed woody material (DWM) to refer broadly to large and small woody materials. We identified additional publications through web searches and references of other articles. To be suitable for use in the meta-analysis, publications needed to include sample sizes, mean responses, a measure of variability, and unmanipulated control units. We conducted 2 separate analyses that examined (1) experimental effects of DWM and snag removal or additions, and (2) effects of precommercial, commercial, and fuels treatment thinning on abundance and diversity of wildlife species. For multiyear studies that presented results annually, we calculated an overall mean effect and standard deviation from the pooled variance. For research that compared multiple treatments to the same control (e.g., multiple levels of DWM retention compared with 100% DWM retention or different thinning intensities compared to unthinned controls), we calculated separate effect sizes for each treatment.

We conducted meta-analyses using MetaWin software (Rosenberg et al. 2000) and calculated the response ratio as the ratio of the experimental manipulation to the control (Hedges et al. 1999). To maintain consistent interpretation of results, we coded the treatment with the least DWM removed as the "control" for meta-analysis when DWM was removed, and unthinned stands as the "control" for the thinning results. Thus, resulting response ratios < 1.00 indicate a negative response to removing woody debris material or thinning, and response ratios > 1.00 indicate a positive response to the DWM removal or thinning treatment. We used bootstrap confidence intervals with 1,000 iterations and considered combined effects as significant when confidence intervals did not overlap 1.00. We examined results separately by taxonomic group and across all taxa for both analyses. Diversity metrics included species richness, diversity measures, and evenness responses; abundance included abundance of taxa, species groups, and guilds; species included abundance of individual species; and invertebrate biomass included biomass of species or taxonomic groupings. Herpetofauna included measures of diversity or guild abundance for amphibians and reptiles combined. For forest thinning, we also examined thinning intensity and type of thinning and described geographic biases in location of studies.

Manipulating Downed Woody Material: Effects on Terrestrial Wildlife

We analyzed 1,083 effect sizes from 33 publications, including 353 additional effect sizes (hereafter, new effect sizes) from 12 articles published since 2010 on effects of woody debris additions or removals on wildlife (Table 2.1). Woody debris manipulation included removals (n=32 studies) or additions (n=9 studies) of snags or DWM. Most (90%) new effect sizes were from studies in the southeastern United States, but they were comprehensive in taxa and included responses of birds (33%), mammals (21%), invertebrates (19%), reptiles (10%), amphibians (10%), or combined herpetofauna (7%). For manipulations of DWM, cumulative effect size across all taxa and metrics was 0.98, and the bootstrapped confidence interval did not include 1, indicating a negative effect from lowered amounts of woody material on forested sites.

For birds, we analyzed 384 effect sizes from 13 publications and, similar to the prior meta-analysis, determined that birds responded negatively to manipulating DWM with a cumulative response of 0.90 (confidence interval: 0.75, 1.00). Bird diversity, taxa abundance, and abundance of specific species all had effect sizes < 1.00, indicating a negative effect of DWM removal on these metrics (Table 2.2, Figure 2.5). We examined effects

of DWM manipulation on mammals from 116 effect sizes derived from 7 publications. We did not detect a significant effect of DWM manipulation on mammal diversity, guild abundance, or abundance of individual species (Table 2.2). Our analyses of amphibian and reptile responses included 106 effect sizes. We did not detect an effect of DWM manipulation on any abundance, diversity, or cumulative metric for amphibians or reptiles (Table 2.2). Finally, we examined effects of DWM manipulation on invertebrates with 359 effect sizes from 8 publications, including diversity, abundance, guild abundance, and biomass. Although the overall effect size was not significant (1.00; Table 2.2), abundance of all invertebrates combined had a positive effect size (1.08), indicating a positive relationship with less DWM.

Forest Thinning: Effects on Terrestrial Wildlife

Twenty-three of the studies included in our meta-analysis related to thinning effects were manipulative experiments, and the remaining

Table 2.2. Summary of effects of forest-based biomass harvesting from snag and downed woody material removal (DWM studies) and forest thinning (Thinning studies) on biodiversity by taxa from 2 meta-analyses. Response ratios < 1.00 indicate a negative response to removing woody debris material or thinning, and response ratios > 1.00 indicate a positive response to the DWM removal or thinning treatment.

	Birds	Mammals	Amphibians	Reptiles	Invertebrates	All Taxa
DWM studies	k = 13	k = 7	k = 4	k = 4	k = 8	k = 33
Diversity	0.98 (n = 30)**	1.02 (n = 12)	1.00 (n = 40)	1.00 (n = 40)	1.00 (n = 90)	1.00 (n = 230)**
Guild abundance	0.87 (n = 49)**	0.98 (n = 19)	1.00 (n = 27)	1.00 (n = 27)	1.08 (n = 231)	1.00 (n = 359)**
Species abundance	0.90 (n = 305)**	0.99 (n = 85)	1.00 (n = 39)	1.00 (n = 26)	1.00 (n = 29)	1.00 (n = 484)**
Cumulative	0.90 (n = 384)**	0.99 (n = 116)	1.00 (n = 106)	1.00 (n = 93)	1.00 (n = 359)	0.98 (n = 1083)**
Thinning studies	k = 16	k = 18	k = 5	k =3	k = 2	k = 39
Diversity	1.12 (n = 10)**	1.06 (n = 10)	0.98 (n = 2)	0.96 (n = 1)	1.11 (n = 4)	1.09 (n = 27)**
Guild abundance	1.07 (n = 43)	1.35 (n = 18)**	0.95 (n = 4)**	0.90 (n = 5)	1.09 (n = 42)**	1.09 (n = 112)**
Species abundance	1.03 (n = 325)**	1.02 (n = 127)	0.94 (n = 14)	1.57 (n = 11)**	—	1.02 (n = 477)**
Cumulative	1.03 (n = 378)**	1.02 (n = 155)	0.94 (n = 20)	1.38 (n = 17)**	1.10 (n = 46)**	1.03 (n = 616)**

**Indicates bootstrap confidence intervals (1,000 iterations) did not include 1.00; k = # of studies, n = # of effect sizes.

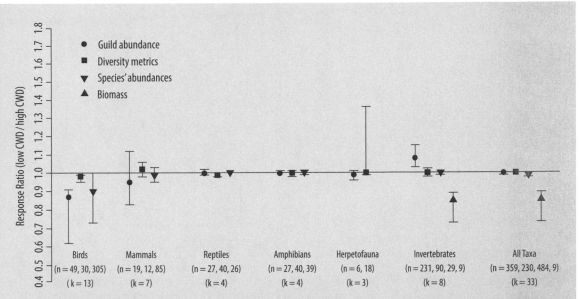

Figure 2.5 Summary of response ratios (95% confidence intervals) from meta-analysis of wildlife responses to manipulation of snags and downed woody material. n = number of individual effect sizes; k = number of publications. *Jessica A. Homyack and Jake Verschuyl.*

16 were observational comparisons of thinned and unthinned areas. Approximately two-thirds (68%) of the effect sizes were from studies in the northwestern United States, and most (94%) were of birds, with small proportions of mammals (5%) and amphibians (1%). Across taxa, cumulative effect size indicated a positive response to thinning.

Like Verschuyl et al. (2011), we determined that birds demonstrated a positive cumulative response to thinning across a diversity of North American forest types. Bird diversity metrics, including reported measures of diversity, evenness and richness, and species abundance also increased in response to thinning (Table 2.2). The effect of forest thinning on bird guild abundance was not significant, but the cumulative effect size was greater than 1 (Table 2.2, Figure 2.6). Unlike Verschuyl et al. (2011), who reported a positive cumulative response by mammals to forest thinning, we identified a neutral

response (i.e., effect size not significantly different than 1). Mammal diversity and species abundance cumulative effect sizes indicated neutral responses for these metrics (Table 2.2). Effect of forest thinning on mammalian guild abundance was significant and positive (1.35). Amphibians had a neutral cumulative response to forest thinning but lower guild abundance with increased thinning (Table 2.2). Responses of amphibian diversity and species abundance were neutral (Table 2.2). We did not include additional studies of reptile responses (published since 2010). Thus, our findings here mirror those of Verschuyl et al. (2011), who reported a positive cumulative reptile response to thinning and a positive relationship between species abundance and thinning. Reptile diversity and species abundance cumulative effect sizes were neutral (Table 2.2). Similarly, we did not include additional studies of invertebrate response to thinning. But

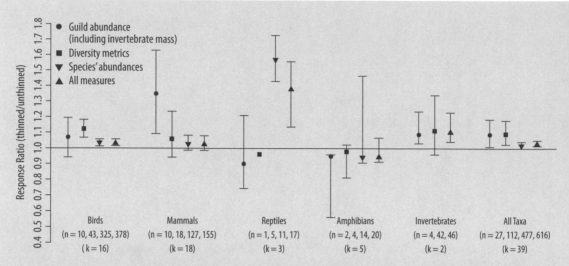

Figure 2.6   Summary of response ratios (95% confidence intervals) from meta-analysis of wildlife responses to forest thinning. n = number of individual effect sizes included in each response ratio; k = number of publications included in each response ratio. *Jessica A. Homyack and Jake Verschuyl.*

Table 2.3.   Summary of effects of forest-based biomass harvesting from forest thinning on overall wildlife responses by proportion of canopy basal area removed. Response ratios < 1.00 indicate a negative response to thinning, and response ratios > 1.00 indicate a positive response to thinning.

	Light (0–33% removed)	Moderate (34–66% removed)	Heavy (> 66% removed)
Thinning studies	k = 18	k = 19	k = 10
Diversity	1.12 (n = 9)**	1.06 (n = 10)	1.11 (n = 8)
Taxa / guild abundance	1.16 (n = 34)**	1.14 (n = 49)**	0.85 (n = 29)
Species abundance	1.07 (n = 157)**	1.01 (n = 228)	1.03 (n = 92)
Cumulative	1.08 (n = 200)**	1.01 (n = 287)	1.02 (n = 129)

**Indicates bootstrap confidence intervals (1,000 iterations) did not include 1.00; k = # of studies, n = # of effect sizes.

Verschuyl et al. (2011) reported a positive cumulative invertebrate response to thinning that was dominated by guild abundance responses. Invertebrate diversity measures were few (*n* = 4) and showed no significant response to thinning (Table 2.2).

Across all taxonomic groups, responses to light-intensity thinning (0 to 33% of the basal area or trees removed) were the most positive (Table 2.3). Species abundance, guild abun-dance, and diversity measures show positive responses to light-intensity forest thinning. Only guild abundance measures indicated a positive response to moderate-intensity (34% to 66% removal) forest thinning, and all other responses were neutral. All responses (i.e., diversity as well as guild and species abun-dance) were neutral for studies reporting effects from heavy-intensity thinning (> 66% removal) (Table 2.3).

Harvesting Woody Biomass: What Have We Learned?

Despite well-documented relationships between numerous wildlife species and DWM, the removal of snags, logs, stumps, or logging residuals does not result in consistent negative effects on biodiversity. Instead, our analysis indicated the most consistent negative effects from both DWM and snag removal occurred for bird diversity and abundance. Why would a mobile taxon with a broad range of habitat requirements be negatively affected by woody debris removal when other groups, such as amphibians that are physiologically linked to microclimatic refugia provided by logs and stumps, be neutrally affected?

The pool of publications incorporated into our meta-analysis included studies of both cavity-nesting and non-cavity-nesting birds and removals of both DWM and snags. Observations of breeding and wintering birds indicate that snags and DWM serve several roles for the avian community. For instance, snags serve as a nesting substrate for cavity nesters, and downed woody debris in managed forest provides cover and perch sites, and may contribute to collection of wind-blown seeds for granivorous winter birds (Grodsky et al. 2016a, 2016b); hence, harvesting biomass may affect bird species through multiple mechanisms. However, across operationally sized clearcut harvests in the southeastern United States, varying levels and arrangement of biomass retention had few effects on either breeding or wintering bird richness or abundance (see Deeper Dive). Although birds used DWM for perching, succession during forest regeneration was a greater driver of avian community dynamics and abundance than was biomass harvesting treatments (Grodsky et al. 2016a, 2016b). Vegetation regrowth following clearcut harvesting influences avian richness and community dynamics in intensively managed forests in the southeastern United States (Lane et al. 2011) and other regions (Jones et al. 2012, Kroll et al. 2016). Studies across several forested regions provide additional evidence that habitat structure of both near-ground (e.g., DWM) and midstory structure drive change in the avian community.

Similar to prior publications, we did not detect responses of mammals, amphibians, and reptiles to manipulations of DWM in forested stands, suggesting that remaining levels of woody debris in managed forests were adequate, at least for the locations, species, and time periods we examined. The pool of studies in the meta-analysis included intense experimental removals of woody material (> 80%) that occurred in simulated (Beauvais 2010) or actual (Fritts et al. 2014) biomass harvests. We note, however, that 90% of the new effect sizes for DWM were from studies that occurred in the southeastern United States, limiting the inferential scope of the analyses. These analysis included the 2 experimental studies of DWM removal in eastern North Carolina, which accounted for 62% of effect sizes for mammals and 63% of effect sizes for herpetofauna. The southeastern United States, particularly upland forests in the Atlantic and Gulf Coastal Plain, historically had relatively low levels of DWM due to frequent wildfires and warm, moist conditions that contribute to rapid decay rates of deadwood. Thus, wildlife in this region may be less sensitive to DWM reductions than species that evolved with greater volumes of woody debris that persist through time. In other geographic regions, such as the Pacific Northwest, exceptionally large trees create logs of greater sizes than trees currently recruited in many intensively managed forest stands. Snag and DWM availability and rates of decay in managed forests under a range of silvicultural regimes across geographic regions are not well studied and warrant future review. Further, studies of both herpetofauna and mammals have been dominated by habitat generalists, such as southern toads (*Anaxyrus terrestris*; 55% of captures in Fritts et al. 2016) or deer mice (*Peromyscus* spp.; 71% of captures in North Carolina in Fritts et al. 2017). Habitat specialists and rare species, particularly those with life-history characteristics tied to woody debris, may display negative effects from DWM removal that are concealed in meta-analyses (Homyack and Kroll 2014).

Finally, the complexity of invertebrate food webs may contribute to the observed contrasting effects of DWM manipulation on invertebrate biomass (negative) as compared with invertebrate abundance (positive). Biomass of invertebrates was negatively associated with DWM removals, but only included the data published in Riffell et al. (2011). In contrast, we included 3 additional publications that examined effects of removing woody debris from midwestern hardwood forests ($n=2$) or mixed pine forest in Quebec ($n=1$) on diversity and abundance of invertebrates (Table 2.1). However, the addition of 66 effect sizes from other forest types and geographic areas did not change the observed positive effect size of DWM removal on invertebrate abundance. Increases in invertebrate abundance from biomass harvesting may be caused by a simplified invertebrate predator community that releases lower trophic levels from effects of predation (Work et al. 2014). Contrary to our overall results, we note that removal of DWM did negatively affect ground beetles and crickets when compared with forest stands without a biomass harvest (Grodsky et al. 2018a). Future research should focus on disentangling the myriad sources of variability in invertebrate food web dynamics from effects of biomass harvesting (e.g., bottom-up effects of fungus on DWM influencing abundance of invertebrate fungivores) and address interactions between invertebrate prey and vertebrate predators (Grodsky et al. 2016b, 2018a). Additionally, evaluating how invasive and noxious invertebrates such as red imported fire ants (*Solenopsis invicta*) influence native species and community composition where DWM is removed may contribute to our understanding of ecosystem-level effects of woody biomass harvesting (Todd et al. 2008, Grodsky et al. 2018b).

Forest Thinning for Biofuels: What Have We Learned?

Traditionally, forest thinning has been used to achieve a variety of management goals, including accelerating diameter growth of crop trees, accelerating old-growth characteristics (Dodson et al. 2012), fuels management and restoration (Fontaine and Kennedy 2012), and improving aesthetics of timber harvesting (Bradley and Kearney 2007). Whole-tree harvesting through forest thinning for bioenergy production is expected to be most profitable with frequent return intervals or when additional trees are harvested at higher prices for solid wood or pulp (Perlack et al. 2011). However, forest thinning for biomass production, without alternate management objectives, is uncommon (Verschuyl et al. 2011). In fact, none of the thinning manipulations undertaken for the studies we reviewed were designed specifically for biomass production. Fuels treatments and forest restoration efforts have been among the more common applications of forest thinning resulting in biomass production (Dorning et al. 2015). Three meta-analyses recently summarized thinning, burning, and fire-surrogate treatment effects on wildlife (Kalies et al. 2010, Fontaine and Kennedy 2012, Willms et al. 2017) and described generally positive effects of restoration and fuels treatment thinning on species abundance across taxonomic groups.

Our results show forest thinning, across all thinning intensities and forest types, generally increases diversity and abundance of a variety of taxa. Recent literature summarizing wildlife diversity responses to forest thinning lends additional support to our findings (Kalies et al. 2010, Kalies and Covington 2012, Neill and Puettmann 2013). The expected mechanism behind these positive responses is increased cover or vigor of understory plant species, which produce important cover and food resources for wildlife, including pollinators (Neill and Puettmann 2013). Remaining tree crowns also respond to increased light and resources following a thinning harvest, meaning that responses of understory vegetation can be ephemeral (Davis et al. 2007). It is important to note that management actions, such as thinning, may also result in a temporary release of non-native plant species (Willms et al. 2017). Disturbance can increase species diversity at stand scales by reducing competitive dominance of overstory species and increasing number of canopy gaps (Lindenmayer et al. 2006). Species adapted to frequently disturbed forests are likely to respond differently to forest thinning than those accustomed to less frequent disturbance, which is more typical of productive forest systems (McWethy et al. 2010, Steen et al. 2010).

Restoration and fuels treatment thinning typically are lighter techniques and exhibited the most positive responses. Although the magnitude of positive responses to thinning across species diversity and abundance measures was generally modest, individual species or taxa, including sensitive or listed species, may not be represented well by these summaries (Steen et al. 2010). When generalist species replace specialists following disturbance, changes are often not captured by richness or diversity metrics but can result in reduced functional diversity (Clavel et al. 2011). Further, temporal considerations are necessary for a holistic understanding of forest thinning. For example, some species, including late-seral specialists that may be negatively affected by management in the short term, will respond positively to thinning treatments in the long term, if key structural features are retained or restored (Kalies et al. 2012).

Several recent publications reporting responses to thinning did not include data necessary for incorporation in the meta-analysis or reported only occupancy modelling or ordination results (e.g., Steen et al. 2010, Kalies et al. 2012, Kelt et al. 2013, Sutton et al. 2013, Stephens et al. 2014), but their findings generally aligned with those reported in our meta-analysis and in Verschuyl et al. (2011). In pine forests of Alabama, reptiles either responded positively or exhibited no cumulative response to thinning, and amphibians had a neutral response to thinning with or without other midstory removal treatments (Steen et al. 2010, Sutton et al. 2013). However, herpetofauna associated with frequent disturbances of the longleaf pine (*Pinus palustris*) ecosystem increased in abundance and diversity after

thinning, burning, and herbicide treatments, illustrating critical needs for examining wildlife responses in an ecological context (Steen et al. 2010).

Small mammal abundance and community composition were unaffected by restoration thinning efforts focused on fuels treatments in the northern Sierra Nevadas (Kelt et al. 2013, Stephens et al. 2014). Bird community responses were similarly unaffected by thinning treatments (Stephens et al. 2014), but territory densities of California spotted owls (*Strix occidentalis occidentalis*) and a few other closed-canopy-associated avian species declined after treatment (Burnett et al. 2013, Stephens et al. 2014). In ponderosa pine (*Pinus ponderosa*) forests of Arizona, small mammals generally responded neutrally or positively to thinning treatments (6 of 8 species), but 2 closed-canopy-associated species had negative responses to thinning, unless large trees, snags, and woody debris were retained (Kalies et al. 2012).

No additional responses, beyond those included in Verschuyl et al. (2011), were added in this effort for reptiles and invertebrates, leaving the response of these taxonomic groups and amphibians to forest thinning still underrepresented in this analysis. Many of the responses included in this meta-analysis were a single abundance measure. Studies focused on entire taxonomic group response to a management action often report abundance measures for more common species and leave out species for which data are sparse. Further, abundance measures, even for common species, may not be a good indicator of habitat quality (Van Horne 1983).

What's Next? Managing for Wildlife and Biomass among Research Gaps

Clearly, much information on wildlife responses to harvesting forest-based biomass has developed through recent years, but we still lack mechanistic understanding of local, landscape, and species-specific factors that contribute to variability in responses among wildlife species, forest types, harvesting systems, and ecoregions. Moreover, some taxonomic groups, such as reptiles, and geographic regions, such as the US Upper Midwest, are poorly represented in the literature. Experimental research that investigates testable hypotheses regarding effects of operationally sized and operationally implemented biomass harvesting treatments will further advance both science-based and empirically based management recommendations for wildlife biologists to implement. Future research on effects of forest-based bioenergy on wildlife should determine whether thresholds exist for both abundance measures and population metrics related to fitness (e.g., reproduction, survival) and whether harvest of woody material above certain volumes (i.e., leaving too little DWM) leads to deleterious ecological effects. For some species and research questions, examining physiological (e.g., energy use or allocation, stress) or behavioral (e.g., space use, foraging patterns, interspecific interactions) responses of individuals to biomass harvest treatments may accelerate understanding of mechanisms that underlie observed patterns at population scales (Homyack 2010, Berger-Tal et al. 2011). We recognize that the analyses we conducted did not distinguish between generalist or specialist life histories or account for species of conservation concern. Therefore, individual species or taxa, including rare, sensitive, and listed species, may not be well represented by these summaries and may warrant consideration when implementing future research.

Without a full understanding of ecological responses to biomass harvesting, land and wildlife managers must still prescribe silvicultural and harvest treatments that meet biological, social, and economic objectives. Land managers must balance these objectives and account for the fine-scale logistics of current economic and market conditions, type of bioenergy being produced (e.g., hardwood chips), transportation logistics, and desired future forest-stand conditions. We suggest that land managers additionally consider biomass harvesting in the context of whether it complements, changes, degrades, or improves habitat conditions for wildlife species of interest. With forethought and ingenuity, there are numerous opportunities where biomass harvests may provide positive effects on wildlife habitat conditions and an economic return. For example, mid-story hardwood removals for bioenergy in southern pine plantations may lead to open-canopy conditions that support endemic species of conservation concern, including gopher tortoises (*Gopherus polyphemus*) (Greene et al. 2016). And, thinning western forests for bioenergy may increase understory diversity (Kalies et al. 2012, Neill and Puettmann 2013) and reduce wildfire risk (Agee and Skinner 2005).

Although this chapter emphasizes biological responses to forest biomass removal, it is important to acknowledge that biomass production from forests, including unmerchantable thinning sources, may provide added economic incentives for landowners. The low economic margin associated with harvest of forest biomass has important implications when considering potential intensity or geographic extent of woody biomass harvesting. Typically, the material being harvested from a site is low enough in value that small increases in transportation or harvesting costs can tip the balance in favor of leaving woody biomass in the stand (US Department of Energy 2011), suggesting that sites with other silvicultural or management actions are the most likely to consider additional biomass removal.

Economic gains that landowners may receive from forest biomass harvesting may help avoid land conversion or development pressures that reduce available habitat types for wildlife (Dorning et al. 2015). Removal of forest biomass during forest restoration can offset treatment costs and provide funding to extend restoration activities across additional forest acreage. Finally, nearly all bioenergy research has occurred at the scale of individual harvest units or at small spatial resolution. However, landscape composition and structure influences biodiversity in managed forests (Kroll et al. 2012, deStefano et al. 2016), and managers are tasked with making management decisions at landscape and ownership scales. Thus, scaling up understanding of bioenergy impacts on wildlife populations and their habitat remains an unfilled knowledge gap and warrants research effort (Riffell et al. 2011).

LITERATURE CITED

Agee, J. K., and C. N. Skinner. 2005. Basic principles of forest fuels reduction treatments. Forest Ecology and Management 211:83–96.

Bagne, K. E., and D. M. Finch. 2010. Response of small mammal populations to fuel treatment and precipitation in a ponderosa pine forest, New Mexico. Restoration Ecology 18:409–417.

Beauvais, C. 2010. Coarse woody debris in a loblolly pine plantation managed for biofuels production. Master's thesis, Duke University, Durham, NC.

Berger-Tal, O., T. Polak, A. Oron, Y. Lubin, B. P. Kotler, and D. Saltz. 2011. Integrating animal behavior and conservation biology: A conceptual framework. Behavioral Ecology 224:236–239.

Bradley, G. A., and A. R. Kearney. 2007. Public and professional responses to the visual effects of timber harvesting: Different ways of seeing. Western Journal of Applied Forestry 22:42–54.

Brooks, R. T., and T. D. Kyker-Snowman. 2008. Forest floor temperature and relative humidity following timber harvesting in southern New England, U.S.A. Forest Ecology and Management 254:65–73.

Cahall, R. E., J. P. Hayes, and M. G. Betts. 2013. Will they come? Long-term response by forest birds to experimental thinning supports the "Field of Dreams" hypothesis. Forest Ecology and Management 304:137–149.

Castro, A., and D. H. Wise. 2010. Influence of fallen coarse woody debris on the diversity and community structure of forest-floor spiders (Arachnida: Araneae). Forest Ecology and Management 260:2088–2101.

Chen, J., S. C. Saunders, T. R. Crow, R. J. Naiman, K. D. Brosofske, G. D. Mroz, B. L. Brookshire, and J. F. Franklin. 1999. Microclimate in forest ecosystem and landscape ecology. Bioscience 49:288–297.

Dale, V. H., K. L. Kline, E. S. Parish, A. L. Cowie, R. Emory, R. W. Malmsheimer, R. Slade, et al. 2017. Status and prospects for renewable energy using wood pellets from the southeastern United States. GCB Bioenergy 8: 1296–1305.

Davis, L. R., K. J. Puettmann, and G. F. Tucker. 2007. Overstory response to alternative thinning treatments in young douglas-fir forest. Northwest Science 81:1–14.

Davis, J. C., S. B. Castleberry, and J. C. Kilgo. 2010. Influence of coarse woody debris on herpetofaunal communities in upland pine stands of the southeastern Coastal Plain. Forest Ecology and Management 259:1111–1117.

Demarais, S., J. P. Verschuyl, G. J. Roloff, D. A. Miller, and T. B. Wigley. 2016. Tamm review: Terrestrial vertebrate biodiversity and intensive forest management in the U.S. Forest Ecology and Management 385:308–330.

Dodson, E. K., A. Ares, and K. J. Puettmann. 2012. Early responses to thinning treatments designed to accelerate late successional forest structure in young coniferous stands of western Oregon, USA. Canadian Journal of Forest Research 42:345–355.

Donner, D. M., T. B. Wigley, and D. A. Miller. 2017. Forest biodiversity and woody biomass harvesting. In 2016 Billion-ton report: Advancing domestic resources for a thriving bioeconomy, Volume 2, Environmental sustainability effects of select scenarios from Volume 1, edited by R. A. Efroymson, M. H. Langholtz, K. E. Johnson, and B. J. Stokes, 397–447. Oak Ridge, TN: Oak Ridge National Laboratory.

Dorning, M. A., J. W. Smith, D. A. Shoemaker, and R. K. Meetenmeyer. 2015. Changing decisions in a changing landscape: How might forest owners in an urbanizing region respond to emerging bioenergy markets? Land Use Policy 49:1–10.

Fontaine, J. B., and P. L. Kennedy. 2012. Meta-analysis of avian and small-mammal response to fire severity and fire surrogate treatments in U.S. fire-prone forests. Ecological Applications 22:1547–1561.

Forest Guild Southeast Working Group. 2012. Forest biomass retention and harvesting guidelines for the Southeast. Santa Fe, NM.

Fox, T. R., E. J. Jokela, and H. L. Allen. 2007. The development of pine plantation silviculture in the southern United States. Journal of Forestry 105:337–347.

Fraver, S., R. Wagner, and M. Day. 2002. Dynamics of coarse woody debris following gap harvesting in the Acadian forest of central Maine, U.S.A. Canadian Journal of Forest Research 32:1–12.

Fritts, S. R., C. E. Moorman, D. W. Hazel, and B. D. Jackson. 2014. Biomass Harvesting Guidelines affect downed woody debris retention. Biomass and Bioenergy 70:382–391.

Fritts, S. R., C. E. Moorman, S. M. Grodsky, D. W. Hazel, J. A. Homyack, C. B. Farrell, and S. B. Castleberry. 2015a. Shrew response to variable woody debris retention: Implications for sustainable forest bioenergy. Forest Ecology and Management 336:35–43.

Fritts, S. R., S. M. Grodsky, D. W. Hazel, J. A. Homyack, S. B. Castleberry, and C. E. Moorman. 2015b. Quantifying multi-scale habitat use of woody biomass by southern toads. Forest Ecology and Management 346:81–88.

Fritts, S. R., C. E. Moorman, D. W. Hazel, J. A. Homyack, C. B. Farrell, S. B. Castleberry, and S. M. Grodsky. 2016. Effects of biomass harvesting guidelines on herpetofauna following harvests of logging residues. Ecological Applications 3:926–939.

Fritts, S. R., C. E. Moorman, S. M. Grodsky, D. W. Hazel, J. A. Homyack, C. B. Farrell, S. B. Castleberry, E. Evans, and D. Greene. 2017. Rodent response to harvesting woody biomass for bioenergy production. Journal of Wildlife Management 81:1170–1178.

Gaines, W. L., M. Haggard, J. Begley, J. F. Lehmkuhl, and A. Lyons. 2010. Short-term effects of thinning and burning restoration treatments on avian community composition, density, and nest survival in the western Cascades dry forests, Washington. Forest Science 56:88–99.

Greene, R. E., R. B. Iglay, K. O. Evans, D. A. Miller, T. B. Wigley, and S. K. Riffell. 2016. A meta-analysis of biodiversity responses to management of southeastern pine forests—opportunities for open pine conservation. Forest Ecology and Management 360:30–39.

Grodsky, S. M., C. E. Moorman, S. R. Fritts, S. B. Castleberry, and T. B. Wigley. 2016a. Breeding, early-successional bird response to forest harvests for bioenergy. PLOS One 11:e0165070.

Grodsky, S. M., C. E. Moorman, S. R. Fritts, D. W. Hazel, J. A. Homyack, S. B. Castleberry, and T. B. Wigley. 2016b. Winter bird use of harvest residues in clearcuts and the implications of forest bioenergy harvest in the southeastern United States. Forest Ecology and Management 379:91–101.

Grodsky, S. M., C. E. Moorman, S. R. Fritts, J. W. Campbell, C. E. Sorenson, M. A. Bertone, S. B. Castleberry, and T. Wigley. 2018a. Invertebrate community response to coarse woody debris removal for bioenergy production from intensively managed forests. Ecological Applications 28:135–148.

Grodsky, S. M., J. W. Campbell, S. R. Fritts, T. B. Wigley, and C. E. Moorman. 2018b. Variable responses of non-native and native ants to coarse woody debris removal following forest bioenergy harvests. Forest Ecology and Management 427:414–422.

Hedges, L. V., J. Gurevitch, and P. S. Curtis. 1999. The meta-analysis of response ratios in experimental ecology. Ecology 80:1150–1156.

Hocking, D. J., G. M. Connette, C. A. Conner, B. R. Scheffers, S. E. Pittman, W. E. Peterman, and R. D. Semlitsch. 2013. Effects of experimental forest management on a terrestrial, woodland salamander in Missouri. Forest Ecology and Management 287:32–39.

Homyack, J. A. 2010. Evaluating habitat quality of vertebrates using conservation physiology tools. Wildlife Research 37:332–342.

Homyack, J., and A. Kroll. 2014. Slow lives in the fast landscape: Conservation and management of plethodontid salamanders in production forests of the United States. Forests 5:2750–2772.

Homyack, J. A., D. J. Harrison, and W. B. Krohn. 2004. Structural differences between precommercial thinned and unthinned conifer stands. Forest Ecology and Management 191:131–143.

Homyack, J. A., C. A. Haas, and W. A. Hopkins. 2011a. Effects of experimental forest harvesting on energetics of terrestrial salamanders. Journal of Wildlife Management 7:1267–1278.

Homyack, J. A., B. J. Paxton, M. D. Wilson, B. D. Watts, and D. A. Miller. 2011b. Snags and cavity-nesting birds within intensively managed pine stands in eastern North Carolina, USA. Southern Journal of Applied Forestry 35:148–154.

Homyack, J. A., K. E. Lucia-Simmons, D. A. Miller, and M. Kalcounis-Rueppell. 2014. Rodent population and community responses to forest-based biofuel production. Journal of Wildlife Management 78:1425–1435.

Jones, J. E., A. J. Kroll, J. Giovanini, S. D. Duke, T. M. Ellis, and M. G. Betts. 2012. Avian species richness in relation to intensive forest management practices in early seral tree plantations. PLOS One 7:e43290.

Kalies, E. L., and W. W. Covington. 2012. Small mammal community maintains stability through compensatory dynamics after restoration of a ponderosa pine forest. Ecosphere 3:art78.

Kalies, E. L., C. L. Chambers, and W. W. Covington. 2010. Wildlife responses to thinning and burning treatments in southwestern conifer forests: A meta-analysis. Forest Ecology and Management 259:333–342.

Kalies, E. L., B. G. Dickson, C. L. Chambers, and W. W. Covington. 2012. Community occupancy responses of small mammals to restoration treatments in ponderosa pine forests, northern Arizona, USA. Ecological Applications 22:204–217.

Kelt, D. A., D. H. Van Vuren, M. L. Johnson, J. A. Wilson, R. J. Innes, B. R. Jesmer, K. P. Ingram, et al. 2013. Small mammals exhibit limited spatiotemporal structure in Sierra Nevada forests. Journal of Mammalogy 94:1197–1213.

Kendrick, S. W., P. A. Porneluzi, F. R. T. Iii, D. L. Morris, J. M. Haslerig, and J. Faaborg. 2015. Stand-level bird response to experimental forest management in the Missouri Ozarks. Journal of Wildlife Management 79:50–59.

Kluber, M. R., D. H. Olson, and K. J. Puettmann. 2009. Downed wood microclimates and their potential impact on plethodontid salamander habitat in the Oregon Coast range. Northwest Science 83:25–34.

Kroll, A. J., S. D. Duke, M. E. Hane, J. R. Johnson, M. Rochelle, M. G. Betts, and E. B. Arnett. 2012. Landscape composition influences avian colonization of experimentally created snags. Biological Conservation 152:145–151.

Kroll, A. J., J. Verschuyl, J. Giovanini, M. G. Betts, and C. Banks-Leite. 2016. Assembly dynamics of a forest bird community depend on disturbance intensity and foraging guild. Journal of Applied Ecology 54:784–793.

Lane, V. R., K. V. Miller, S. B. Castleberry, R. J. Cooper, D. A. Miller, T. B. Wigley, G. M. Marsh, and R. L. Mihalco. 2011. Bird community responses to a gradient of site preparation intensities in pine plantations in the Coastal Plain of North Carolina. Forest Ecology and Management 262:1668–1678.

Loman, Z. G., S. K. Riffell, D. A. Miller, J. A. Martin, and F. J. Vilella. 2013. Site preparation for switchgrass intercropping in loblolly pine plantations reduces retained trees and snags, but maintains downed woody debris. Forestry 86:353–360.

Manning, T., J. C. Hagar, and B. C. McComb. 2012. Thinning of young Douglas-fir forests decreases density of northern flying squirrels in the Oregon Cascades. Forest Ecology and Management 264:115–124.

McWethy, D. B., A. J. Hansen, and J. P. Verschuyl. 2010. Bird response to disturbance varies with forest productivity in the northwestern United States. Landscape Ecology 25:533–549.

Moorman, C. E., K. R. Russell, G. R. Sabin, and D. C. Guynn. 1999. Snag dynamics and cavity occurrence in the South Carolina piedmont. Forest Ecology and Management 118:37–48.

NARA. First commercial flight using biojet fuel made from wood. https://nararenewables.org/first-flight-using-biojet-made-from-wood/. Accessed November 2017.

Neill, A. R., and K. J. Puettmann. 2013. Managing for adaptive capacity: Thinning improves food availability for wildlife and insect pollinators under climate change conditions. Canadian Journal of Forest Research 43:428–440.

Otto, C. R. V., A. J. Kroll, and H. C. McKenny. 2013. Amphibian response to downed wood retention in managed forests: A prospectus for future biomass harvest in North America. Forest Ecology and Management 304:275–285.

Parrish, C., and K. Summerville. 2015. Effects of logging and coarse woody debris harvest on lepidopteran communities in the eastern deciduous forest of North America. Agricultural and Forest Entomology 17:317–324.

Phalan, B., M. Onial, A. Balmford, and R. E. Green. 2011. Reconciling food production and biodiversity conservation: Land sharing and land sparing compared. Science 333:1289–1291.

Riffell, S., J. Verschuyl, D. Miller, and T. B. Wigley. 2011. Biofuel harvests, coarse woody debris, and biodiversity: A meta-analysis. Forest Ecology and Management 261:878–887.

Rosenberg, R. W., D. C. Adams, and J. Gurevitch. 2000. MetaWin: Statistical software for meta-analysis. Sinauer Associates Version 2.0.Sunderland, MA.

Rothermel, B. B., and T. M. Luhring. 2005. Burrow availability and desiccation risk of mole salamanders (Ambystoma talpoideum) in harvested versus unharvested forest stands. Journal of Herpetology 39:619–626.

Rupp, S. P., L. Biels, A. Glaser, C. Kowaleski, T. McCoy, T. Rentz, S. A. M. Riffell, J. Sibbing, J. Verschuyl, and T. B. Wigley. 2012. Effects of bioenergy production on wildlife and wildlife habitat. Wildlife Society Technical Review 12-03. Bethesda, MD: Wildlife Society.

Scott, V. E., K. E. Evans, D. R. Patton, and C. P. Stone. 1977. Cavity-nesting birds of North American forests. Handbook 511. US Department of Agriculture, Forest Service.

Spies, T. A., J. F. Franklin, and T. B. Thomas. 1988. Coarse woody debris in douglas-fir forests of western Oregon and Washington. Ecology 69:1689–1702.

Steen, D. A., A. E. Rall McGee, S. M. Hermann, J. A. Stiles, S. H. Stiles, and C. Guyer. 2010. Effects of forest management on amphibians and reptiles: Generalist species obscure trends among native forest associates. Open Environmental Sciences 4:24–30.

Stephens, S. L., S. W. Bigelow, R. D. Burnett, B. M. Collins, C. V. Gallagher, J. Keane, D. A. Kelt, et al. 2014. California spotted owl, songbird, and small mammal responses to landscape fuel treatments. Bioscience 64:893–906.

Sullivan, T. P., and D. S. Sullivan. 2012. Woody debris, voles, and trees: Influence of habitat structures (piles and windrows) on long-tailed vole populations and feeding damage. Forest Ecology and Management 263:189–198.

Sutton, W. B., Y. Wang, and C. J. Schweitzer. 2013. Amphibian and reptile responses to thinning and prescribed burning in mixed pine-hardwood forests of northwestern Alabama, USA. Forest Ecology and Management 295:213–227.

Talbert, C., and D. Marshall. 2005. Plantation productivity in the douglas-fir region under intensive silvicultural practices: Results from research and operations. Journal of Forestry 103:65–70.

Todd, B. D., B. B. Rothermel, R. N. Reed, T. M. Luhring, K. Schlatter, L. Trenkamp, and J. W. Gibbons. 2008. Habitat alteration increases invasive fire ant abundance to the detriment of amphibians and reptiles. Biological Invasions 10:539–546.

US Energy Information Administration. 2016. Southern states lead growth in biomass electricity generation. Today in Energy. https://www.eia.gov/todayinenergy /detail.php?id=26392.

Van Hook, R. I., D. W. Johnson, D. C. West, and L. K. Mann. 1982. Environmental effects of harvesting forests for energy. Forest Ecology and Management 4:79–84.

Van Horne, B. 1983. Density as a misleading indicator of habitat quality. Journal of Wildlife Management 47:893–901.

Wear, D. N., and J. G. Greis, eds. 2013. The southern forests futures project. Asheville, NC: USDA-Forest Service, Southern Research Station.

Willms, J., A. Bartuszevige, D. W. Schwilk, and P. L. Kennedy. 2017. The effects of thinning and burning on understory vegetation in North America: A meta-analysis. Forest Ecology and Management 392:184–194.

Work, T. T., S. Brais, and B. D. Harvey. 2014. Reductions in downed deadwood from biomass harvesting alter composition of spiders and ground beetle assemblages in jack-pine forests of Western Quebec. Forest Ecology and Management 321:19–28.

Impacts on Wildlife of Annual Crops for Biofuel Production

Introduction

In the United States, human use of natural resources for food, fiber, and fuel has a disproportionate effect on ecosystems and wildlife. Whereas the United States supports only 5% of the Earth's human population, it is responsible for over 30% of global natural resource consumption (Primack 2002). Current US energy policy has generated significant societal concerns due to overexploitation of natural resources and the threats it poses to human welfare, ecosystem function, global climate change, and biodiversity. These concerns, coupled with rising petroleum costs and the desire for energy independence, have generated interest in pursuing alternative and more sustainable fuel sources. Use of agricultural feedstocks as alternative fuel sources first received national attention during the 1970s because of rising petroleum costs and again during the 1990s and 2000s as a means to curb greenhouse gas (GHG) emissions and secure energy independence (Tyner 2008). Since the passage of the US Energy Independence and Security Act of 2007 (US Congress 2007), liquid biofuel production has grown considerably, and it is now a staple fuel source for the US transportation sector. In 2006, the United States produced over 18.5 billion liters (L) (4.9 billion gallons) of ethanol and 946 mil-lion L (250 million gallons) of biodiesel, derived largely from corn kernels and soybean oil, respectively (USDA-ERS 2017). In 2016, production levels of ethanol and biodiesel increased to 57.9 billion L (15.3 billion gallons) and 5.7 billion L (1.5 billion gallons), respectively. Today, liquid biofuels accounts for 7.1% of total US transportation fuel consumption, with corn being the most common feedstock (USDA-ERS 2017).

Rising demand for biofuel crop production has engendered, and will continue to engender, significant changes in agriculture and wildlife habitat. Among these changes, conversion of natural land covers such as grassland and forest to cropland presents the greatest threats to wildlife and ecosystem function, both nationally and globally (Fearnside 2001). The US Renewable Fuel Standard (RFS) program, authorized by Congress under the Energy Policy Act of 2005 (42 U.S.C. ch. 149 § 15801 et seq) and greatly expanded under the Energy Independence and Security Act of 2007, calls for the production of 136 billion L (36 billion gallons) of renewable biofuels annually by 2022, of which 57 billion L (15 billion gallons) will be conventional biofuels produced from corn starch. Fargione et al. (2008) estimated that meeting this volume goal would require about 13.1 million ha (32.4 million ac) of land to be planted in corn. Growing demand for corn- and soybean-based biofuels is being met through improved crop yields, improved crop varieties, agricultural intensification, and conversion of natural land and pasture to cropland (Wright and Wimberly 2013, Lin and Henry 2016). Growth in biofuel feedstocks has sparked significant national debate over the risk of allocating more land to growing biofuel feedstocks, which leaves less land to support global food supply (Tenenbaum 2008). Rapid growth in corn and soybean production, spurred by the Energy Independence and Security Act of 2007, represents a unique situation for US agriculture and renewable biofuels policy.

Production, Energetics, and Use

The United States produces 36% and 34% of the world's corn and soybeans, respectively—the highest production among all countries (USDA-OCE 2017). Much of that production occurs in the midwestern states, including Iowa, Nebraska, Illinois, Indiana, Ohio, western Kentucky, southern Wisconsin, and southern Minnesota. Though Iowa and Illinois are typically the top producers of corn and soybeans, corn production has pushed farther west and north into the Great Plains region in western Nebraska, South Dakota, and North Dakota, in part due to increased demand for biofuel feedstocks (Figure 3.1). Even states with historically high acreages planted in corn (e.g., Nebraska) have seen significant gains in corn acreage since the Energy Independence and Security Act (https://quickstats.nass.usda.gov/; Figure 3.1). In the 2015–2016 growing season, the United States produced 13.6 billion bushels of corn, of which 5.2 billion bushels were used for ethanol production (USDA-ERS 2017). In recent years, the amount of corn used for ethanol production is nearly equal to, or sometimes greater than, that of corn used for livestock feed—typically the top use of corn (5.1 billion bushels in 2015).

Corn-based ethanol is produced from corn-seed feedstocks through starch fermentation and distilla-

tion, leaving distillers grain as a cereal byproduct. In the United States, ethanol is often blended with petroleum gasoline to make E10, E15, or E85 fuel for automobiles. Soybean-based biodiesel is produced from soybean oil through transesterification, in which fatty acids from soybeans are mixed with alcohol to produce biodiesel and glycerol. Biodiesel is typically blended at 5% to 20% with petroleum diesel fuel for use in diesel engines. Biodiesel derived from soybeans uses only the oil, leaving the meal protein for other uses, such as livestock feed. One advantage of soybean biodiesel over corn ethanol is that it produces 93% more usable energy than the fossil fuel energy needed for its production, whereas corn ethanol results in just a 25% net energy gain (Hill et al. 2006).

Scientists and the popular press have debated the net benefits of using row crops as biofuel feedstocks to reduce GHGs (http://science.sciencemag.org /content/319/5867/1238/tab-e-letters). Biofuel crop proponents suggest a primary benefit of biofuels derived from row crops is reduction in GHG emissions relative to conventional petroleum-based fuels (Wang et al. 1999, Farrell et al. 2006). However, earlier estimates of GHGs from row crop–based biofuels did not take into account potential land-use conversion (Searchinger et al. 2008). If previously uncultivated lands (e.g., native prairie, restored grasslands) were tilled to plant a biofuel row crop, these land conversions would release significant amounts of sequestered carbon into the atmosphere, thereby negating any GHG reductions from ethanol or biodiesel (Searchinger et al. 2008, Piñeiro et al. 2009, Gelfand et al. 2011, Johnson et al. 2014). Research has estimated the resulting "carbon debt" from land-use change may take several decades before reverting to carbon-neutral or positive. For example, Fargione et al. (2008) estimated that corn produced on converted grasslands in the central United States would result in a carbon debt that would take roughly 93 years to repay. Searchinger et al. (2008) calculated that corn-based ethanol grown on converted lands would nearly double

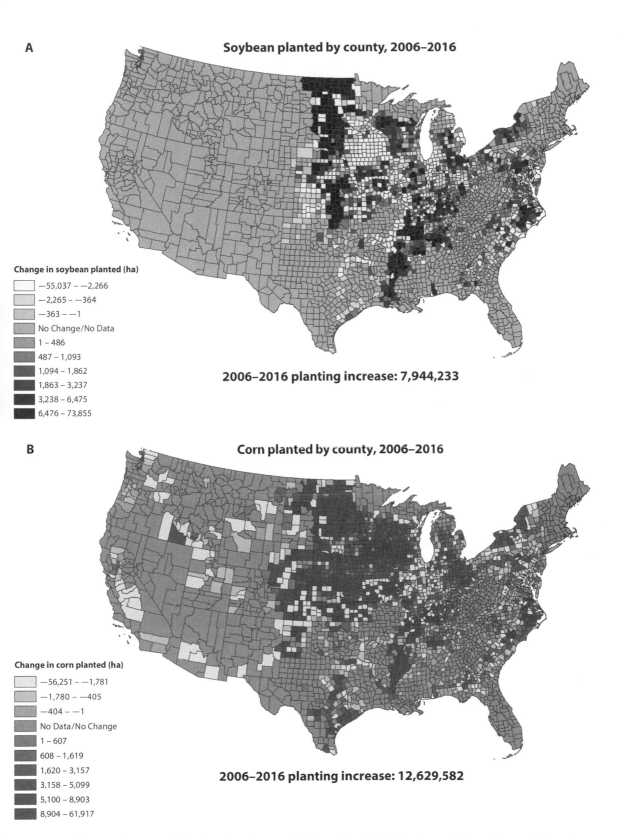

Figure 3.1 Change in hectares of (*A*) planted soybeans, and (*B*) corn from 2006 (pre-Energy Independence and Security Act) to 2016 across the contiguous United States. *Clint Otto.*

GHGs over a 30-year period and increase GHGs for 167 years; yet, these estimates were in stark contrast to the net savings published in earlier studies. These studies and ensuing debates fuel a critical discussion about potential effects of biofuels derived from row crops and unintended consequences for GHGs if increased cropping area leads to land-use conversion.

The Expanding Biofuel Frontier

Continued loss of native prairie and other grasslands caused by an expanding biofuel frontier threatens wildlife habitat and impairs ecosystem function. Temperate grassland is one of the most imperiled ecosystems on Earth, where habitat conversion exceeds habitat protection by an 8:1 ratio (Hoekstra et al. 2005). In the United States, 70% of the Great Plains region has been converted from grassland to agriculture, and in some areas, the rate of grassland loss is increasing (Samson et al. 2004, Stephens et al. 2008, Claassen et al. 2011). The rise of commodity crop prices, renewable fuel mandates, and industry subsidies for biofuel production have contributed to conversion of grassland to cropland during the late 2000s (Lark et al. 2015). The most dramatic rates of grassland conversion are occurring in the US Midwest and Great Plains regions. Wright and Wimberly (2013) estimated that nearly 530,000 ha (1.3 million ac) of grassland in North Dakota, South Dakota, Nebraska, Minnesota, and Iowa were converted to corn or soybeans from 2006 to 2011. Much of that conversion took place in close proximity to wetlands, presenting increased risk to waterfowl populations and other wetland-dependent wildlife (Drum et al. 2015). Lin and Henry (2016) identified a 1.6 million ha (3.9 million ac) loss of grassland from land conversion in 9 states in the Midwest from 2007 to 2012. Johnston (2014) showed a 27% increase in corn and soybean acreage from 2010 to 2012 in the Prairie Pothole Region of North and South Dakota, largely due to acreage loss in small grains, pasture, rangeland, and remnant prairies. Stephens et al. (2008) showed that 5.2% of previously untilled prairie along the Co-

teau du Missouri in North and South Dakota was cultivated from 1984 to 2003. In general, these investigations revealed that agricultural expansion has pushed westward into more arid environments and marginal lands prone to soil erosion. Grassland-to-cropland conversion rates from 2008 to 2012 were greatest in proximity to ethanol refineries in the Great Plains (Wright et al. 2017); 1.5 million ha (3.6 million ac) of grassland were converted to cropland within 161 km (100 mi) of ethanol refineries. Commodity crop prices for corn and soybeans have dropped since 2015, but it is unclear whether the rapid rate of land-use conversion will continue in this region under more modest crop prices.

Often, wildlife species are underrepresented in economic-based research that quantifies trade-offs between biofuel production and its projected outcomes for land-use change. This underrepresentation stems from the fact that "wildlife production" is an ecosystem service that is external from biofuel market considerations and incentives, thereby making wildlife more vulnerable to unanticipated consequences of biofuel policy (Fargione et al. 2009). Furthermore, with the exception of grassland birds, there is a paucity of data regarding the direct effect of biofuel row crop production on most wildlife species, thereby limiting understanding of how current and future biofuel scenarios will affect wildlife. Review articles written on the impacts of biofuel crops on "biodiversity" or "wildlife" tend to focus on grassland birds (e.g., Fargione et al. 2009, Fletcher et al. 2011, Landis et al. 2017). For example, Fargione et al. (2009) focused their wildlife-biofuel review on grassland birds, in part because of limitations in the primary literature on potential impacts on other species. This apparent lack of research beyond grassland birds is especially problematic, considering the most significant rates of land-use change brought on by biofuel production are occurring along major migration corridors and in prime nesting areas for waterfowl and shorebirds in the United States Additionally, it is unclear how large-scale changes in cropping patterns, including switching from small grains to corn and soybeans, affect migratory bird energetics during the fall migration along the US Central Flyway. Residual corn grains remaining in the field after harvest can provide important nutritional resources to migratory bird species (Krapu et al. 1995, Abraham et al. 2005). Thus, increased production of corn may be beneficial to some migratory species, such as geese (*Anser* spp.); however, this association needs to be tested for additional migratory species. Furthermore, continued reduction in waste grain associated with improved harvest techniques could reduce potential benefits of increased biofuel crops (Krapu et al. 2004).

Over 60% of all grassland bird species in North America are in decline (Ziolkowski et al. 2010); hence, there is a pressing need for research on their response to biofuels development and agricultural practices associated with growing biofuel feedstocks. Within the United States, landscapes experiencing the highest conversion of grasslands to corn and soybeans also represent the core ranges of many grassland bird species (Figure 3.2). Biofuels development poses multiple direct and indirect threats to grassland birds, including conversion of habitat to biofuel row crops, fragmentation of grassland patches, intensive agricultural practices, and crop genetic varieties that are resistant to pesticides. Among these

threats, conversion of grassland to agriculture poses the greatest risk to grassland birds. In his review of grassland bird use of Conservation Reserve Program areas (CRP; see below), Johnson (2000) summarized several studies demonstrating the benefits of conserving grasslands to breeding birds relative to the croplands they replace, noting that grassland patch size and landscape context have a strong influence on whether grassland birds use particular fields. Thus, future grassland loss and fragmentation of existing grassland patches are expected to further reduce landscape suitability to support grassland birds, other wildlife, and (indirectly) fisheries (Fargione et al. 2009). Biofuel forecast models developed for the Upper Midwest suggest that continued expansion of corn and soybean production on marginal lands will lead to declines in avian richness between 7% and 65% across nearly a quarter of the region and will particularly threaten "at-risk" species (Meehan et al. 2010).

In general, corn and soybean in the Midwest are grown as large monocultures with significant amounts of agrochemicals applied, providing little habitat for grassland birds. Although the relative contribution of habitat loss, industrial agriculture, and agrochemical use to grassland bird declines has been debated in the literature (Murphy and Moore 2003, Mineau and Whiteside 2013, Hill et al. 2014), it is clear that large-scale agriculture, including growing corn and soybeans as biofuel feedstocks, poses multiple threats to grassland birds. These studies did not directly implicate development of biofuel crops to the decline of grassland birds, but rather showed how avian populations may respond to intensive agricultural practices that are often associated with large-scale biofuel crop production. In a meta-analysis of multiple biofuel feedstock impacts on wildlife, Fletcher et al. (2011) showed that biofuels derived from row crops had the largest negative impact on birds, relative to reference habitat types and other types of biofuel production derived from poplar (*Populus* spp.) and pine (*Pinus* spp.). Furthermore, bird species of conservation concern were projected to

Figure 3.2 Grasslands provide critical habitat for grassland birds, such as Baird's sparrow (*Ammodramus bairdii*). *Photo courtesy of Rich Bohn.*

have the strongest negative response to agricultural practices to produce corn-based ethanol. These findings led Fletcher et al. (2011) to conclude that structural and compositional homogenization resulting from large-scale biofuel crop plantings does not meet habitat requirements for many wildlife species, including grassland birds.

Research has demonstrated that commodity crop prices have a substantial influence on landowner decisions on whether to participate in the CRP (Atkinson et al. 2011), which has been the largest agricultural land retirement program in the United States since it was initiated in 1985. The CRP is a voluntary federal program that offers landowners annual payments in exchange for taking agricultural lands out of production and establishing a conservation cover that provides known environmental benefits. Federal

subsidies for biofuel production and high commodity prices for corn and soybeans have contributed to conversion of CRP lands to row crops over the past decade and, in particular, from 2008 to 2012 (Morefield et al. 2016). The most significant losses of CRP areas were concentrated in the Great Plains region of the United States (Figure 3.3). Over 530,000 ha (1.3 million ac) of expiring CRP land throughout the Midwest and Great Plains was put back into production from 2010 to 2013 (Morefield et al. 2016), with over 70% of the converted acreage being planted in corn and soybeans. Of the CRP land converted, 360,000 ha (889,600 ac), 76,000 ha (187,800 ac), and 53,000 ha (131,000 ac) previously were in grasslands, wildlife habitat, and wetland cover practices, respectively. Considerable research has demonstrated the importance of the CRP to grassland-nesting

CRP change by county, 2006–2016

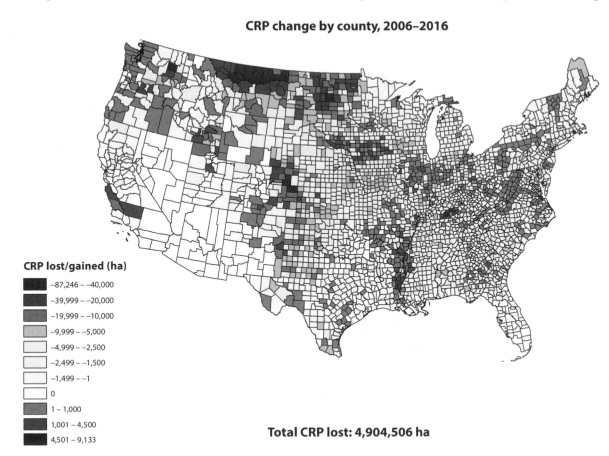

CRP lost/gained (ha)

■	−87,246 − −40,000
■	−39,999 − −20,000
■	−19,999 − −10,000
■	−9,999 − −5,000
■	−4,999 − −2,500
■	−2,499 − −1,500
□	−1,499 − −1
□	0
■	1 − 1,000
■	1,001 − 4,500
■	4,501 − 9,133

Total CRP lost: 4,904,506 ha

Figure 3.3 Change in Conservation Reserve Program area from 2006 to 2016 across the contiguous United States. *Clint Otto.*

Table 3.1.   Estimated mean values of number of breeding pairs (1,000s) of birds in Conservation Reserve Program (CRP) fields in North Dakota, estimated number if CRP lands were cropland, predicted change in number of breeding pairs, statewide number of breeding pairs, and predicted change in number of breeding pairs as percentage of statewide population for 1992–1993. Reproduced with author permission from Johnson and Igl (1995).

Species	Population			Statewide population	Percent change
	CRP	Crop	Change		
Lark bunting	211	21	−190	1,113	−17
Grasshopper sparrow	206	12	−193	945	−20
Red-winged blackbird	187	18	−170	1,421	−12
Savannah sparrow	94	10	−84	445	−19
Western meadowlark	80	16	−64	1,260	−5
Brown-headed cowbird	74	33	−41	1,380	−3
Bobolink	73	31	−42	388	−11
Clay-colored sparrow	54	0	−54	593	−9
Common yellowthroat	27	0	−27	286	−9
Horned lark	20	316	296	3,042	10
Sedge wren	16	0	−16	61	−26
Baird's sparrow	10	2	−8	225	−4
Dickcissel	10	1	−9	52	−17
Ring-necked pheasant	8	1	−8	84	−9
Sharp-tailed grouse	7	1	−6	88	−7

species (Johnson and Schwartz 1993, Johnson and Igl 1995, Ryan et al. 1998). Of the 18 grassland bird species surveyed by Johnson and Igl (1995), 12 species were less common in cropland than on CRP fields. Johnson and Igl (1995) showed that CRP supported >20% of the populations of many grassland species in North Dakota, even though CRP land amounted to just 7% of the total land area. The authors estimated that, if all CRP fields in North Dakota were returned to cropland, numbers of sedge wrens (*Cistothorus platensis*), grasshopper sparrows (*Ammodramus savannarum*), and savannah sparrows (*Passerculus sandwichensis*) would be reduced by 26%, 20%, and 19%, respectively (Table 3.1). Drum et al. (2015) estimated that CRP areas supported 6% to 34% of populations of grassland passerines in the Prairie Pothole Region.

In addition to supporting grassland birds, CRP land provides habitat for numerous wildlife taxa. Reynolds et al. (2001) estimated that CRP areas con-

tributed to the recruitment of 12.4 million additional ducks to the Prairie Pothole Region from 1992 to 1997. Nielson et al. (2008) showed that counts of ring-necked pheasants (*Phasianus colchicus*) were positively correlated with the amount of herbaceous vegetation in CRP fields within 1 km of Breeding Bird Survey routes across 9 states. The CRP also provides valuable habitat and dispersal corridors for amphibians (Mushet et al. 2014). Loss of CRP areas in the Prairie Pothole Region resulted in a decline of amphibian habitat by 22% (1.1 million ha) from 2007 to 2012, with an additional forecasted loss of 26% (Mushet et al. 2014). In addition to the direct loss of habitat for wildlife, conversion of CRP land to biofuel production can also affect the delivery of multiple ecosystem services, such as reduced soil erosion, sequestered carbon, and retained soil nutrients provided by the CRP and other federally funded conservation programs that target private lands (Randall et al. 1997, Gleason et al. 2011, Johnson et al. 2016).

Deep Dive: Grassland Birds and Biofuel Development in the Northern Great Plains

Biofuel development has the potential to negatively affect grassland bird populations, especially if it results in conversion of grassland to row crops. Although scientific investigations of land-use change and wildlife generally take multiple decades of data to elucidate systematic patterns, there is significant concern that the rapid rate and spatial extent of land-use change brought on by biofuel development in the Northern Great Plains of the United States may negatively affect wildlife populations over a relatively short timescale. Experimental research designed to investigate population-level effects of biofuel development on wildlife are severely lacking. Grassland birds are one of the most well-studied groups of organisms, with respect to the impacts of agriculture and the expansion of bioenergy crops into natural systems. Nevertheless, it is not well understood how grassland bird populations have responded to the observed changes in land use brought on by biofuel development over the past decade, particularly in the Northern Great Plains. Many of the studies reviewed herein are limited by short timescales or provide forecasts of how grassland birds are expected to respond to hypothetical biofuel scenarios. Here, I used observational data collected for savannah sparrows along 32 North American Breeding Bird Survey (BBS) routes and land-cover data from 2006 to 2016 to investigate concurrent trends in grassland bird populations and land-cover change in the Prairie Pothole Region of North Dakota. I selected this region because of the rapid land-use changes that have taken place here since the 2007 Energy Independence and Security Act.

The goals of this Deep Dive were to (1) investigate associations between biofuel crop and CRP areas with savannah sparrow counts collected along BBS routes; (2) estimate trends in savannah sparrow counts; (3) estimate land-cover trends for biofuel crops (i.e., corn and soybeans) and CRP lands around BBS routes; and (4) determine whether modeled trends in savannah sparrow counts are correlated with modeled land-cover trends along each BBS route. This Deep Dive provides a working example of how future researchers can investigate systematic changes in corn and soybean production and estimate how these changes may affect wildlife population trends.

Methodology

The study area consisted of BBS routes distributed throughout the Prairie Pothole Region of North Dakota, a region characterized by nutrient-rich glacial soils. Prior to European settlement, the Prairie Pothole Region was dominated by mixed-grass prairie and isolated wetlands; however, during the past 150 years, much of this area has been converted to cropland. I consulted the previous literature to identify a bird species that would likely exhibit a negative response to row crops and positive response to CRP lands. Savannah sparrows are a grassland species that show strong preference for CRP grasslands (Johnson and Igl 1995, Johnson and Igl 2001). I obtained georeferenced route locations and annual count data for savannah sparrows collected at each BBS route from 2006 to 2016 (Sauer et al. 2014). A BBS route consists of a 40 km road transect with 50 stop locations spaced 0.8 km apart. A single BBS observer surveys each route annually, primarily in late May–June. At each stop location, the observer conducts a 3-minute point count, recording all birds seen or heard within a ~400 m radius. Avian count data for the BBS are best viewed as an index of true abundance because of variation in observer detection probability and roadside biases. In a geographic information system, I created a 400 m buffer (50.2 ha) around each BBS route. The Cropland Data Layer (https://nassgeodata.gmu.edu/Crop Scape/), available annually from 2006 to 2016, was overlaid on each buffer to quantify land-cover composition within 400 m of each route. For this analysis, I reclassified the Cropland Data Layer into 6 land-use and land-cover categories and calculated the annual areas of each. For the purposes of this Deep Dive, I reported the results for only corn and soybean. I also calculated the annual area of CRP lands within each route buffer from 2006 to 2016 (minus 2007) from CRP data provided via a data-sharing agreement with the US Department of Agriculture—Farm Service Agency. I removed 2 BBS routes from the analysis because of missing avian count data for most years.

I graphically displayed savannah sparrow counts and land-cover data to explore potential linear trends for the 32 BBS routes included in the final analysis. For Objective 1, I fit generalized linear models to determine whether spatial variation in savannah sparrow counts can be explained as a function of the area of corn, soybeans, or CRP land surrounding BBS routes. I performed this analysis using data from all years, but report only the results for the earliest and latest survey years (i.e., 2006 and 2016). I assumed a Poisson distribution for the count data and used the raw land-cover area data as the independent variable. To assess trends in savannah sparrows, CRP land, and corn or soybean crops (hereafter biofuel crops) from 2006 to 2016

(Objectives 2 and 3), I fit simple linear regression models for each BBS route. I accommodated the repeated surveys on each route across years by assuming an autoregressive lag of one. I used the mixed linear model procedure of SAS (PROC MIXED, SAS 2013) with the AR(1) option to compute trend coefficients. To compare the trends among savannah sparrows, CRP, and biofuel crops (Objective 4), I first standardized all the data prior to running regression models, resulting in standardized trend coefficients. I used the Pearson correlation coefficient to compare trends across all 32 routes. In addition, bivariate plots of CRP versus biofuel crops were used to group each route into 1 of 4 categories: low CRP-low biofuels, low CRP-high biofuels, high CRP-low biofuels, and high CRP-high biofuels. I computed correlation coefficients among counts, biofuels, and CRP trends separately for each grouping. Grouping of the routes was subjective, as some routes had large changes in CRP and/or biofuels, while other routes realized little changes, but in general the routes could be broadly categorized.

Key Findings and What They Mean

Variation in savannah sparrow counts was negatively related to the area of biofuel row crops within 400 m of BBS routes in 2006 (z value $= -9.69$, $p < 0.001$) and 2016 (z value $= -3.25$, $p = 0.001$) and positively related to the area of CRP land in 2006 (z value $= 3.66$, $p < 0.001$), but not 2016 (z value $= -1.77$, $p = 0.077$; Figure 3.4). In 2016, the relationship between savannah sparrow counts and CRP area was slightly negative; however, the slope parameter had overlapping 95% confidence intervals. The distribution of CRP along BBS routes in 2016 was right-skewed, and the apparent lack of BBS routes with more than

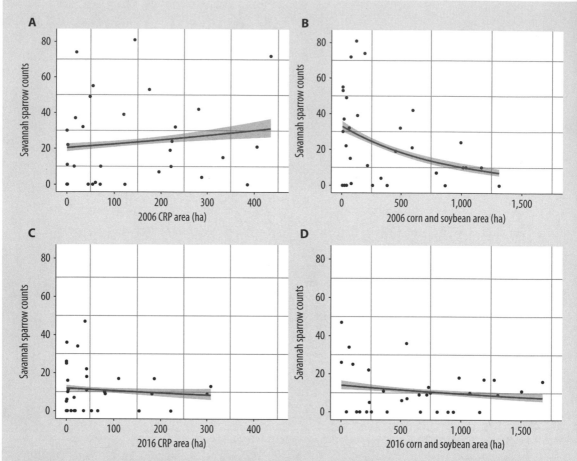

Figure 3.4   Associations between savannah sparrow (*Passerculus sandwichensis*) counts and land-cover area quantified within 400 m of Breeding Bird Survey routes. (*A*) 2006 Conservation Research Program area (CRP); (*B*) 2006 corn and soybeans; (*C*) 2016 CRP area; and (*D*) 2016 corn and soybeans. Shaded region denotes 95% CI for each generalized linear model. *Clint Otto.*

100 ha (245 ac) of CRP land in 2016 made modeling an association between CRP area and savannah sparrow counts challenging (Figure 3.4D).

In general, land-cover trends for corn and soybeans increased from 2006 to 2016, while CRP area and savannah sparrow counts decreased (Figure 3.5). For individual BBS routes, standardized trend coefficients for savannah sparrows ranged from −0.85 to 0.70, with a mean of −0.17 (SD = 0.36), suggesting that counts decreased during the study period.

Mean savannah sparrow counts were highest in 2008 (29.6+/− 4.2 1SE) and lowest in 2016 (11.2 +/− 2.1; Figure 3.5A). During this same period, trend coefficients for corn and soybeans ranged from −0.29 to 0.97, with a mean of 0.71 (SD = 0.28). The most substantial gains in corn and soybeans occurred after 2010 (Figure 3.5B). Concurrent with the increase in corn and soybeans, coefficient estimates for CRP area tended to decrease (x = −0.58 +/− 0.61), with a range of −1.16 to 0.95. Correlation coefficients between trends

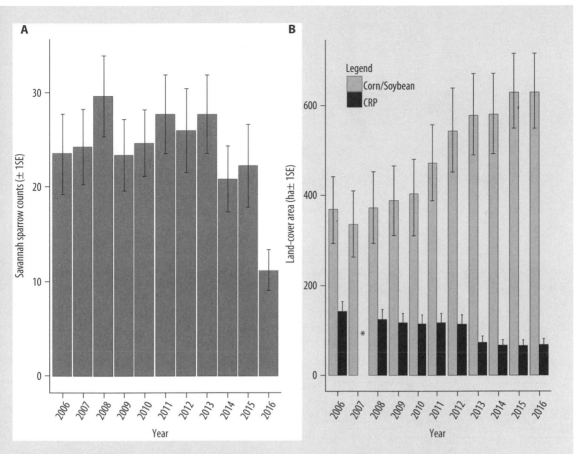

Figure 3.5   (*A*) Average (+/− 1SE) number of savannah sparrows (*Passerculus sandwichensis*) detected along 32 Breeding Bird Survey (BBS) routes in North Dakota from 2006 to 2016. (*B*) Average (+/− 1SE) area of corn and soybeans (gray) or Conservation Reserve Program (CRP) (black) within 400 m of BBS routes. Asterisks denote lack of 2007 CRP data.

in CRP and corn/soybeans, CRP and savannah sparrows, and corn/soybeans and savannah sparrows were 0.10 (p = 0.60), 0.26 (p = 0.17), and −0.01 (p = 0.94), respectively, suggesting a lack of associations. Grouping routes by land cover revealed correlations between CRP area and savannah sparrow counts among routes with high CRP; however, these groups also had low sample size (Table 3.2).

Collectively, these results demonstrate how annual biological and land-cover data can be used to investigate effects of biofuel development on species of interest. Although there

were no significant correlations among temporal trends in savannah sparrow counts, CRP area, and biofuel crop area, I believe this is an effective approach to be applied to other species over larger spatial and temporal scales in areas with preexisting BBS data. My analysis showed that BBS routes surrounded by more CRP land in 2006 and fewer biofuel crops in 2006 and 2016 had higher savannah sparrow counts in those years. This finding is supported by much of the literature reviewed in this chapter describing the benefits of CRP areas for wildlife as well as the threats posed by

Table 3.2. Correlation coefficients for trend estimates between savannah sparrows (*Passerculus sandwichensis*, SAVS), Conservation Reserve Program (CRP) areas and biofuels (i.e., corn and soybeans) for 32 North American Breeding Bird Survey routes. Each route was assigned to one of 4 groups: low CRP-low biofuels; low CRP-high biofuels; high CRP-low biofuels; and high CRP-high biofuels. Two routes had no CRP area for all years so trends were inestimable.

CRP group	Biofuel crop group	SAVS vs. CRP	SAVS vs. biofuels	CRP vs. biofuels
Low	Low	0.27 (n = 8)	0.15 (n = 10)	0.02 (n = 8)
Low	High	0.18 (n = 11)	−0.01 (n = 11)	0.37 (n = 11)
High	Low	0.9 (n = 5)	−0.01 (n = 5)	0.26 (n = 5)
High	High	0.8 (n = 6)	0.03 (n = 6)	0.03 (n = 6)

annual row crops. The analysis also showed how savannah sparrow counts are trending down, concurrent with a negative trend in CRP area and positive trend in biofuel crops across all BBS routes in the study region. However, count and land-cover trends observed at the BBS route level were not correlated, suggesting that savannah sparrow count trends were independent of the land-cover trends I estimated. The analysis revealed a tendency for savannah sparrows to decline with decreasing CRP area but the correlation was not statistically significant. Given the substantial body of literature written about biofuel development and its negative consequences for grassland birds, and the rate of land-use change taking place in the Northern Great Plains, it seemed reasonable to test whether observed declines in savannah sparrow counts were related to recent loss of CRP grasslands or growth in biofuel crops. My analysis was limited by a relatively small spatial and temporal sample size, thereby making it difficult to correlate trends in land cover with trends in savannah sparrow counts. The baseline year for this analysis was 2006 because that is the first year for which CRP spatial data were available. Subsequent analyses can be improved by using data over a larger study region, additional land-cover types, and longer time series as data become available.

It is important to note that land-use change in the Northern Great Plains occurs at the scale of individual fields and farmsteads— land-use choices that culminate in differential patterns in land use through time. Thus, future researchers investigating trends in biofuel production and grassland bird populations should consider using more fine-scale data, such as BBS stop-level data, to investigate potential impacts, a finding supported by other grassland bird and land-use research (Scholtz et al. 2017).

Pollinators and Biofuel Development

Insect pollinators are responsible for facilitating sexual reproduction in 85% of angiosperms globally (Ollerton et al. 2011). Plant-pollinator interactions play an important role in maintaining ecosystem function and terrestrial food webs. Furthermore, over one-third of global crop production by volume depends on animal pollination (Klein et al. 2007). A comprehensive analysis of US food crop production estimated the value of pollinator-dependent crops at $15.1 billion in 2009 (Calderone 2012). Thus, pollinator populations are intimately intertwined with human health and ecosystem function.

As human population growth continues, our dependence on pollinators and pollination services for food production is growing, even as pollinator populations exhibit global declines (Allen-Wardell et al. 1998). Population models estimate a 23% decline in wild bee abundance from 2008 to 2013 throughout the contiguous United States (Koh et al. 2016). Similar to many vertebrate wildlife species, wild insect pollinators are exhibiting significant range contractions and negative population trends. The rusty patched bumble bee (*Bombus affinis*), once widely distributed across the Midwest, was officially listed under the Endangered Species Act in 2017. Concurrent with wild bee declines, beekeepers have experienced a 30%–40% loss of their honey bee (*Apis mellifera*) colonies in recent years (Lee et al. 2015). Global declines in pollinator populations have raised considerable societal concern over food security, human health, and loss of ecosystem function. Although scientists are still investigating causes of pollinator declines, current evidence suggests that habitat loss, pesticide use, lack of forage, parasites, and diseases all play a significant role in pollinator health (Goulson et al. 2015, Spivak et al. 2017). Societal outcry over pollinator declines led to the development of the US Pollinator Health Task Force (2015) to improve the health of honey bees and native pollinators. This task force established 3 key goals, including the establishment of 2.8 million ha (7 million ac) of pollinator habitat within the United States by 2020.

The Northern Great Plains provides summer foraging ground to over 40% (~1 million) of US honey bee colonies and hundreds of native bee species. States within the region are routinely among the top honey producers, with North Dakota being the top producing state since 2004 (USDA Honey 2014). For multiple generations, beekeepers have brought honey bee colonies to the Dakotas and surrounding areas to make a valuable honey crop and to improve the health of their honey bee colonies following the

winter and spring crop-pollination season. Honey bees in the Northern Great Plains are transported across the United States to provide pollination services for a variety of agricultural crops, including almonds, apples, cherries, and melons, during the winter months.

Changes in land use and biofuel development have reduced habitat suitability for native bees and honey bees in the Northern Great Plains and Upper Midwest (Koh et al. 2016, Otto et al. 2016). Pollinator health, nutrition, and annual survival are lower in landscapes supporting more row crop agriculture in the region (Smart et al. 2016a, Smart et al. 2016b). In recent years, conservation of grasslands and native prairies that provide forbs needed to support native bees and honey bees have become increasingly less common and have been replaced by biofuel crops (Wright and Wimberly 2013, Morefield et al. 2016, Otto et al. 2018). Landscape composition also has a significant effect on bee abundance, community composition, and diversity, with the greatest diversity and abundance occurring in perennial grasslands. At small scales, wild bee abundance can be 2 to 3 times greater in grasslands than in fields of corn or soybeans (Gardiner et al. 2010). At larger scales, wild bee richness and abundance of specialized bees is often greater in fields surrounded by more grasslands relative to row crops (Bennett and Isaacs 2014). In turn, increased bee abundance has a positive effect on pollination service of some annual crops (Bennett and Isaacs 2014). Forecast models developed by Bennett et al. (2014) indicate that expansion of biofuel row crops into 600,000 ha (1.5 million ac) of marginal land could reduce wild bee abundance by 0% to 71% and bee diversity by 0% to 28% in southern Michigan.

Agrochemical Use, Biodiversity, and Ecosystem Function

In the United States, producers apply multiple agrochemicals to biofuel crops such as corn and soybeans in attempts to increase yield and reduce pest populations. Increased use of agrochemicals is not a trend unique to biofuel crops, but rather represents a larger, global trend occurring across most sectors of agriculture. A wealth of literature exists exploring direct and indirect effects of agrochemicals on vertebrate and invertebrate wildlife (Relyea 2005, Relyea 2009, Köhler and Triebskorn 2013, Main et al. 2014). In this section, I chose to focus on particular classes of agrochemicals that have specific relevance to biofuel crops and, as such, have seen dramatic increases in use over the past 10 years. I highlighted a few pertinent examples of how key classes of chemicals used on corn and soybeans affect wildlife and ecosystem function, but I do not consider this an exhaustive review.

Producers annually spend over $4 billion on insecticides in the United States, totaling 45 million kg (100 million lb) of active ingredients (Grube et al. 2011). Neonicotinoids, a class of neurotoxic chemical that is highly effective at killing undesirable insect pests, is the most commonly used class of insecticide globally (Jeschke et al. 2010, Douglas and Tooker 2015). Neonicotinoids are systemic, meaning they are transferred to all parts of the plant, including roots, stems, leaves, pollen, nectar, and fruiting bodies produced by the plant. Neonicotinoids were implemented in large-scale agriculture because of their acute toxicity to insect pests and relatively low toxicity to mammals, including humans. Although the Environmental Protection Agency has enacted guidelines to curtail use of neonicotinoids in croplands, overall use of neonicotinoids has increased substantially in the United States since 2003 (Douglas and Tooker 2015). In 2011, 34%–44% of US soybeans and 79%–100% of corn were treated with neonicotinoids; virtually all the treatments were applied as seed coatings. Douglas and Tooker (2015) reported that neonicotinoid seed treatments accounted for 43% and 21%–23% of the insecticide mass applied to corn and soybeans, respectively. Neonicotinoids are also applied as foliar sprays later in the growing season. Although neonicotinoids are intended to remain in crop fields, they persist in the

local environment and have been detected in drainage areas, such as isolated wetlands, streams, and wildlife conservation plantings (Main et al. 2014, Hladik and Kolpin 2016, Mogren and Lundgren 2016, Hladik et al. 2017).

The rapid rise in neonicotinoid use across the United States suggests a pressing need to understand how these chemicals travel throughout ecosystems and quantify their impacts on vertebrate wildlife, pollinators, and other beneficial insects. Although information gaps exist, a significant body of evidence demonstrates that large-scale use of neonicotinoids can have negative effects on vertebrate wildlife (Gibbons et al. 2015), wild bees (Stanley et al. 2015), honey bees (Henry et al. 2012), and aquatic invertebrates (Morrissey et al. 2015). Pollinating bees have been particularly harmed by widespread use of neonicotinoids and can be exposed to these chemicals via multiple pathways. Krupke et al. (2012) showed that foraging honey bees can be exposed to neonicotinoids via the following pathways: (1) during the planting of seed-treated corn and soybeans; (2) by visiting wildflowers adjacent to seed-treated fields; and (3) by collecting pollen from seed-treated corn that has reached anthesis. Conservation strips along field margins, which are often considered beneficial for wildlife in agroecosystems, can be contaminated with neonicotinoids; bees foraging on wildflowers in these conservation strips consume neonicotinoids, resulting in adverse health effects (Mogren and Lundgren 2016). Exposure to neonicotinoids by bees via contaminated nectar and pollen can lead to both lethal and sublethal effects, including behavioral abnormalities and reproductive deficiencies (Henry et al. 2012, Whitehorn et al. 2012, Baron et al. 2017), which can decrease a bee's ability to pollinate crops (Stanley et al. 2015). For example, Baron et al. (2017) demonstrated that field-relevant exposure of the neonicotinoid thiamethoxam on wild bumblebee queens caused a reduction in feeding and impeded ovary development in multiple bumblebee species. Although additional field-level research is needed to quantify effects of neonicotinoid on pollinators, emerging evidence suggests that neonicotinoids commonly used on biofuel crops pose significant threats to pollinators and other beneficial insects.

Concurrent with the increase in neonicotinoid use, global use of genetically engineered, glyphosate-resistant corn and soybeans has increased 15-fold since 1974 (Benbrook 2016). Glyphosate (i.e., Roundup) is a chemical herbicide developed to kill undesirable broadleaf plants and grasses and is commonly used on agricultural fields. Genetic engineering of glyphosate-resistant corn and soybeans has contributed to a significant increase in glyphosate applications in US agriculture. In the United States, 8.6 billion kg (19 billion lb) of glyphosate has been applied since 1974, with 66% of the total volume having been applied from 2004 to 2014 (Benbrook 2016). Although glyphosate-resistant corn and soybeans provide producers with the ability to prophylactically manage weeds and reduce crop injury in and adjacent to crop fields (Carpenter et al. 2002), growing these crops also eliminates an important, yet often underrepresented, forage base for beneficial insects and pollinators, including monarch butterflies (*Danaus plexippus*; Figure 3.6).

The monarch butterfly was proposed for listing under the Endangered Species Act (16 U.S.C. ch. 35 § 1531 et seq) in 2014 because of significant population declines and extinction risk (Semmens et al. 2016). Widespread use of glyphosate-resistant corn and soybeans in the United States has been implicated in the decline of eastern populations of monarch butterflies through the widespread elimination of milkweed (*Asclepias* spp.), the essential host plant for monarch larvae (Stenoien et al. 2016, Thogmartin et al. 2017). In their review of threats to monarch butterflies, Stenoien et al. (2016) reported that widespread use of herbicide-resistant crops has become more prevalent in the core summer breeding range of monarchs and that use of glyphosate has likely contributed to a landscape-level reduction in milkweed. This is of particular concern in the Midwest and Northern Great Plains regions, which represent a major migratory pathway of the monarch. A better

Figure 3.6   The monarch butterfly (*Danaus plexippus*) is an iconic insect that exhibits a unique, multigenerational migration across North America. Monarch populations are threatened by habitat loss and reduction in milkweed (*Asclepias* spp.), an essential host plant, throughout the United States. *Photo courtesy of Savannah Adams.*

understanding of how changes in agricultural practices and use of glyphosate-resistant crops have affected the distribution of milkweed, at multiple scales, would improve population viability analyses for monarchs. Recent scenario modeling conducted by Thogmartin et al. (2017) showed that restoration efforts on agricultural lands are essential for achieving monarch conservation goals and called for restoration of over 1.3 billion milkweed plants across the Midwest and Northern Great Plains. Similar to reductions in milkweed from agricultural fields, removal of flowers deemed undesirable by landowners and land managers may also reduce the forage base for adult monarchs, as well as for native bees and honey bees in agricultural landscapes (Bretagnolle and Gaba 2015, Otto et al. 2017).

Virtually no information exists on how the rapid increase in glyphosate-resistant corn and soybeans, which constitute the primary landscape matrix in many parts of the United States, may have altered landscape suitability for supporting pollinators and other beneficial insects in agroecosystems. Reduction in biodiversity on farms is correlated with increased pest populations (Lundgren and Fausti 2015), presumably due to loss of predatory insects (Gardiner et al. 2009). Large-scale expansion of corn production has been directly linked with loss of arthropod diversity and the crop pest control services they provide (Landis et al. 2008). This suggests that agronomic practices that result in loss of biological diversity may require additional agrochemical inputs from agricultural producers to compensate for the loss in arthropod-mediated pest control services, perpetuating the continued use of agrochemicals for treating crop pests and further loss of beneficial insects.

Multifunctioning Landscapes

Given the rapid conversion of natural areas to agriculture across the globe and a burgeoning human population, there has never been a more pressing need to understand how land-use changes affect wildlife, in tandem with the multiple ecosystem services wildlife provide to human society. Viewing the impacts of biofuel crop production on wildlife alone is not the most effective approach for infusing wildlife conservation goals into national policy and decision making. Rather, I believe what is needed is research that quantifies how land-use change affects multiple ecosystem services concurrently, and, in turn, how these changes affect the multiple human stakeholders. Given the complexities of modeling how multiple ecosystem services and stakeholders are affected by land-use change, this area of research has only recently been developed (Nelson et al. 2009, Polasky et al. 2011, Zheng et al. 2016, Schulte et al. 2017). Polasky et al. (2011) applied a spatially explicit modeling tool (InVEST) to quantify how alternative land-use change scenarios in Minnesota might affect

multiple ecosystem services. Results showed significant agricultural expansion generated the largest returns for landowners (i.e., market commodities) but the fewest net social benefits, such as carbon sequestration and water quality. In a review of over 35 studies on biomass cropping systems in Michigan and Wisconsin, Landis et al. (2017) determined that perennial grass cropping systems have the potential to increase (relative to feedstocks derived from annual row crops) multiple ecosystem services, including pest suppression, decreased GHG emissions, and increased grassland bird diversity, but at the expense of biomass yield. This line of research clearly demonstrates how trade-offs between crop production profits (i.e., landowner benefits) and multiple societal benefits can be evaluated within a unified framework.

Because of the global footprint and far-reaching effects of biofuel crop expansion, ecosystems service models can be developed that use a life-cycle assessment approach to determine how environmental perturbations in one area affect stakeholders and ecosystem service delivery in other parts of the country (Zhang et al. 2010). For example, Smart et al. (2018) demonstrated how honey bee colonies surrounded by larger areas of row crops in the Northern Great Plains were smaller in size at the end of the growing season, leading to reduced pollination services the subsequent spring in Californian almond orchards. This analysis could be further extended to investigate how biofuel development in the Northern Great Plains affects pollinator habitat, which, in turn, affects agricultural producers in California and ultimately consumer cost for almonds at the grocery store. Ecosystem service analyses such as these present significant challenges and would require increased interdisciplinary collaboration; however, they would provide a method for monetizing land-use decisions and ecosystem service outcomes and identifying unintended consequences of biofuel development.

As important as evaluation of multiple services is, this alone will not lead to land-use decisions that have beneficial results for wildlife. Effective communication of research results to policy makers and decision makers is of paramount importance for enacting policy that maximizes services provided to society. For example, ecosystem service–based research can assist policy makers with development of programs that incentivize landowners to make land-use decisions that have greater societal benefits (Polasky et al. 2011). Developing ecosystem service models will require increased collaboration between economists and ecologists to ensure that "ecological production functions" are based on sound science. This is particularly true for ecosystems services models that include wildlife, a group that has been traditionally absent from service-based research because their market value can be more difficult to ascertain.

It is important to recognize that many of the issues surrounding wildlife and biofuel development are social, economic, and political in nature. Science will always play an important role in documenting and quantifying threats to biodiversity, informing decision makers, and evaluating costs and benefits of conservation actions. However, the premier societal challenge at hand is promoting a fundamental shift in human perception of natural systems and their benefits to humankind (Ehrlich and Pringle 2008). With the growing global human population and rising standard of living, there is a pressing need to develop sustainable fuel sources that can meet growing demand for energy with minimal environmental degradation. We face the challenge of balancing the immediate human needs for food, fiber, and fuel with long-term maintenance of ecosystem function and natural resources for future generations (Foley et al. 2005). Given the significant impacts that increased corn and soybean production have had on natural systems in the United States since the 2007 US Energy Independence and Security Act, it does not appear that annual biofuel crops represent an effective long-term solution to environmentally sound energy production. Second-generation feedstocks (see Chapter 4, Rupp and Ribic) provide potential alternatives that may lessen concerns over habitat loss

and biodiversity (Werling et al. 2014), but current technological, political, and transportation barriers limit their expansion.

Key Research Needs

There are numerous research needs to identify strategies for minimizing the impact of biofuel row crops on wildlife and ecosystem function. Here, I outline a few key priority areas:

- Identify and monetize consequences of a broad range of land-management actions and policy decisions, including those that optimize ecosystem service benefits to society and individual producers. This will provide useful information to policy makers for incentivizing these services through federal policy and programs (Polasky et al. 2011).
- Conduct research at larger spatial and temporal scales to identify population-level effects and assist with landscape planning for wildlife and pollinators. Small-scale research can be useful for elucidating mechanisms of how populations respond to biofuel practices and inform specific management actions, but its application to informing national policy and landscape planning is limited. Meehan et al. (2010) provided an example of how biological data can be used to forecast outcomes of biofuel development on grassland birds across a six-state ecoregion.
- Research neonicotinoids and glyphosate used on biofuel crops to better understand how these agrochemicals move through terrestrial and aquatic environments and their long-term impacts on wildlife populations in terrestrial, palustrine, and riverine systems.
- Conduct proactive research that evaluates potential impacts of novel agricultural practices or policies before they are implemented at a national scale. For example, RNA interference (RNAi) is a biotechnology that is being considered as a novel pest-control strategy; however, its effects on non-target beneficial insects have not been thoroughly

evaluated in working landscapes (Huvenne and Smagghe 2010).

- Develop national land-cover data sets that track changes in natural vegetation on an annual basis. Lack of these data sets makes it challenging to study effects of land-use change and casts doubt on wildlife and habitat studies, particularly in areas of recent biofuel expansion (Reitsma et al. 2016, Flather et al. 2017).

Best Management Practices

Below, I outline several best management practices that can be implemented on biofuel crop farms to maintain agricultural productivity and reduce impacts to wildlife and pollinators.

- Precision agriculture is a crop-management concept based on observing, measuring, and responding to intra-field variation in crop production. New precision agriculture technology can be used to identify locations within farm fields where producers are losing money because input costs are greater than revenue generated. Land use in these areas could be modified by planting cover crops, perennial conservation strips, and associated covers or by allowing the land to remain fallow, to improve ecosystem service delivery and benefit both the producer and society.
- Agricultural producers can improve wildlife habitat and reduce soil erosion by leaving crop residues within fields and using less-intensive weed control practices (Krapu et al. 2004).
- At larger spatial and temporal scales, heterogeneous land covers, including a diversity of crop types, can increase wildlife abundance and wildlife use of row crop fields in agroecosystems (Best et al. 2001).
- Producers can maintain or enhance habitat for butterflies and pollinators by establishing filter strips and natural areas beside agricultural fields or unproductive areas with agricultural fields (Reeder et al. 2005, Park et al. 2015, Schulte et al.

2017). In addition to providing forage for pollinators, these areas can also enhance habitat for wildlife (Moorman et al. 2013, Plush et al. 2013) and facilitate pollination of crops grown in adjacent fields (Morandin and Winston 2006).

- Producers can implement best management practices to minimize drift of insecticides and herbicides into non-crop areas to reduce harmful effects to beneficial insects and pollinators. Examples of practices that reduce drift include proper calibration of application equipment and being cognizant of optimal weather conditions for chemical applications and seed treatments (Spivak et al. 2017).
- Crop consultants and producers can reevaluate the prophylactic use of pesticides on biofuel crops to ensure these greater input costs result in higher profit margins for the producer, rather than just increased yield. Integrated pest management may provide producers with a framework for reducing crop pests and thereby the prophylactic use of pesticides on agricultural fields.

ACKNOWLEDGMENTS
I thank Wes Newton and Deb Buhl for providing assistance with statistical analyses and Alec Boyd for conducting GIS work and developing figures. Aaron Pearse, Rich Iovanna, Melanie Colon, and Skip Hyberg provided thoughtful comments that improved this chapter. Conservation Reserve Program data were provided via a data-sharing agreement with USDA—Farm Service Agency. Funding for this chapter was provided by the USGS Ecosystems Mission Area. Any use of trade, firm, or product names is for descriptive purposes only and does not imply endorsement by the US government.

LITERATURE CITED
Abraham, K. F., R. L. Jefferies, and R. T. Alisauskas. 2005. The dynamics of landscape change and snow geese in mid-continent North America. Global Change Biology 11:841–855.
Allen-Wardell, G., P. Bernhardt, R. Bitner, A. Burquez, S. Buchmann, J. Cane, P. Allen Cox, V. Dalton, P. Feinsinger, and M. Ingram. 1998. The potential consequences

of pollinator declines on the conservation of biodiversity and stability of food crop yields. Conservation Biology 12:8–17.

Atkinson, L., R. J. Romsdahl, and M. J. Hill. 2011. Future participation in the Conservation Reserve Program of North Dakota. Great Plains Research 21:203–214.

Baron, G. L., N. E. Raine, and M. J. Brown. 2017. General and species-specific impacts of a neonicotinoid insecticide on the ovary development and feeding of wild bumblebee queens. Proceedings of the Royal Society B: Biological Sciences 284:20170123.

Benbrook, C. M. 2016. Trends in glyphosate herbicide use in the United States and globally. Environmental Sciences Europe 28:3.

Bennett, A. B., and R. Isaacs. 2014. Landscape composition influences pollinators and pollination services in perennial biofuel plantings. Agriculture, Ecosystems and Environment 193:1–8.

Bennett, A. B., T. D. Meehan, C. Gratton, and R. Isaacs. 2014. Modeling pollinator community response to contrasting bioenergy scenarios. PLOS One 9:e110676.

Best, L. B., T. M. Bergin, and K. E. Freemark. 2001. Influence of landscape composition on bird use of rowcrop fields. Journal of Wildlife Management 65:442–449.

Bretagnolle, V., and S. Gaba. 2015. Weeds for bees? A review. Agronomy for Sustainable Development 35:891–909.

Calderone, N. W. 2012. Insect pollinated crops, insect pollinators and U.S. agriculture: Trend analysis of aggregate data for the period 1992–2009. PLOS One 7:e37235.

Carpenter, J., A. Felsot, T. Goode, M. Hammig, D. Onstad, and S. Sankula. 2002. Comparative environmental impacts of biotechnology-derived and traditional soybean, corn, and cotton crops. Council for Agricultural Science and Technology 2002:3.

Claassen, R., F. Carriazo, J. Cooper, D. Hellerstein, and K. Ueda. 2011. Grassland to cropland conversion in the Northern Plains: The role of crop insurance, commodity, and disaster programs. In Economic Research Report Number 120, 1–85. Washington, DC: US Department of Agriculture Economic Research Service.

Douglas, M. R., and J. F. Tooker. 2015. Large-scale deployment of seed treatments has driven rapid increase in use of neonicotinoid insecticides and preemptive pest management in U.S. field crops. Environmental Science and Technology 49:5088–5097.

Drum, R. G., C. R. Loesch, K. M. Carrlson, K. E. Doherty, and B. C. Fedy. 2015. Assessing the biological benefits of the USDA-Conservation Reserve Program (CRP) for waterfowl and grassland passerines in the Prairie Pothole Region of the United States: Spatial analyses for targeting CRP to maximize benefits for migratory birds. Prairie Pothole Joint Venture. Final report for USDA-FSA Agreement: 12-IA-MRE-CRP-TA. https://www.fsa.usda.gov/Assets/USDA-FSA-Public/usdafiles/EPAS/PDF/drumetal2015_crp_prr_final.pdf.

Ehrlich, P. R., and R. M. Pringle. 2008. Where does biodiversity go from here? A grim business-as-usual forecast and a hopeful portfolio of partial solutions. Proceedings of the National Academy of Sciences 105:11579–11586.

Fahrig, L. 2003. Effects of habitat fragmentation on biodiversity. Annual Review of Ecology, Evolution, and Systematics 34:487–515.

Fargione, J., J. Hill, D. Tilman, S. Polasky, and P. Hawthorne. 2008. Land clearing and the biofuel carbon debt. Science 319:1235–1238.

Fargione, J. E., T. R. Cooper, D. J. Flaspohler, J. Hill, C. Lehman, T. McCoy, S. McLeod, E. J. Nelson, K. S. Oberhauser, and D. Tilman. 2009. Bioenergy and wildlife: Threats and opportunities for grassland conservation. Bioscience 59:767–777.

Farrell, A. E., R. J. Plevin, B. T. Turner, A. D. Jones, M. O'Hare, and D. M. Kammen. 2006. Ethanol can contribute to energy and environmental goals. Science 311:506–508.

Fearnside, P. M. 2001. Soybean cultivation as a threat to the environment in Brazil. Environmental Conservation 28:23–38.

Flather, C., M. Knowles, and L. Baggett. 2017. Land cover dynamics across the Great Plains and their influence on breeding birds: Potential artefact of data and analysis limitations. Biological Conservation 213:243–244.

Fletcher, R. J., B. A. Robertson, J. Evans, P. J. Doran, J. R. R. Alavalapati, and D. W. Schemske. 2011. Biodiversity conservation in the era of biofuels: Risks and opportunities. Frontiers in Ecology and the Environment 9:161–168.

Foley, J. A., R. DeFries, G. P. Asner, C. Barford, G. Bonan, S. R. Carpenter, F. S. Chapin, et al. 2005. Global consequences of land use. Science 309:570–574.

Gardiner, M., D. Landis, C. Gratton, C. DiFonzo, M. O'Neal, J. Chacon, M. Wayo, N. Schmidt, E. Mueller, and G. Heimpel. 2009. Landscape diversity enhances biological control of an introduced crop pest in the north-central USA. Ecological Applications 19:143–154.

Gardiner, M. A., J. K. Tuell, R. Isaacs, J. Gibbs, J. S. Ascher, and D. A. Landis. 2010. Implications of three biofuel crops for beneficial arthropods in agricultural landscapes. Bioenergy Research 3:6–19.

Gelfand, I., T. Zenone, P. Jasrotia, J. Chen, S. K. Hamilton, and G. P. Robertson. 2011. Carbon debt of Conservation

Reserve Program (CRP) grasslands converted to bioenergy production. Proceedings of the National Academy of Sciences 108:13864–13869.

Gibbons, D., C. Morrissey, and P. Mineau. 2015. A review of the direct and indirect effects of neonicotinoids and fipronil on vertebrate wildlife. Environmental Science and Pollution Research 22:103–118.

Gleason, R. A., N. H. Euliss Jr., B. A. Tangen, M. K. Laubhan, and B. A. Browne. 2011. USDA conservation program and practice effects on wetland ecosystem services in the Prairie Pothole Region. Ecological Applications 21:S65–S81.

Goulson, D., E. Nicholls, C. Botías, and E. L. Rotheray. 2015. Bee declines driven by combined stress from parasites, pesticides, and lack of flowers. Science 347:1255957.

Grube, A., D. Donaldson, T. Kiely, and L. Wu. 2011. EPA pesticide industry sales and usage report: 2006 and 2007 market estimates. Washington, DC: US Environmental Protection Agency.

Henry, M., M. Beguin, F. Requier, O. Rollin, J.-F. Odoux, P. Aupinel, J. Aptel, S. Tchamitchian, and A. Decourtye. 2012. A common pesticide decreases foraging success and survival in honey bees. Science 336:348–350.

Hill, J., E. Nelson, D. Tilman, S. Polasky, and D. Tiffany. 2006. Environmental, economic, and energetic costs and benefits of biodiesel and ethanol biofuels. Proceedings of the National Academy of Sciences 103:11206–11210.

Hill, J. M., J. F. Egan, G. E. Stauffer, and D. R. Diefenbach. 2014. Habitat availability is a more plausible explanation than insecticide acute toxicity for U.S. grassland bird species declines. PLOS One 9:e98064.

Hladik, M. L., and D. W. Kolpin. 2016. First national-scale reconnaissance of neonicotinoid insecticides in streams across the USA. Environmental Chemistry 13:12–20.

Hladik, M. L., S. Bradbury, L. A. Schulte, M. Helmers, C. Witte, D. W. Kolpin, J. D. Garrett, and M. Harris. 2017. Neonicotinoid insecticide removal by prairie strips in row-cropped watersheds with historical seed coating use. Agriculture, Ecosystems and Environment 241:160–167.

Hoekstra, J. M., T. M. Boucher, T. H. Ricketts, and C. Roberts. 2005. Confronting a biome crisis: Global disparities of habitat loss and protection. Ecology Letters 8:23–29.

Huvenne, H., and G. Smagghe. 2010. Mechanisms of dsRNA uptake in insects and potential of RNAi for pest control: A review. Journal of Insect Physiology 56:227–235.

Jeschke, P., R. Nauen, M. Schindler, and A. Elbert. 2010. Overview of the status and global strategy for neonicoti-noids. Journal of Agricultural and Food Chemistry 59:2897–2908.

Johnston, C. A. 2014. Agricultural expansion: Land use shell game in the U.S. Northern Plains. Landscape Ecology 29:81–95.

Johnson, D. H. 2000. Grassland bird use of Conservation Reserve Program fields in the Great Plains. USGS Publication 29. Reston, VA: US Geological Survey.

Johnson, D. H., and L. D. Igl. 1995. Contributions of the Conservation Reserve Program to populations of breeding birds in North Dakota. Wilson Bulletin 107:709–718.

Johnson, D. H., and L. D. Igl. 2001. Area requirements of grassland birds: A regional perspective. Auk 118:24–34.

Johnson, D. H., and M. D. Schwartz. 1993. The Conservation Reserve Program: Habitat for grassland birds. Great Plains Research 3:273–295.

Johnson, J. A., C. F. Runge, B. Senauer, J. Foley, and S. Polasky. 2014. Global agriculture and carbon trade-offs. Proceedings of the National Academy of Sciences 111:12342–12347.

Johnson, K. A., B. J. Dalzell, M. Donahue, J. Gourevitch, D. L. Johnson, G. S. Karlovits, B. Keeler, and J. T. Smith. 2016. Conservation Reserve Program (CRP) lands provide ecosystem service benefits that exceed land rental payment costs. Ecosystem Services 18:175–185.

Klein, A. M., B. E. Vaissiére, J. H. Cane, I. Steffan-Dewenter, S. A. Cunningham, C. Kremen, and T. Tschamtke. 2007. Importance of pollinators in changing landscapes for world crops. Proceedings of the Royal Society B 274:303–313.

Koh, I., E. V. Lonsdorf, N. M. Williams, C. Brittain, R. Isaacs, J. Gibbs, and T. H. Ricketts. 2016. Modeling the status, trends, and impacts of wild bee abundance in the United States. Proceedings of the National Academy of Sciences 113:140–145.

Köhler, H.-R., and R. Triebskorn. 2013. Wildlife ecotoxicology of pesticides: Can we track effects to the population level and beyond? Science 341:759–765.

Krapu, G. L., K. J. Reinecke, D. G. Jorde, and S. G. Simpson. 1995. Spring-staging ecology of midcontinent greater white-fronted geese. Journal of Wildlife Management 59:736–746.

Krapu, G. L., D. A. Brandt, and R. R. Cox Jr. 2004. Less waste corn, more land in soybeans, and the switch to genetically modified crops: Trends with important implications for wildlife management. Wildlife Society Bulletin 32:127–136.

Krupke, C. H., G. J. Hunt, B. D. Eitzer, G. Andino, and K. Given. 2012. Multiple routes of pesticide exposure for honey bees living near agricultural fields. PLOS One 7:e29268.

Landis, D. A., M. M. Gardiner, W. van der Werf, and S. M. Swinton. 2008. Increasing corn for biofuel production reduces biocontrol services in agricultural landscapes. Proceedings of the National Academy of Sciences 105:20552–20557.

Landis, D. A., C. Gratton, R. D. Jackson, K. L. Gross, D. S. Duncan, C. Liang, T. D. Meehan, et al. 2018. Biomass and biofuel crop effects on biodiversity and ecosystem services in the North Central U.S. Biomass and Bioenergy 114:18–29.

Lark, T. J., J. M. Salmon, and H. K. Gibbs. 2015. Cropland expansion outpaces agricultural and biofuel policies in the United States. Environmental Research Letters 10:044003.

Lee, K. V., N. Steinhauer, K. Rennich, M. E. Wilson, D. R. Tarpy, D. M. Caron, R. Rose, K. S. Delaplane, K. Baylis, and E. J. Lengerich. 2015. A national survey of managed honey bee 2013–2014 annual colony losses in the USA. Apidologie 46:292–305.

Lin, M., and M. C. Henry. 2016. Grassland and wheat loss affected by corn and soybean expansion in the Midwest corn belt region, 2006–2013. Sustainability 8:1177.

Lundgren, J. G., and S. W. Fausti. 2015. Trading biodiversity for pest problems. Science Advances 1:e1500558.

Main, A. R., J. V. Headley, K. M. Peru, N. L. Michel, A. J. Cessna, and C. A. Morrissey. 2014. Widespread use and frequent detection of neonicotinoid insecticides in wetlands of Canada's Prairie Pothole Region. PLOS One 9:e92821.

Meehan, T. D., A. H. Hurlbert, and C. Gratton. 2010. Bird communities in future bioenergy landscapes of the Upper Midwest. Proceedings of the National Academy of Sciences 107:18533–18538.

Mineau, P., and M. Whiteside. 2013. Pesticide acute toxicity is a better correlate of U.S. grassland bird declines than agricultural intensification. PLOS One 8:e57457.

Mogren, C. L., and J. G. Lundgren. 2016. Neonicotinoid-contaminated pollinator strips adjacent to cropland reduce honey bee nutritional status. Scientific Reports 6:29608.

Moorman, C. E., C. J. Plush, D. Orr, and C. Reberg-Horton. 2013. Beneficial insect borders provide northern bobwhite brood habitat. PLOS One 12:e83815.

Morandin, L. A., and M. L. Winston. 2006. Pollinators provide economic incentive to preserve natural land in agroecosystems. Agriculture, Ecosystems and Environment 116:289–292.

Morefield, P. E., S. D. LeDuc, C. M. Clark, and R. Iovanna. 2016. Grasslands, wetlands, and agriculture: The fate of land expiring from the Conservation Reserve Program in the midwestern United States. Environmental Research Letters 11:094005.

Morrissey, C. A., P. Mineau, J. H. Devries, F. Sanchez-Bayo, M. Liess, M. C. Cavallaro, and K. Liber. 2015. Neonicotinoid contamination of global surface waters and associated risk to aquatic invertebrates: A review. Environment International 74:291–303.

Murphy, M. T., and F. Moore. 2003. Avian population trends within the evolving agricultural landscape of eastern and central United States. Auk 120:20–34.

Mushet, D. M., J. L. Neau, and N. H. Euliss Jr. 2014. Modeling effects of conservation grassland losses on amphibian habitat. Biological Conservation 174:93–100.

Nelson, E., G. Mendoza, J. Regetz, S. Polasky, H. Tallis, D. Cameron, K. Chan, G. C. Daily, J. Goldstein, and P. M. Kareiva. 2009. Modeling multiple ecosystem services, biodiversity conservation, commodity production, and tradeoffs at landscape scales. Frontiers in Ecology and the Environment 7:4–11.

Nielson, R. M., L. L. McDonald, J. P. Sullivan, C. Burgess, D. S. Johnson, D. H. Johnson, S. Bucholtz, S. Hyberg, and S. Howlin. 2008. Estimating the response of ring-necked pheasants (Phasianus colchicus) to the Conservation Reserve Program. Auk 125:434–444.

Ollerton, J., R. Winfree, and S. Tarrant. 2011. How many flowering plants are pollinated by animals? Oikos 120:321–326.

Otto, C. R., C. L. Roth, B. L. Carlson, and M. D. Smart. 2016. Land-use change reduces habitat suitability for supporting managed honey bee colonies in the Northern Great Plains. Proceedings of the National Academy of Sciences 113:10430–10435.

Otto, C., S. O'Dell, R. Bryant, N. Euliss, R. Bush, and M. Smart. 2017. Using publicly available data to quantify plant-pollinator interactions and evaluate conservation seeding mixes in the Northern Great Plains. Environmental Entomology 46:565–578.

Otto, C. R. V., H. Zheng, A. L. Gallant, R. Iovanna, B. L. Carlson, M. D. Smart, and S. Hyberg. 2018. Past role and future outlook of the Conservation Reserve Program for supporting honey bees in the Great Plains. Proceedings of the National Academy of Sciences, 115:7629–7634.

Park, M. G., E. Blitzer, J. Gibbs, J. E. Losey, and B. N. Danforth. 2015. Negative effects of pesticides on wild bee communities can be buffered by landscape context. Proceedings of the Royal Society B 282:20150299.

Piñeiro, G., E. G. Jobbágy, J. Baker, B. C. Murray, and R. B. Jackson. 2009. Set-asides can be better climate investment than corn ethanol. Ecological Applications 19:277–282.

Plush, C. J., C. E. Moorman, D. Orr, and C. Reberg-Horton. 2013. Overwintering sparrow use of field borders planted as beneficial insect habitat. Journal of Wildlife Management 77:200–206.

Polasky, S., E. Nelson, D. Pennington, and K. A. Johnson. 2011. The impact of land-use change on ecosystem services, biodiversity and returns to landowners: A case study in the state of Minnesota. Environmental and Resource Economics 48:219–242.

Pollinator Health Task Force. 2015. National strategy to promote the health of honey bees and other pollinators. Washington, DC: The White House.

Primack, R. B. 2002. Habitat destruction fragmentation, degradation, and global climate change. In Essentials of Conservation Biology, 213–263. Sunderland, MA: Sinauer.

Randall, G., D. Huggins, M. Russelle, D. Fuchs, W. Nelson, and J. Anderson. 1997. Nitrate losses through subsurface tile drainage in conservation reserve program, alfalfa, and row crop systems. Journal of Environmental Quality 26:1240–1247.

Reeder, K. F., D. M. Debinski, and B. J. Danielson. 2005. Factors affecting butterfly use of filter strips in Midwestern USA. Agriculture, Ecosystems and Environment 109:40–47.

Reitsma, K. D., D. E. Clay, S. A. Clay, B. H. Dunn, and C. Reese. 2016. Does the U.S. cropland data layer provide an accurate benchmark for land-use change estimates? Agronomy Journal 108:266–272.

Relyea, R. A. 2005. The lethal impact of Roundup® on aquatic and terrestrial amphibians. Ecological Applications 15:1118–1124.

Relyea, R. A. 2009. A cocktail of contaminants: How mixtures of pesticides at low concentrations affect aquatic communities. Oecologia 159:363–376.

Reynolds, R. E., T. L. Shaffer, R. W. Renner, W. E. Newton, and B. D. Batt. 2001. Impact of the Conservation Reserve Program on duck recruitment in the U.S. Prairie Pothole Region. Journal of Wildlife Management 65:765–780.

Ryan, M. R., L. W. Burger, and E. W. Kurzejeski. 1998. The impact of CRP on avian wildlife: A review. Journal of Production Agriculture 11:61–66.

Samson, F., F. Knopf, and W. Ostlie. 2004. Great Plains ecosystems: Past, present, and future. Wildlife Society Bulletin 32:6–15.

SAS Institute Inc. 2013. SAS/STAT 13.1 User's guide. Cary, NC: SAS Institute.

Sauer, J., J. Hines, J. Fallon, K. Pardieck, D. Ziolkowski Jr., and W. Link. 2014. The North American Breeding Bird Survey, results and analysis 1966–2013. Laurel, MD: US Geological Survey, Patuxent Wildlife Research Center.

Scholtz, R., J. Polo, S. Fuhlendorf, and G. Duckworth. 2017. Land cover dynamics influence distribution of breeding birds in the Great Plains, USA. Biological Conservation 209:323–331.

Schulte, L. A., J. Niemi, M. J. Helmers, M. Liebman, J. G. Arbuckle, D. E. James, R. K. Kolka, et al. 2017. Prairie strips improve biodiversity and the delivery of multiple ecosystem services from corn-soybean croplands. Proceedings of the National Academy of Sciences 114:11247–11252.

Searchinger, T., R. Heimlich, R. A. Houghton, F. Dong, A. Elobeid, J. Fabiosa, S. Tokgoz, D. Hayes, and T.-H. Yu. 2008. Use of U.S. croplands for biofuels increases greenhouse gases through emissions from land-use change. Science 319:1238–1240.

Semmens, B. X., D. J. Semmens, W. E. Thogmartin, R. Wiederholt, L. López-Hoffman, J. E. Diffendorfer, J. M. Pleasants, K. S. Oberhauser, and O. R. Taylor. 2016. Quasi-extinction risk and population targets for the eastern, migratory population of monarch butterflies (*Danaus plexippus*). Scientific Reports 6:23265.

Smart, M., J. Pettis, N. Rice, Z. Browning, and M. Spivak. 2016a. Linking measures of colony and individual honey bee health to survival among apiaries exposed to varying agricultural land use. PLOS One 11:e0152685.

Smart, M. D., J. S. Pettis, N. H. Euliss Jr., and M. Spivak. 2016b. Land use in the Northern Great Plains region of the U.S. influences the survival and productivity of honey bee colonies. Agriculture Ecosystems and Environment 230:139–149.

Smart, M. D., C. R. V. Otto, B. L. Carlson, and C. L. Roth 2018. The influence of spatiotemporally decoupled land use on honey bee colony health and pollination service delivery. Environmental Research Letters 13:84016.

Spivak, M., Z. Browning, M. Goblirsch, K. Lee, C. Otto, M. Smart, and J. Wu-Smart. 2017. Why does bee health matter? The science surrounding honey bee health concerns and what we can do about it. Ames, IA: Council for Agricultural Science and Technology.

Stanley, D. A., M. P. D. Garratt, J. B. Wickens, V. J. Wickens, S. G. Potts, and N. E. Raine. 2015. Neonicotinoid pesticide exposure impairs crop pollination services provided by bumblebees. Nature 528:548–550.

Stenoien, C., K. R. Nail, J. M. Zalucki, H. Parry, K. S. Oberhauser, and M. P. Zalucki. 2016. Monarchs in decline: A collateral landscape-level effect of modern agriculture. Insect Science 23.

Stephens, S. E., J. A. Walker, D. R. Blunck, A. Jayaraman, D. E. Naugle, J. K. Ringelman, and A. J. Smith. 2008. Predicting risk of habitat conversion in native temperate grasslands. Conservation Biology 22:1320–1330.

Tenenbaum, D. J. 2008. Food vs. fuel: Diversion of crops could cause more hunger. Environmental Health Perspectives 116:A254.

Thogmartin, W. E., L. López-Hoffman, J. Rohweder, J. Diffendorfer, R. Drum, D. Semmens, S. Black, et al.

2017. Restoring monarch butterfly habitat in the Midwestern US: "All hands on deck." Environmental Research Letters 12:074005.

Tyner, W. E. 2008. The U.S. ethanol and biofuels boom: Its origins, current status, and future prospects. America Institute of Biological Sciences Bulletin 58:646–653.

US Congress. 2007. Energy Independence and Security Act of 2007. http://www.gpo.gov/fdsys/pkg/BILLS-110hr6enr/pdf/BILLS-110hr6enr.pdf.

USDA-ERS (US Department of Agriculture, Economics Research Service). 2017. U.S. bioenergy statistics. https://www.ers.usda.gov/data-products/us-bioenergy-statistics/us-bioenergy-statistics/#Feedstocks.

USDA-OCE (US Department of Agriculture, Office of the Chief Economist). 2017. World agricultural supply and demand estimates. ISSN:1554-9089. https://www.usda.gov/oce/commodity/wasde/.

USDA-NASS (US Department of Agriculture, National Agricultural Statistics Service). 2014. Honey. ISSN: 1949-1492. https://www.nass.usda.gov/Publications/Todays_Reports/reports/hony0318.pdf.

Wang, M., C. Saricks, and D. Santini. 1999. Effects of fuel ethanol use on fuel-cycle energy and greenhouse gas emissions. Center for Transportation Research, Energy Systems Division, Argonne National Laboratory. ANL/ESD-38:1–39.

Werling, B. P., T. L. Dickson, R. Isaacs, H. Gaines, C. Gratton, K. L. Gross, H. Liere, C. M. Malmstrom, T. D. Meehan, and L. Ruan. 2014. Perennial grasslands enhance biodiversity and multiple ecosystem services in bioenergy landscapes. Proceedings of the National Academy of Sciences 111:1652–1657.

Whitehorn, P. R., S. O'Connor, F. L. Wackers, and D. Goulson. 2012. Neonicotinoid pesticide reduces bumble bee colony growth and queen production. Science 336:351–352.

Wright, C. K., and M. C. Wimberly. 2013. Recent land use change in the western corn belt threatens grasslands and wetlands. Proceedings of the National Academy of Sciences 110:4134–4139.

Wright, C. K., B. Larson, T. J. Lark, and H. K. Gibbs. 2017. Recent grassland losses are concentrated around U.S. ethanol refineries. Environmental Research Letters 12:044001.

Zhang, Y., S. Singh, and B. R. Bakshi. 2010. Accounting for ecosystem services in life cycle assessment, Part I: A critical review. Environmental Science and Technology 44:2232–2242.

Zheng, H., Y. Li, B. E. Robinson, G. Liu, D. Ma, F. Wang, F. Lu, Z. Ouyang, and G. C. Daily. 2016. Using ecosystem service trade-offs to inform water conservation policies and management practices. Frontiers in Ecology and the Environment 14:527–532.

Ziolkowski, D., K. Pardieck, and J. R. Sauer. 2010. On the road again for a bird survey that counts. Birding 42:32–41.

4

Susan P. Rupp and
Christine A. Ribic

Second-Generation Feedstocks from Dedicated Energy Crops

Implications for Wildlife and Wildlife Habitat

Introduction

Concerns about the cost of energy, US dependence on foreign oil, greenhouse gas emissions, and increasing global energy demand are bringing attention to bioenergy (i.e., heat, fuel, and power produced from organic material). Biomass constitutes almost 80% of the global renewable energy portfolio (Immerzeel et al. 2014). In the United States, renewable energy accounts for roughly 11% of the energy portfolio, of which roughly 45% is supplied by biomass (Energy Information Administration 2018). Pelletized biomass (Figure 4.1) to generate thermal and electric energy is the most active current market (see Chapter 2, Homyack and Verschuyl), but there is much interest in liquid biofuels from cellulosic sources.

Conventional biofuels (i.e., first-generation biofuels) are those produced from corn-kernel starch, soybeans (e.g., biomass-based diesel), and other row crops (see Chapter 3, Otto). Because first-generation biofuels use crops also grown for food, concerns have been voiced about the diversion of food crops to fuel purposes (McGuire 2012, Manning et al. 2015), as well as the potential environmental impacts of biofuel crop production due to increased water and fertilizer use and soil erosion caused by conventional tilling methods (Alshawaf et al. 2016, Liu et al. 2017,

Hoekman et al. 2018). In contrast, second-generation biofuels, or cellulosic fuels, are derived from *nonfood* feedstocks, such as field crop residues (e.g., corn stover), forest product residues (e.g., chips, mulch), organic waste (e.g., manure, municipal solid waste), or fast-growing dedicated energy crops such as switchgrass (*Panicum virgatum*) or giant miscanthus (*Miscanthus* x *giganteus*; Marquis 2016). Because second-generation biofuels are not used for food, their use for fuel may cause fewer social and environmental concerns.

Perennial grasses are considered prime candidates for cellulosic bioenergy production because of their high biomass production, ability to sequester carbon, tolerance to variable climatic conditions, compatibility with conventional farming practices, and minimal maintenance requirements (Graham et al. 1995, Tolbert and Schiller 1995, McLaughlin et al. 1999, Rinehart 2006, Kibit et al. 2016). Additionally, they can be derived as a coproduct of row crops grown for first-generation biofuels or food (e.g., cornfields with switchgrass borders, prairie STRIPS [Schulte et al. 2017], or use of cover crops) or grown as dedicated energy crops on lands too poor (e.g., steep, rocky, nutrient-poor) for food production. In 2011, the US Department of Energy (DOE) estimated that monocultures of perennial grasses (e.g., switchgrass, mis-

Figure 4.1   Pelletized biomass can be made out of any plant material and is widely used to generate thermal and electric energy. *Photo courtesy of Bill McGuire.*

canthus) alone could produce between 90 and 154 million dry tons (82–140 million Mg) by 2017 and 255 to 462 million tons (231–419 million Mg) by 2030 (US Department of Energy 2011); however, yields may be affected by soil quality, previous land use, and use of soil amendments (e.g., fertilizers).

Using perennial plants instead of annual row crops as feedstocks may also tend to have lower fertilizer and pesticide requirements than many first-generation feedstocks, which reduces cost to producers, limits nutrient runoff to waterways, and reduces greenhouse gas emissions (Adler et al. 2009, Davis et al. 2010)—especially compared with corn, for which greater fertilizer application rates are correlated with emissions of nitrous oxide (N_2O), a greenhouse gas with 310 times the warming potential of CO_2 (Davis et al. 2012, Smith and Scarchinger 2012). The dense rooting systems of many second-generation feedstocks also improve the physical characteristics of soil and reduce soil erosion that can prevent downstream flooding and phosphorous transport by changing surface hydrology (Blanco-Canqui 2010, Tufekcioglu et al. 2003). Wildlife populations can benefit from the increased vertical and horizontal structure these feedstocks create, which may provide foraging and nesting substrates and protective cover, while reduced use of pesticides and

herbicides means better habitat for supporting pollinators (Fargione et al. 2008, Gardiner et al. 2009, Fletcher et al. 2011, Wiens et al. 2011).

Second-generation feedstock cultivars continue to be evaluated, and varieties are genetically modified to increase yield while minimizing land requirements and need for external inputs. However, several factors need to be considered to make these new feedstocks socially, environmentally, and economically feasible. As with first-generation feedstock crops, growers emphasize maximizing production and yield while minimizing up-front costs in establishment (i.e., seed cost and availability, site preparation), reducing maintenance (i.e., fertilizers, pesticides, herbicides, water requirements), and increasing ease of harvest (i.e., using existing equipment as opposed to developing new technology specific to individual crops). Growers also want feedstocks that are reliable, drought and pest resistant, and rapidly established. If there are alternative uses for a feedstock or its by products (i.e., "added-value"), such as feed for livestock, bedding for poultry, fertilizer (in the case of animal wastes), or bio-based chemicals that can be used in cosmetics, pharmaceuticals, or plastics, that is even more desirable.

However, native wildlife species abundance and diversity as well as other ecosystem services (e.g., air, water, and soil quality/quantity, and reduced greenhouse gas emissions) may be affected by increasing bioenergy development regardless of the feedstock used because the characteristics of ideal bioenergy crops can be in direct conflict with characteristics that make habitat attractive to wildlife (Table 4.1). Potential effects on wildlife depend on the type of land that is replaced, planting density, stand management, and surrounding land uses (Bakker and Higgins 2009, McGuire 2012, Rupp et al. 2012, McGuire and Rupp 2013, Pedroli et al. 2013, Robertson et al. 2013). In addition, bioenergy crop production can raise concerns about agricultural intensification, potential for species' invasiveness, feedstock location and geographic context within the landscape, and crop diversity (Fletcher et al. 2010, Wiens et al. 2011,

Table 4.1. Characteristic traits of dedicated bioenergy crops versus desired wildlife habitat. Items listed in bold represent traits for which there is a commonality (from McGuire 2012 and Rupp et al. 2012).

Dedicated Bioenergy Crops	Desired Wildlife Habitat
Monoculture crop	Diverse plant community
Highly productive crops	Non-commodity emphasis
Aggressive, often non-native, hybrid, or genetically modified plants	Native and nonaggressive plants
Higher plant density	Lower plant density
Harvest after dormancy[a]	**Avoid nesting disturbance**
Remove all useable biomass	Leave some residual biomass
Irrigate if needed	Conserve water
Use fertilizer/pesticides when needed	Avoid fertilizers and pesticides

[a] Fall-season harvest common for some bioenergy crops (e.g., switchgrass and miscanthus), but not others. Ratooning crops (e.g., sorghum) may be harvested several times per year.

Rupp et al. 2012, Immerzeel et al. 2014, Tarr et al. 2017). Ultimately, effects of bioenergy development will vary across spatial and temporal scales, geographic regions, wildlife species considered, feedstock types, feedstock management practices, and land-use histories (McGuire 2012, Rupp et al. 2012, Immerzeel et al. 2014, Tarr et al. 2017, Hoekman and Broch 2018, Hoekman et al. 2018).

Introduction to Common Feedstocks

Of the many species of perennial forage crops available, only a few have been researched intensively as biomass feedstock for commercial-scale bioenergy production as part of regional Coordinated Agricultural Projects (CAPs) awarded by the USDA National Institute of Food and Agriculture (NIFA) Agriculture and Food Research Initiative (AFRI; Table 4.2). The primary purpose of CAPs is to enhance rural prosperity and national energy security through the development of regional systems for sustainable production of advanced biofuels and industrial chemicals (i.e., bio-based products). A brief overview of the more common feedstocks investigated by CAPs is provided below, although it should be noted that ad-

ditional feedstocks continually are evaluated to determine their potential for large-scale bioenergy production.

Switchgrass—a native perennial grass—was chosen as the model herbaceous bioenergy crop in DOE study trials because of its endemic nature, relative abundance, high biomass production, tolerance to climatic conditions, compatibility with conventional farming practices, and minimal maintenance requirements (McLaughlin et al. 1999, Adler et al. 2006, Rinehart 2006, Wright 2007, Mitchell et al. 2008, Sanderson and Adler 2008, Lee et al. 2018). Switchgrass can be used as a source of cellulosic ethanol or combined with other energy sources (e.g., coal) to produce heat or electricity. In recent years, there has been increased use of switchgrass for direct combustion (e.g., switchgrass pellets or briquettes used to generate heat) and as a feedstock co-fired with coal for heat or electricity production to lower emissions associated with burning coal alone (Rinehart 2006). Though biomass yields of switchgrass cultivars are not as bountiful as those of some other available feedstocks, its other benefits still make it attractive as a bioenergy feedstock. This has led to several commercial-scale, pre-operational testing facilities using switchgrass as a primary feedstock over the past several years; however, political uncertainty and costs of converting cellulosic material to useable energy have caused several of those facilities to close (e.g., Abengoa's project in Hugoton, Kansas; Lane 2016).

Miscanthus (*Miscanthus* spp.), a species native to Asia, has been grown for many years in the United States as a landscape plant, but it was not proposed for evaluation in early bioenergy crop screening studies (Wright 2007). However, given the high yields that have been achieved in Europe (up to 40 Mg ha^{-1} in some regions; Lee et al. 2018) and on small-scale field plots in the United States with a single sterile hybrid cultivar, giant miscanthus, questions are being asked about how it compares with switchgrass as a model bioenergy crop (Heaton et al. 2004a, Wright 2007, Lee et al. 2018). Yields from US studies typically

Table 4.2.   USDA Coordinated Agricultural Projects and their focal bioenergy feedstock (USDA National Institute of Food and Agriculture 2017, 2018)

US Region	Coordinated Agricultural Project	Focal Feedstock	Website
Pacific Northwest	Advanced Hardwood Biofuels Northwest	poplar	http://hardwoodbiofuels.org/
Pacific Northwest	Advanced Renewables Alliance	forest residues	https://nararenewables.org/
Rocky Mountains	Bioenergy Alliance Network of the Rockies	beetle-killed pine	http://banr.nrel.colostate.edu/
Southwest	Sustainable Bio-economy for Arid Regions	guayule (flowering shrub), guar (legume)	https://energy.arizona.edu/sbar
Midwest	CenUSA Bioenergy	switchgrass, big bluestem (*Andropogon gerardii*), indiangrass (*Sorghastrum nutans*)	https://cenusa.iastate.edu/
Northeast	Woody/Warm-season Biomass Consortium	willow, switchgrass, miscanthus	http://www.newbio.psu.edu/
Southeast	Partnership for Advanced Renewables	carinata (oilseed)	https://sparc-cap.org/
Southeast	Partnership for Integrated Biomass Supply Systems	switchgrass, poplar, eucalyptus, pine	http://www.se-ibss.org/
South	Sustainable Bioproduct Initiative Project	sorghum, energy cane	https://www.facebook.com /agcentersubi/

average about 23 Mg ha^{-1}, but much lower values also have been reported, bringing into question the ability of the feedstock to reliably reach the higher yields reported overseas (Lee et al. 2018). Giant miscanthus has been bred to be sterile and produce no seeds. It is established by planting rhizome cuttings started in greenhouses, which delays its full production for 2 to 3 years (Heaton et al. 2004a). Reports from Europe indicate that, once established, plants can live in excess of 30 years without a significant loss in yield (Schill 2007). However, studies in the United States documented declines in yield over time that were proportionately greater in giant miscanthus than in switchgrass, suggesting a stronger effect of stand age on giant miscanthus (e.g., Arundale et al. 2014).

Reed canarygrass (*Phalaris arundinacea*) is a sod-forming, potentially invasive rhizomatous perennial grass native to portions of North America, Europe, and Asia that grows in wet soils across a wide pH range, making it a suitable contender for bioenergy production in areas where feedstocks such as switchgrass may not grow (Lewandowski et al. 2003, Sanderson and Adler 2008, Tahir et al. 2010, Glaser and Glick 2012, Rancane et al. 2012). In suitable con-

ditions (i.e., aerated humus and nutrient-rich soils), reed canarygrass yields of over 8.16 Mg ha^{-1} yr^{-1} of dry matter can be expected (Rancane et al. 2012). A unique characteristic of reed canarygrass is that its biomass increases linearly with applied nitrogen, making it a potentially useful feedstock to use in situations where the ability to take up high nutrient levels is necessary (e.g., disposal of manure from intensive industrial livestock and poultry farms or at municipal wastewater facilities). It also can improve the structure of clay-based soil (Cherney et al. 1991, Tahir et al. 2010).

Giant reed (*Arundo donax*) is a non-native, perennial, cool-season grass from the Mediterranean region that has become a serious pest in tropical and temperate parts of the world; it is now on the list of 100 of the World's Worst Alien Species (Boland 2006, Glaser and Glick 2012, EDDMapS 2018, Global Invasive Species Database 2018). It reproduces vegetatively through rhizomes and vegetative fragments (i.e., rooting of nodes), making it particularly prone to invasion via flooding, and it severely degrades wildlands by altering native vegetation structure, displacing native plant species, reducing habitat quality, and increasing fire frequencies (Boland 2006,

Glaser and Glick 2012). Despite these concerns, it can produce over 30 Mg ha^{-1} yr^{-1} and was approved as a renewable feedstock by the US Environmental Protection Agency (EPA; Ingwell et al. 2014).

Napiergrass (*Pennisetum purpureum*), also known as elephant grass, is an invasive, non-native, clump-forming grass indigenous to Africa. It is ideally suited as a bioenergy crop because it is established with just one seeding, requires little water or added nutrients, and develops deep root systems that improve soil fertility and minimize erosion (USDA 2016). However, unlike switchgrass, the thick-stemmed napiergrass requires specialized harvesting equipment and needs to be left in the field longer to remove moisture content, which may delay harvesting (Lewandowski et al. 2003). In the 1980s, napiergrass was used as a feedstock for methane generation and also was co-fired with coal to produce electricity (Glaser and Glick 2012). Today, it is being considered as a feedstock for several kinds of biofuels, including renewable jet fuel (USDA 2016). It is attractive as a feedstock because it can yield up to 20 tons ac^{-1} yr^{-1} (44.8 Mg ha^{-1} yr^{-1}) of biomass and can be harvested twice per year (Glaser and Glick 2012).

Energycane (sugarcane, *Saccharum* spp., x wild cane, *Saccharum spontaneum*) is being researched to supplement sugarcane production by providing an economically viable and yearlong sustainable biorefinery in the southeastern United States (Kim and Day 2011, Han et al. 2012). Energycane, like sugarcane, is a tropical perennial that is vegetatively propagated. Unlike most other summer crops, energycane is established in the fall from mature canes of existing plants (Lee et al. 2018). Energycane has been bred for higher fiber to sugar ratios, thinner stalks, elevated plant density population traits (compared with sugarcane), disease and pest resistance, ratooning performance (i.e., ability to resprout/stubble crop), cold tolerance, and vigor (Matsuoka et al. 2014, Lee et al. 2018). Five to 6 harvests within a year for the L79-1002 variety of energycane are common (Kim and Day 2011, Han et al. 2012). In the United States, yields range from 14.0 Mg ha^{-1} in Georgia to 39.7 Mg ha^{-1} in Hawai'i, and stems contain substantial amounts of sugar that can be exploited through extraction (pressing) or via in situ fermentation (Lee et al. 2018).

Sorghum (*Sorghum bicolor*) is unique among energy crops because different sorghum varieties (e.g., grain, energy, or sweet) can produce economic quantities of different products in differing proportions (e.g., starch, sugar, and lignocellulosic biomass), all of which can be used for biofuel or bioproduct production (Almodares and Hadi 2009, Monge et al. 2014, Viator et al. 2014, Lee et al. 2018). Sorghum can produce high biomass yields (~40 Mg ha^{-1}) and it is widely adapted and highly amenable to US production and cultivation systems, though yields are highly sensitive to environmental conditions (Lee et al. 2018). In addition, the bagasse (i.e., dry pulpy residue left after the extraction of juice) can be used to make bioproducts from the remaining cellulose, hemicellulose, and lignin or burned for power generation (Kim and Day 2011, Lee et al. 2018).

Invasive Potential

Crops bred for biofuels or other bioenergy production inherently have several characteristics associated with high risk of invasiveness: these crops are robust, outcompete other plants, thrive in a variety of conditions, and are resistant to pests and diseases (Glaser and Glick 2012; Table 4.3). Negative effects of invasive species on native wildlife and the environment are well documented (e.g., McGuire 2012, Glaser and Glick 2012, National Invasive Species Council 2016), and damage control can reach into the billions of dollars annually (National Invasive Species Council 2016). Researchers estimate that nearly half of the species that are listed as threatened or endangered under the US Endangered Species Act of 1973 (16 U.S.C. § 1531 et seq.) are at risk due, at least in part, to the impacts of invasive species (Glaser and Glick 2012).

Several studies have assessed the potential invasive qualities of feedstocks grown for bioenergy pro-

from the vertical position) in the field after senescence (Heaton et al. 2004b). Such characteristics suggest it may offer winter thermal cover for wildlife in more northerly latitudes, assuming some stubble remains in the field.

Though we are not aware of any studies on wildlife use of miscanthus fields in the United States, studies in the United Kingdom have shown that farmland birds use miscanthus stands during the establishment phase, early in the growing season, or when weeds are present (i.e., in failed plantings; Semere and Slater 2007, Bellamy et al. 2009, Sage et al. 2010, Bright et al. 2013). Small mammals appear to use margins of miscanthus fields that are next to natural areas (Clapham and Slater 2008, Semere and Slater 2007). Bellamy et al. (2009) concluded that managing miscanthus fields for wildlife was not compatible with managing miscanthus fields for maximum yield (i.e., bioenergy production). However, McCalmont et al. (2017) suggested there may be some indirect wildlife benefits, such as increased relative habitat quality of natural areas adjacent to miscanthus fields (e.g., due to reduction in pesticide use and soil erosion), as well as increased winter cover provided by miscanthus.

Stand Management

Stand management for different biomass crops (e.g., cutting height, timing of harvest, seeding density) affects stand quality for use by specific wildlife species (Maves 2011, Rupp et al. 2012, Uden et al. 2015, Moorman et al 2017). Seeding rates vary for different feedstocks to encourage optimum density for biomass production, although attractiveness to most species of wildlife declines with increasing stand density (Rupp et al. 2012, McGuire and Rupp 2013). This is an especially important factor for the young of ground-nesting birds that have trouble moving through dense vegetation. Many wildlife species prefer open ground conditions that are prominent during the establishment period of perennial biomass crops when stands have the greatest plant and structural diversity; as stands mature, the wildlife species composition changes (McGuire and Rupp 2013). Periodic disturbance of perennial grasslands—through harvesting or mechanical manipulations such as discing—helps maintain plant diversity and thereby benefits wildlife species that use early successional vegetation (Rupp et al. 2012, McGuire and Rupp 2013, Uden et al. 2015, Tarr et al. 2017).

Harvest timing and intensity is feedstock-dependent and affects both the quality of the biomass produced and potential wildlife responses (Rupp et al. 2012, Tarr et al. 2017). Dormant-season harvesting maximizes cellulosic content of standing biomass, minimizes process contaminants such as metals, reduces moisture, and reduces fertilizer inputs needed to maintain a healthy stand (Adler et al. 2006, Harper and Keyser 2008, Caslin et al. 2010, Lee et al. 2018). Whereas removal of standing biomass in the fall is beneficial to avoid impacts to wildlife during the nesting, fawning, or brood-rearing seasons, it may negatively impact wildlife if fields are used for winter thermal or escape cover (Harper and Keyser 2008, Rupp et al. 2012, McGuire and Rupp 2013). Implementing a spring harvest instead may maintain winter cover for resident wildlife species while simultaneously increasing biomass quality for certain bioenergy applications (Murray and Best 2003, Adler et al. 2006); however, yields may decrease as much as 40% because of lodging in the field caused by snow in northern latitudes (Adler et al. 2006).

Following a 1- to 3-year establishment phase, switchgrass and miscanthus are typically harvested in the fall after the first hard frost, when plants have moved the bulk of aboveground nutrients back to their roots (Lee et al. 2007, Mitchell et al. 2008, Caslin et al. 2010, Lee et al. 2018). In contrast, reed canarygrass—a cool-season grass—can be harvested either in the fall after the first hard frost or in early summer when warm-season grass biomass is not available, facilitating a constant feedstock flow to bioreactors or power plants (Sanderson and Adler 2008, Tahir et al. 2010, Glaser and Glick 2012). Potential negative effects on nesting, fawning, or brood rearing for wildlife species may become a concern during early harvest periods. Napiergrass and energycane can be harvested twice annually. However, recent studies on harvest frequency and timing for napiergrass and energycane grown in Florida reported that 2 annual harvests negatively affected long-term biomass yield and plant persistence (Na et al. 2015a, 2015b); therefore, a single fall harvest appears to maximize the concentration of cellulose in total biomass for both feedstocks. Delaying harvest from early November to early December increases nonstructural carbohydrates and soluble sugars in energycane but not in napiergrass (Na et al. 2016).

As for sweet sorghum, research indicates harvest should be delayed until seeds reach the "hard dough" stage of development, when sugar content and fresh yields are maximized; early maturing sweet sorghums, which are photoperiod insensitive, typically mature in approximately 90 days and can be harvested multiple times each year, whereas certain full-season varieties or hybrids can take more than 150 days to mature (Viator et al. 2014). However, ratoon growth in sweet sorghum is inconsistent because it is dependent on late-season rainfall, and varying the planting date to provide biomass or fermentable sugar over a longer period of time appears to be a better alternative than depending on ratoon cropping of sweet sorghum (Viator et al. 2014).

Biomass harvesting in the fall or winter may have indirect effects on wildlife species that use fields the following spring/summer, and reproductive success might change because of changes in vegetation structure caused by harvesting. For example, Murray and Best (2003) reported that abundances of different grassland bird species changed based on differences in vegetation structure present in totally harvested, partially harvested, and unharvested switchgrass fields, although total bird abundance and species richness did not. Maves (2011) determined that the most important factor affecting grassland songbird use of switchgrass-dominated fields after a fall harvest was structural diversity the following spring determined by harvest intensity. These results are not surprising, given the established importance of vegetation structure to grassland birds (Sample and Mossman 1997, Askins et al. 2007). Ultimately, producers that can vary harvest intensities can help create a mosaic of diverse vegetative structures and grassland bird communities on the landscape (Sample and Mossman 1997).

For pheasants and waterfowl, Bender (2012) reported no effect of residual stubble height on diversity, abundance, or nest success during summer surveys following a fall harvest of the same switchgrass-dominated fields studied by Maves (2011). However, Bender (2012) documented a significant difference in nest initiation dates for hen mallards (*Anas platyrhynchos*); nests were initiated a full month later on harvested sites than on nonharvested sites. In contrast, there was no effect of a single fall biomass harvest on nest initiation or nest success the following breeding season for ducks and pheasants on switchgrass-dominated US Waterfowl Production Areas in southwestern Minnesota, though nest density was greater in the uncut fields (Jungers et al. 2015). Given such conflicting results, long-term implications of a fall harvest on game bird production, maturation prior to migration from more northerly latitudes, and overall fitness are unclear (Bender 2012).

Land-Use Change and Landscape Composition

With changes in land use potentially shifting to more monoculture crops for bioenergy production, there is increased concern over effects of bioenergy development on terrestrial and aquatic biodiversity (Gasparatos et al. 2011, Joly et al. 2015). In a review of biodiversity impacts from 53 studies of bioenergy crop production, Immerzeel et al. (2014) concluded that land-use change (i.e., conversion from natural plant communities or shifts in land use from one crop type to another) appeared to be the key driver of biodiversity change in both temperate and tropical regions. Whereas terrestrial biodiversity is more likely to be affected by land-use change associated with expanded biofuel development, aquatic biodiversity is at risk from impacts to water quality, including increased sedimentation, nutrient, and pesticide runoff that typically accompany land conversion (Hoekman and Broch 2018).

Additionally, landscape heterogeneity contributes to greater species abundance and species diversity, increases the quality of soil, and provides greater genetic variation within communities (Hartman et al. 2011). Hartman et al. (2011) concluded that one of the central tenets associated with maximizing structural and functional characteristics of grassland ecosystems following switchgrass cultivation is the maintenance of landscape heterogeneity, which can be maximized by altered harvest rotations, no-till farming, and mixed species composition. In a meta-analysis of crops being considered for feedstock production in the United States, Fletcher et al. (2011) concluded that vertebrate densities and abundances were typically lower in biofuel crops than in the non-biofuel crops they replaced, with effects generally greater for corn than pine (*Pinus* spp.) or poplar (*Poplar* spp.); they predicted birds of conservation concern would experience greater negative effects from corn production for biofuel than would more generalist species. Fletcher et al. (2011) also determined that conversion of row crop fields to grasslands dedicated for bioenergy production could have positive effects on local vertebrate diversity and abundance of birds, especially if grass bioenergy systems are polycultures (see Monoculture Compared to Polyculture Feedstocks, below).

If there were a shift to more cellulose-based feedstocks, the general consensus is that most of that development would occur on "marginal lands" (i.e., lands not highly suited for row crop production; Kang et al. 2013, Stoof et al. 2015). These lands, however, are highly valuable to wildlife, given that agricultural production and other forms of human land use have relegated wildlife into what habitat still remains (Samson and Knopf 1994, Askins et al.

2007). An example of this is Conservation Reserve Program (CRP) lands, which have been suggested for bioenergy production because they are often perceived as underutilized land that could be put into other uses while at the same time not competing with food production (Fargione et al. 2009, Lee et al. 2018). However, the importance of CRP lands to wildlife is well documented (e.g., Johnson 2005, Reynolds 2005, Niemuth et al. 2007), suggesting caution if CRP areas are targeted for bioenergy production.

Monoculture Compared to Polyculture Feedstocks

There is growing interest in the possible use of multispecies crops as feedstocks for bioenergy production. Polyculture plantings are likely to be more beneficial to wildlife communities than monotypic stands of other feedstocks, because polyculture plantings have more compositional and structural (i.e., horizontal and vertical) diversity (Rupp et al. 2012). This greater plant diversity, in turn, increases biological diversity and ecosystem resilience. However, landscape context and land-use change are important drivers that need to be taken into consideration (Joly et al. 2015, Manning et al. 2015). Landscape composition can have an important effect on wildlife. Grassland bird use of switchgrass and other perennial grass fields during the breeding season (reviewed in Robertson et al. 2012b) and migratory periods (Robertson et al. 2011, 2013) is affected by landscape composition (e.g., increased bird density in fields). If traditional row crops such as soybeans and corn are replaced with polyculture plantings, then habitat for row crop–averse wildlife groups will likely increase (Rupp et al. 2012, McGuire and Rupp 2013, Klimstra et al. 2015, Moorman et al. 2017). However, if bioenergy crops (including polyculture crops) replace natural areas and are improperly managed (e.g., intensive harvest schedule, planted to pure grass stands lacking forbs), additional habitat could be lost and populations of native wildlife may be further diminished (McGuire and Rupp 2013, Klimstra et al. 2015, Moorman et al. 2017).

When compared with monoculture plantings, polyculture stands of native grass species benefit wildlife. Studies indicate replanted native grass mixes used for bioenergy support more arthropod diversity (including pollinators when forbs are present) than monocultures of crops or bioenergy plantings, but not as much as unbroken native prairies (Landis and Werling 2010, Robertson et al. 2012a). Robertson et al. (2012a) report mixed-grass–forb prairie plantings were associated with a 324% increase in arthropod family diversity and a 2,700% increase in arthropod biomass compared with annually planted corn. Several grassland bird species may also benefit from use of mixed native grass polycultures for bioenergy production (unless they replace native prairie), given extensive losses of native grasslands across the United States. However, management practices used to maintain or harvest polycultures will likely play an important role in determining habitat quality for birds and other wildlife (e.g., Birckhead et al. 2014, Klimstra et al. 2015, Moorman et al. 2017). For example, crop placement within the landscape may affect diversity and biomass of terrestrial arthropod communities and provisioning of arthropod functional groups responsible for important ecosystem services (Robertson et al. 2012a). Because mixed-grass polyculture feedstocks would be treated much the same as any other agricultural crop, issues caused by agricultural intensification—even with polyculture stands—may be a concern as bioenergy develops (Askins et al. 2007). These include direct and indirect mortality resulting from potential use of fertilizers or pesticides, harvest intensity and frequency, season of harvest, plant density, and use of agricultural machinery for planting, maintenance, and harvest.

Also, use of polyculture feedstocks can benefit bioenergy feedstock producers. Diverse mixtures tend to be more resilient to environmental changes (e.g., disease, pests, shifts in precipitation) and have reduced year-to-year variability of aboveground bio-mass compared with monoculture bioenergy crops (Jarchow and Liebman 2010, Picasso et al. 2011, Hong et al. 2013, McGuire and Rupp 2013). This, in turn, may lead to reduced dependence on external inputs, such as pesticides or herbicides, and lower overall risk to producers. Monoculture bioenergy crops have shown some susceptibility to insect pests, including corn leaf aphids (*Rhopalosiphum maidis*) in miscanthus (Huggett et al. 1999), switchgrass moths (*Blastobasis repartella*) in switchgrass (Torrez et al. 2013), and cereal leaf beetles (*Oulema melanopus*) in reed canarygrass (Wilson and Shade 1966), each of which can reduce yields and potentially transmit disease (Huggett et al. 1999). However, introducing additional perennial grass species into bioenergy crops also increases the complexity of technology needed to convert such crops into fuels or other bio-based products, thus increasing production costs (Griffith et al. 2011; Figure 4.2).

Inconsistent or reduced yields may be problematic when considering the economic, social, and environmental benefits of perennial grass mixtures as a bioenergy feedstock (Mulkey et al. 2008, Griffith et al. 2011, Gamble et al. 2015, Anderson et al. 2016, Lee et al. 2018). In US South Dakota total biomass yield on marginal land of mixed-grass fields with >50% switchgrass averaged 0.8 short tons/ha

Figure 4.2   A side-by-side comparison of a native warm-season grass mixture (*left*) and giant miscanthus (*right*). *Photo courtesy of Bill McGuire.*

(0.73 Mg ha^{-1}) on sites with a residual stubble height of 10 cm, whereas sites harvested at 30 cm averaged 0.4 short tons/ha (0.36 Mg ha^{-1}; Bender 2012)—well below the target goal of 3.2 to 4 dry tons ha^{-1} yr^{-1} (2.9 to 3.6 Mg ha^{-1}) established by the DOE (English et al. 2006). However, in another South Dakota study, yields peaked at 5.46 Mg ha^{-1} in switchgrass-dominated fields and varied substantially with precipitation (Mulkey et al. 2006). The greatest effects on seasonal biomass production and changes in vegetation composition for polycultures may be caused by where the crops are grown (e.g., marginal versus "prime" lands), location-specific precipitation (especially below 50% of average), and use of supplemental nitrogen to maximize biomass yields (Mulkey et al. 2006, Anderson et al. 2016, Lee et al. 2018).

Current Status of Bioenergy Production and Cellulosic Biofuels

Lee et al. (2018) emphasized that herbaceous dedicated energy crops, including switchgrass, miscanthus, energycane, and sorghum, will play an important role in future sustainable bioenergy feedstock production, as first outlined in the 2005 Billion-Ton Study (US DOE 2005) followed by the 2011 Billion Ton Update (US DOE 2011) and, most recently, in the 2016 Billion Ton Report (US DOE 2017). However, converting perennial grasses into energy—especially diverse mixtures of prairie grasses—is more complicated and less consistent than converting monoculture row crops to fuel. Therefore, the desirable qualities often sought in feedstocks, including their capacity for high-yielding biomass, ease of conversion, and reliable production through time (Casler et al. 2009, Griffith et al. 2011), may be difficult to achieve in perennial grass and polyculture plantings. Such complications in technology generally have stagnated advancements in the cellulosic biofuel/bioenergy industry.

Though potential for greater commercial production of second-generation biofuels at a larger scale still exists, their expansion and advancement depend on the sociopolitical environment of renewable energy development. As part of the American Recovery and Reinvestment Act of 2009 (ARRA), renewed emphasis was put on renewable energy development (Mundaca and Richter 2015). The ARRA provided more than $90 billion in strategic clean energy investments and tax incentives to promote job creation and deployment of low-carbon technologies. On November 30, 2017, the EPA finalized a rule that established the required renewable fuel volumes under the Renewable Fuel Standard program for 2018 and biomass-based diesel for 2019 (Table 4.4). The established volumes were similar to those for 2017, but at the time of this writing, the 2018 Farm Bill is being

Table 4.4. Final volume requirements for Renewable Fuel Standard Program: Standards for 2018 and biomass-based diesel volume for 2019 compared to 2016 and 2017 (US Environmental Protection Agency 2017a, 2017b)

	Proposed Volume Requirements[a]			
	2016	2017	2018	2019
Cellulosic biofuel (million gallons)	230	311	288	n/a
Biomass-based diesel (billion gallons)	1.90	2.00	2.10[b]	2.10
Advanced biofuel (billion gallons)	3.61	4.28	4.29	n/a
Renewable fuel (billion gallons)	18.11	19.28	19.29	n/a

[a] All values are ethanol-equivalent on an energy content basis, except for BBD, which is biodiesel-equivalent.

[b] The 2018 BBD volume requirement was established in the 2017 final rule (81 FR 89746, December 12, 2016).

negotiated in Congress and it is unknown what changes will be made in the Energy Title (Title IX) (US Senate Committee on Agriculture, Nutrition and Forestry 2018). Originally developed as part of the 2002 Farm Bill, the Energy Title contains several programs developed to reduce energy costs in rural America, increase use and production of renewable energy in rural and farming communities, and spur development and innovation of a domestic bio-based economy (Congress.gov 2018; see Appendix 4.1 for details).

Although government policies and related industry activity are important for creating a market niche for cellulosic biofuels (Chen and Smith 2017), geographic, technical, and institutional barriers may limit industry development and regional branching (Kedron and Bagchi-Sen 2016). Another commonly cited barrier to development and expansion of both first- and second-generation biofuels is the "blend wall" (Peplow 2014)—an artificial barrier to the amount of 10% ethanol (i.e., E10) to be produced in the United States, according to the volumetric requirements set as part of the Renewable Fuel Standard program established by the Energy Independence and Security Act of 2007. The Renewable Fuel Standard program targeted the amount of conventional ethanol to be produced at 15 billion gallons/yr (56.8 billion liters/yr)—enough to supply a nationwide supply of E10 fuel—which was surpassed by production of corn ethanol in 2018. The ability to break the "blend wall" is there; the EPA allows use of 15% ethanol (E15) in cars, light trucks, and medium-duty vehicles produced in 2001 or later (US EPA 2017c), and the production of vehicles in the United States that can run on as much as 85% ethanol (E85) has increased over time (US DOE 2016). Another obstacle is that it still costs more to convert second-generation feedstocks to ethanol than it does to convert first-generation feedstocks. Recent estimates of biofuel production costs show that second-generation biofuels are 2 to 3 times more expensive than petroleum fuels on an energy-equivalent basis (Balan 2014).

Emergence of Bio-based Products

The emergence and proliferation of bio-based products has complemented the expansion of the biofuels industry, and they are now a significant component of the overall US bioeconomy. Bio-based products—which include diverse categories such as lubricants, cleaning products, inks, fertilizers, and bioplastics (Figure 4.3)—can be derived from the by-products of first- or second-generation feedstocks, as well as more advanced biofuels (e.g., algae; Golden et al. 2016). Established by the Farm Security and Rural Investment Act of 2002 (2002 Farm Bill) and strengthened by the Food, Conservation, and Energy Act of 2008 (2008 Farm Bill) and the Agriculture Act of 2014 (H.R. 2642, the 2014 Farm Bill), the USDA Bio-Preferred Program elevated attention to and support for bio-based products and is credited with transforming the marketplace for bio-based products and creating jobs in rural America (Golden et al. 2016). The USDA BioPreferred Program is transforming the marketplace for bio-based products through 2 initiatives: (1) mandatory purchasing requirements for federal agencies and federal contractors; and (2) voluntary product certification and labeling.

Though only a few life-cycle analyses of the production of bio-based products have been conducted, key environmental benefits of manufacturing and using bio-based products include reducing use of fossil fuels and reducing associated greenhouse gas emissions (Golden et al. 2016). Using a modified life-cycle assessment, Golden et al. (2016) estimated the petroleum saved by the bio-based products industry may have been as much as 6.8 million barrels (1.1 million kl) of oil, and greenhouse gas emission reductions may have been be as much as 10 million metric tons of CO_2 equivalents in 2014. However, they also stated that, given the increasing interest in and use of bio-based products, it is essential to conduct additional analyses of their potential impacts on water quality, water use, land use, and other ecosystem

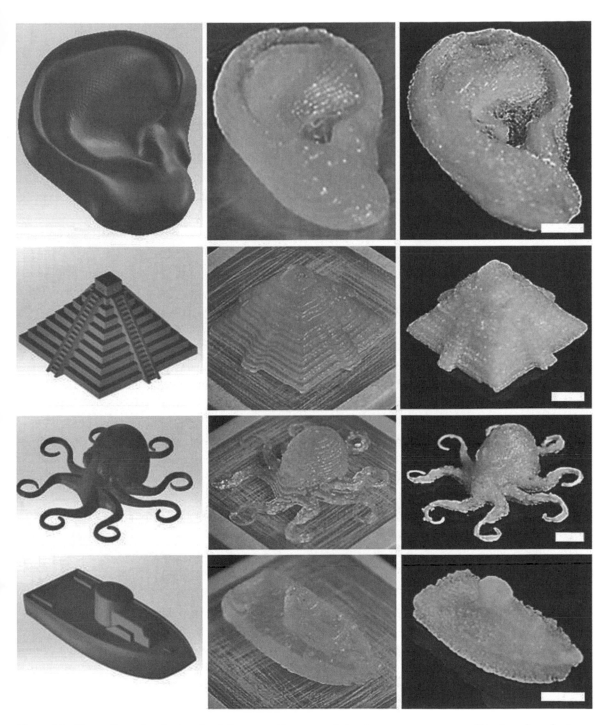

Figure 4.3 Three-dimensional printed cellulose nanofiber aerogel structures are among several bio-based products being developed. These structures could be incorporated with metals and metal oxides to develop electromechanically responsive properties or integrated/cultured with biological cells for use in medical applications. The first column (*left*) is a SolidWorks model; the second column (*center*) is the 3-D printed gel structure; and the third column (*right*) is resultant structures after freeze-drying. Displayed scale bars are 1 cm. *Reprinted with permission from Li et al. 2018. Copyright 2018. American Chemical Society.*

services. With second-generation feedstocks potentially supplying the base ingredient for bio-based products, impacts to ecosystem services, including effects on wildlife habitat discussed throughout this chapter, will be relevant to the emergence of this new industry.

Mitigating Impacts of Bioenergy Production on Wildlife

Because the specific feedstocks used for bioenergy are changing, more efforts should be made to include wildlife resources—especially native wildlife and species of greatest conservation need (SGCN)—as a pre-implementation component of economic sustainability in the discussion of bioenergy production, rather than having to assess impacts after the fact. Second-generation feedstocks could benefit wildlife and other aspects of biodiversity by improving on many of the disadvantages of first-generation feedstocks (Rupp et al. 2012). Effects of second-generation feedstocks on terrestrial biodiversity are largely the result of direct changes in landscape composition, habitat fragmentation, and shifting structural characteristics of vegetation that lead to changes in biological communities, whereas changes to aquatic biodiversity result from indirect effects of bioenergy-induced landscape change, including nutrient and pesticide runoff, changes in hydrology, sedimentation of waterways, and eutrophication of aquatic ecosystems. However, conserving biodiversity in bioenergy landscapes will depend on the specific crops grown, the lands brought into production, the management practices adopted, and the ecological communities encountered (McGuire 2012, McGuire and Rupp 2013, Pedroli et al. 2013, Evans et al. 2015, Joly et al. 2015, Manning et al. 2015, Tarr et al. 2017).

Several studies have suggested that changes in crop management could help mitigate some of the potential negative effects of bioenergy production on wildlife (Fletcher et al. 2011, Uden et al. 2015, Hoekman and Broch 2018, Hoekman et al. 2018). As of this writing (2018), Wisconsin is the only US state to have approved sustainable planting and harvest guidelines for non-forest biomass (Hull et al. 2011, Ventura et al. 2012). In the Prairie Pothole Region of the US Northern Great Plains, best management guidelines for perennial herbaceous biomass production and harvest that incorporated sustainability of wildlife resources were developed using an advisory group of natural resource professionals with expertise in agronomy, production aspects of energy crops, wildlife (e.g., amphibians, birds, insects, mammals, reptiles), and native ecosystems (McGuire and Rupp 2013). Some cooperative extension offices (e.g., University of Tennessee Extension, Harper and Keyser 2008; University of Missouri Extension, Pierce et al. 2014) have short publications that address bioenergy crop management in relation to wildlife sustainability. Efforts have also been made by groups like the Association of Fish and Wildlife Agencies (AFWA; newsletter excerpt on pp. 79–80) to reflect the viewpoints of state fish and wildlife agencies on how fish, wildlife, and native ecosystem sustainability needs can be integrated with bioenergy production.

These resources are good starting points for discussion about how to minimize negative effects of perennial biomass production for bioenergy on wildlife and wildlife habitat, but more efforts could be made. Additional research specific to wildlife responses to bioenergy production is needed, and advanced planning using the collaborative efforts of wildlife experts can benefit both the wildlife and bioenergy communities.

Opportunities exist to improve potential wildlife habitat with strategic placement of bioenergy crops and holistic landscape-scale management. Joly et al. (2015) suggested the following 3 guiding principles for reducing negative effects and enhancing positive effects of bioenergy production on biodiversity and ecosystem services: (1) identification and conservation of priority biodiversity areas is critical, because biodiversity is the foundation for sustainable development; (2) effects of feedstock produc-

FISH, WILDLIFE & BIOENERGY

Perspectives from the Association of Fish & Wildlife Agencies on Integrating Fish & Wildlife Conservation with Bioenergy Production

©NRCS

A native grass field border.

BEST MANAGEMENT GUIDELINES

These Best Management Guidelines (BMGs) reflect the viewpoint of state fish and wildlife agencies on how fish, wildlife, and native ecosystem sustainability needs can be integrated with bioenergy production. Use these BMGs as a template, and customize them to address the needs of your project and priority fish and wildlife species with the help of your state fish and wildlife agency and other conservation partners.

NATIVE HABITATS

Avoid the conversion of native habitats to establish energy crops. Harvest biomass so as to maintain as much as possible of the plant species and structural diversity of the native habitat.

When planting a bioenergy crop, try to use site-appropriate species that match the native habitat, like prairie or forest. Avoid introducing, or prevent the escape of, aggressive plant species that could become invasive. Nonnative, hybrid, or genetically modified plants that become invasive can have significant and costly impacts on ecosystems. Work with your state fish and wildlife agency and other experts to develop a containment plan.

BIOENERGY PLANTINGS

Using native plants as bioenergy crops will generally provide better wildlife habitat than nonnative plants. Allow native plants to grow between rows of bioenergy crops (such as short-rotation woody plantings). Mixed species plantings, especially those with wildflowers and noninvasive legumes, provide better habitat than monocultures. Breaking up expansive single-species plantings into a polyculture with smaller blocks of diversified crops can offer more for wildlife. Aggressive, potentially invasive bioenergy crops should not be planted next to native habitats. Buffers or other safeguards can be used to reduce risk of escape.

WATER QUALITY & QUANTITY

Minimize the use of water for bioenergy production so we can meet the needs for drinking water, other agriculture, and aquatic species. Select bioenergy crops that use water efficiently.

Founded in 1902, the Association of Fish & Wildlife Agencies represents North America's fish and wildlife agencies to advance science-based management and conservation of species and their habitats for the public's long-term benefit and use. www.fishwildlife.org @fishwildlife

ASSOCIATION *of* FISH & WILDLIFE AGENCIES

(continued)

Residual Cover

After bioenergy or other crops have been harvested, wildlife—like the greater prairie-chicken, above—can be left exposed, vulnerable to predators and harsh weather. Leaving some crop stubble on the field and planting native field borders can help reduce the impact.

Riparian Buffers

Aquatic habitats are often impacted by invasive plant species; and invasive plants can be spread by the flowing waters of rivers and streams. Establishing buffers of native plants between bioenergy crops and aquatic habitats can reduce the risks.

To protect water quality, use containment measures to prevent migration of sediments, nutrients, pesticides, herbicides, or seed or other vegetative materials into aquatic ecosystems. Minimize fertilizer, herbicide, and pesticide inputs, and avoid spraying near streams or water bodies. Separate energy crop plantings from streams and other bodies of water with wide vegetative buffers of native plants.

Wetlands, rivers, streams, or other natural aquatic habitats should not be used for production of bioenergy crops such as algae or other cultured or cultivated aquatic organisms. Produce algal biomass only with containment and backup measures in place to prevent escape of aggressive algal strains to areas outside the production area.

Avoid introduction of invasive or aggressive species into aquatic habitats. Harvest invasive aquatic plants as a bioenergy crop only when harvest helps reduce or eliminate the invasive species.

HARVEST GUIDELINES

Harvest bioenergy crops in late summer or fall to protect ground-nesting species and the fawns and calves of large herbivores. At the same time, consider harvesting early enough in the fall to allow regrowth of cover—ideally 10-12 inches for native grasses. Consider leaving a portion of the bioenergy crop unharvested each year

to provide winter cover and spring nesting habitat. Stiff species like switchgrass that resist bending under snow make good winter cover.

Harvesting in blocks rather than strips lessens the risk from predators for species nesting in unharvested areas. Planting native field borders helps wildlife move through harvested areas to suitable cover. Avoid placing corridors so they connect to local roadways to help reduce wildlife/auto collisions.

Develop and enforce a sanitation and containment protocol that ensures that transportation of bioenergy crops and movement of harvesting equipment does not spread potentially invasive species offsite.

Contact your state fish and wildlife agency to customize these Best Management Guidelines to meet the needs of your bioenergy project and priority fish and wildlife species in your area.

Learn more at:
bit.ly/FishWildlifeBioenergy

Reference:

Bill McGuire. 2012. Assessment of the Bioenergy Provisions in the 2008 Farm Bill. Association of Fish and Wildlife Agencies, Washington, D.C., pages 17-23.

This project is made possible by funding from the Sport Fish and Wildlife Restoration Programs of the USFWS, pursuant to the Stevens Amendment to P.L. 11-463.

tion on biodiversity and ecosystem services are context-specific and need to consider the values of stakeholders; and (3) location-specific management of feedstock production systems should recognize the trade-offs between environmental resources and energy production when developing management plans. Additionally, understanding and incorporating social aspects of bioenergy issues (e.g., Lehrer 2010, Kulcsar et al. 2016) is an important part of the equation that has been acknowledged for some eco-system services but has yet to be specifically addressed for wildlife, as far as we know. Difficulties linking social issues and wildlife conservation may be exacerbated by the inherent difficulties of quantifying and qualifying the value of wildlife resources. Creative partnerships that incorporate wildlife values and other ecosystem services into large-scale agricultural production systems are essential; one such case is exemplified by the accompanying Deep Dive.

Deep Dive: Hog Waste to Hog Heaven (For Wildlife, That Is!)

Concentrated animal feeding operations (CAFOs) help meet the food demands of a growing human population by providing a low-cost source of meat, milk, and eggs, facilitated by efficient feeding and housing of animals, increased facility size, and animal specialization (i.e., focus on a single species of livestock). However, when CAFOs are not properly located, managed, and monitored, they can pose public health and environmental risks, including contamination (e.g., from excess nutrients, antibiotics, pathogens, growth hormones, and cleaning chemicals), decreased air quality (i.e., odors, greenhouse gas emissions, and particulates), point and nonpoint sources of water pollution, and decreased native plant density and diversity—all of which have negative impacts on fish and wildlife populations (Burkholder et al. 2007, Hribar 2010). As the human population continues to grow, we may not be able to completely eradicate risks associated with CAFOs, but what if there were ways to take these existing waste streams and convert them into useable energy while simultaneously conserving natural resources?

This Deep Dive illustrates how a creative partnership worked at the nexus of energy production, environmental stewardship, and wildlife conservation to create a successful program that turned a waste stream, which can be a liability, into a source of renewable energy.

Anaerobic digestion of animal manure and slurry offers several environmental, agricultural, and socioeconomic benefits, including improved fertilizer quality, considerable reduction of odors, inactivation of pathogens, and the possibility for production of biogas as clean, renewable fuel for multiple utilizations (Holm-Nielsen et al. 2009). Biogas is primarily a mixture of methane and carbon dioxide produced by the bacterial decomposition of organic materials in the absence of oxygen (i.e., anaerobic digestion). Biogas can be used to produce heat, electricity, plastics, and chemicals. It can also be upgraded to a renewable natural gas that can be used to fuel vehicles or be directly injected into the natural gas grid. Other by-products of anaerobic digestion can include non-energy products, such as nutrient-rich soil amendments, biochar, and fertilizers.

In 2014, Roeslein Alternative Energy (RAE) and Premium Standard Farms—a subsidiary of

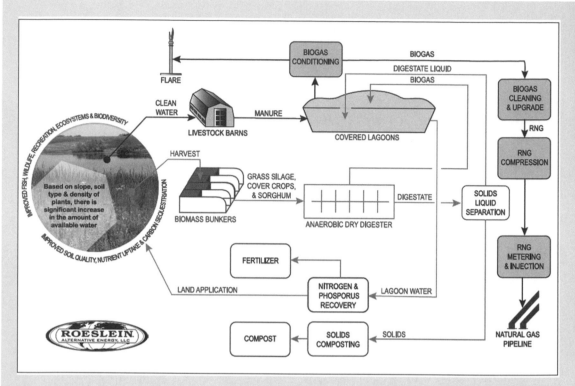

Figure 4.4   Conceptual diagram of a model system integrating mixed native prairie plants into existing concentrated animal feeding operations as a method for producing biogas while simultaneously reducing greenhouse gas emissions, improving water quality/quantity, stabilizing soils and soil health, improving native wildlife habitat, and providing socioeconomic benefits across the greater landscape. *Source: Rudi Roeslein, Roeslein Alternative Energy 2018.*

Smithfield Foods, Inc. (the world's largest pork producer)—announced plans to develop a $100 million renewable biogas project in the United States, in northern Missouri. The aim of the project was to combine biogas generated from waste lagoons at hog CAFO facilities in northern Missouri with biogas produced from digestion of grassy biomass composed of a mixture of native warm-season grass prairie, cover crops, and sorghum. The combined and purified biogas would then be delivered to most major US markets for conversion to renewable compressed or liquefied natural gas (R-CNG or R-LNG, respectively) for use in the transportation and passenger vehicles fuel market (Figure 4.4).

The first phase of the project (Horizon 1; Figure 4.5A) involved capturing biogas by covering 88 manure lagoons situated at 9 hog finishing farms with impermeable covers, flares, and biogas upgrade systems (Roeslein Alternative Energy 2018). The simple act of covering the lagoons provided several immediate benefits. First, it reduced noxious odors that had been the source of multiple complaints over the years. Second, it initiated the capture of the methane-rich biogas. Third, the covers prevented rainwater from infiltrating the lagoon systems, preventing the lagoons from overflowing and contaminating surrounding soils and waterways.

Figure 4.5   Stages of the Smithfield Foods / Roeslein Alternative Energy project: (*A*) Horizon 1 includes covering 88 manure lagoons at nine hog finishing farms in northern Missouri, US, with impermeable covers and flare systems used to destroy methane emissions, thereby obtaining carbon credits from the California carbon market. (*B*) Horizon 2 began with converting 202 ha (500 ac) of cool-season grass and corn spray fields to native prairie for use as a co-digestate in the anaerobic digestion system. *Photos courtesy of Rudi Roeslein, Roeslein Alternative Energy.*

The second phase (Horizon 2; Figure 4.5B) began with conversion of 500 acres (202 ha) of cool-season grass and corn spray fields to mixed native warm-season grass prairie for use in the anaerobic digestion system (Roeslein Alternative Energy 2018). Unlike high-yielding monoculture crops typically considered for bioenergy production (e.g., corn, miscanthus, switchgrass), use of native warm-season grass prairie mixtures as a feedstock in the anaerobic digestion process provides multiple ecosystem service benefits. The fibrous root systems of grasses used in these prairie mixes increase soil organic matter, sequester carbon, and change soil surface hydrology to offer a natural water accumulation system to reduce the environmental impacts of major floods. The use of multispecies forb and grass mixtures creates new habitat for wildlife and pollinators by offering structural and compositional diversity across the landscape. Furthermore, the seed can be harvested and sold or used in additional native prairie plantings.

At completion, the project hopes to restore ~250,000 acres (101,000 ha) of marginal lands in northern Missouri to native prairie, which, when combined with the manure from over 2 million finishing hogs, would be used as a feedstock in the anaerobic digestion process to produce up to 60 million diesel-gallon-equivalents (227 million diesel-liter-equivalents) per year. Smithfield and RAE are using the US EPA's cellulosic biofuel renewable identification numbers (RINs) system and the California Offset Program to improve economic viability of the project. RINs are credits used for compliance and are the "currency" of the EPA's Renewable Fuel Standard program. Similarly, anaerobic digestion of livestock manure has been adopted by California as an eligible project type for the generation of carbon credit offsets under its statewide cap-and-trade program, providing a developing market demand for dairy and swine manure digester projects (US DOE 2014). Smithfield and RAE capitalized on these programs by injecting the upgraded renewable natural gas

ENERGY
- Yield Per Acre
- Methane Production
- Compost / Organic Fertilizer

ECOSYSTEM
- Clean Water
- Water Storage
- Soil Health
- CO2 Sequestration
- Carbon Rebuilding

WILDLIFE
- Pollinators
- Grassland Birds
- Food & Shelter

Figure 4.6 An intersection of energy production, environmental stewardship, and wildlife conservation. *Source: Roeslein Alternative Energy 2018.*

into the interstate natural gas pipeline to obtain carbon credits from the California Carbon market.

The United States currently has more than 2,000 sites (239 of which are manure-based) producing biogas (US DOE 2014). With future political support, more than 11,000 additional biogas systems could be deployed in the United States, producing enough energy to power more than 3 million American homes and reduce methane emissions up to 54 million metric ton equivalents of carbon dioxide emissions by 2030, roughly equal to 11 million passenger vehicles (US DOE 2014). Biogas projects similar to the one described in this Deep Dive are unique in that they occur at the interface of energy production, environmental stewardship, and wildlife protection (Figure 4.6). When used as a model, creative partnerships such as these could help restore millions of acres of grassland prairie around the globe while also creating wildlife habitat and producing energy for a growing human population.

ACKNOWLEDGMENTS

We thank C. Moorman, S. Grodsky, M. Ahlering, and K. Bakker for their comments on previous versions of this manuscript. Their thorough reviews were greatly appreciated and resulted in a cleaner, more concise finished product. We would also like to extend our appreciation to Brent Bailey for his assistance in pulling together the information provided in Appendix 4.1. Any use of trade, firm, or product names is for descriptive purposes only and does not imply endorsement by the US government.

Appendix 4.1.   Farm Bill Energy Title programs' funding levels as prescribed by the 2014 Farm Bill as compared with actual appropriations provided by Congress and approved by the president each year. Based on the Consolidated Appropriations Acts of 2014–2018 (available at www.congress.gov).

2014 Farm Bill	Funding (Millions $)	FY14	FY15	FY16	FY17	FY18	5 Yr Total
REAP	Mandatory	50	50	50	50	50	250
	Discretionary	20	20	20	20	20	100
	Actual	53.5	47.7[†]	48[†]	50.352	50.293	249.85

Description: Provides funding to farmers, ranchers, and small businesses to install renewable energy (wind, solar, biomass, geothermal, etc.) and energy efficiency programs.

BAP	Mandatory	100	50	50	0	0	200
	Discretionary	75	75	75	75	75	375
	Actual	59.306[^]	30	27	0	0	116.3

Description: Provides loan guarantees for the development of facilities that produce advanced biofuels, renewable chemicals, and bio-based products.

BCAP	Mandatory	25	25	25	25	25	125
	Discretionary	0	0	0	0	0	0
	Actual	23.2[†]	21.2[†]	3	3	0	50.4

Description: Provides cost-share up to 50 percent of costs for the establishment of cellulosic crops, and/or provides matching payments for the harvesting of eligible materials, such as forestry and agricultural residues.

BPAP	Mandatory	15	15	15	15	15	75
	Discretionary	20	20	20	20	20	100
	Actual	7[^]	< 15	< 15	15	15	< 67

Description: Pays advanced biofuels producers to expand their production levels.

BRDI	Mandatory	3	3	3	3	3	15
	Discretionary	20	20	20	20	20	100
	Actual	3	3	3	3	3	15

Description: Provides funding in the following areas: (A) feedstocks development; (B) biofuels and bio-based products development; and (C) biofuels development analysis.

BMP	Mandatory	3	3	3	3	3	15
	Discretionary	2	2	2	2	2	10
	Actual	3	3	3	3	3	15

Description: Designed to increase the purchase and use of bio-based products (i.e., USDA BioPreferred Program).

[^] Balance rescinded via the Consolidated Appropriations Act of 2014 (H.R.3547).

[†] After Sequestration cuts.

REAP = Rural Energy for America Program; BAP = Biorefinery, Renewable Chemical and Biobased Product Manufacturing Assistance Program; BCAP = Biomass Crop Assistance Program; BPAB = Bioenergy Program for Advanced Biofuels; BRDI = Biomass Research and Development Initiative; BMP = Biobased Markets (BioPreferred) Program.

LITERATURE CITED

Able, K. W., and S. M. Hagan. 2000. Effects of common reed (*Phragmites australis*) invasion on marsh surface macrofauna: Response of fishes and decapod crustaceans. Estuaries 23:633–646.

Adler, P. R., M. A. Sanderson, A. A. Boateng, P. J. Weimer, and H. G. Jung. 2006. Biomass yield and biofuel quality of switchgrass harvested in fall or spring. Agronomy Journal 98:1518–1525.

Adler, P. R., M. A. Sanderson, P. J. Weimer, and K. P. Vogel. 2009. Plant species composition and biofuel yields of conservation grasslands. Ecological Applications 19:2202–2209.

Almodares, A., and M. R. Hadi. 2009. Production of bioethanol from sweet sorghum: A review. African Journal of Agricultural Research 4:772–778.

Alshawaf, M., E. Douglas, and K. Ricciardi. 2016. Estimating nitrogen load resulting from biofuel mandates.

International Journal of Environmental Research and Public Health 13:478.

Anderson, E. K., E. Aberle, C. Chen, J. Egenolf, K. Harmoney, V. G. Kakani, R. Kallenbach, M. Khanna, W. Wang, and D. Lee. 2016. Impacts of management practices on bioenergy feedstock yield and economic feasibility on Conservation Reserve Program (CRP) grasslands. Global Change Biology Bioenergy 8:1178–1190.

Arundale, R. A., F. G. Dohleman, E. A. Heaton, J. M. Mcgrath, T. B. Voigt, and S. P. Long. 2014. Yields of Miscanthus × giganteus and Panicum virgatum decline with stand age in the midwestern USA. Global Change Biology Bioenergy 6:1–13.

Askins, R. A., F. Chavez Ramirez, B. C. Dale, C. A. Haas, J. R. Herkert, F. L. Knopf, and P. D. Vickery. 2007. Conservation of grassland birds in North America: Understanding ecological processes in different regions. Ornithological Monograph 64. Report of the AOU Committee on Conservation. doi:10.2307/40166905.

Bakker, K. K., and K. F. Higgins. 2009. Planted grasslands and native sod prairie: Equivalent habitat for grassland birds? Western North American Naturalist 69:235–242.

Balan, V. 2014. Current challenges in commercially producing biofuels from lignocellulosic biomass. ISRN Biotechnology. Article 463074. doi:10.1155/2014/463074.

Barney, J. N. 2014. Bioenergy and invasive plants: Quantifying and mitigating future risks. Invasive Plant Science and Management 7:199–209.

Barney, J. N., and J. M. DiTomaso. 2008. Nonnative species and bioenergy: Are we cultivating the next invader? BioScience 58:64–70.

Bellamy P. E., P. J. Croxton, M. S. Heard, S. A. Hinsley, L. Hulmes, S. Hulmes, P. Nuttall, R. F. Pywell, and P. Rothery. 2009. The impact of growing Miscanthus for biomass on farmland bird populations. Biomass and Bioenergy 33:191–199.

Bender, J. J. 2012. Development of sustainable harvest strategies for cellulose-based biofuels: The effect of intensity and season of harvest on cellulosic feedstock and upland nesting game bird production. Thesis, South Dakota State University, Brookings, SD.

Birckhead, J. L., C. A. Harper, P. D. Keyser, D. McIntosh, E. D. Holcomb, G. E. Bates, and J. C. Waller. 2014. Structure of avian habitat following hay and biofuels production in native warm-season grass stands in the Mid-South. Journal of the Southeastern Association of Fish and Wildlife Agencies 1:115–121.

Bischoff, K. P., K. A. Gravois, T. E. Reagan, J. W. Hoy, C. A. Kimbeng, C. M. LaBorde, and G. L. Hawkins. 2008. Registration of "L 79-1002" sugarcane. Journal of Plant Registrations 2:211–217.

Blanco-Canqui, H. 2010. Energy crops and their implications on soil and environment. Agronomy Journal 102:403–419.

Boland, J. M. 2006. The importance of layering in the rapid spread of Arundo donax (giant reed). Madroño 53:303–312.

Bright, J. A., G. Q. A. Anderson, T. McArthur, R. Sage, J. Stockdale, P. V. Grice, and R. B. Bradbury. 2013. Bird use of establishment-stage Miscanthus biomass crops during the breeding season in England. Bird Study 60:357–369.

Burkholder, J., B. Libra, P. Weyer, S. Heathcote, D. Kolpin, P. S. Thorne, and M. Wichman. 2007. Impacts of waste from concentrated animal feeding operations on water quality. Environmental Health Perspectives 115:308–312.

Casler, M. D., E. Heaton, K. J. Shinners, H. G. Jung, P. J. Weimer, M. A. Liebig, R. B. Mitchell, and M. F. Digman. 2009. Grasses and legumes for cellulosic bioenergy. In Grassland: Quietness and Strength for a New American Agriculture, edited by W. F. Wedin and S. L. Fales, 205–219. Madison, WI: American Society of Agronomy, Crop Science Society of America, Soil Science Society of America.

Caslin, B., J. Finnan, and L. Easson. 2010. Miscanthus best practice guidelines. Dublin: Teagasc and the Agri-Food and Bioscience Institute.

Chen, M., and P. M. Smith. 2017. The U.S. cellulosic biofuels industry: Expert views on commercialization drivers and barriers. Biomass and Bioenergy 102:52–61.

Cherney, J. H., K. D. Johnson, J. J. Volenec, and D. K. Greene. 1991. Biomass potential of selected grass and legume crops. Energy Sources 13:283–292.

Clapham, S. J., and F. M. Slater. 2008. The biodiversity of established biomass grass crops. Aspects of Applied Biology 90:325–330.

Congress. Gov. 2018. H.R.2646—Farm Security and Rural Investment Act of 2002. https://www.congress.gov/bill/107th-congress/house-bill/2646. Accessed July 30, 2018.

Cox, G. W. 1999. Alien species in North America and Hawaii: Impacts on natural ecosystems. Washington, DC: Island Press.

Davis, S. C., W. J. Parton, F. G. Dohleman, C. M. Smith, S. Del Grosso, A. D. Kent, and E. H. DeLucia. 2010. Comparative biogeochemical cycles of bioenergy crops reveal nitrogen-fixation and low greenhouse gas emissions in a Miscanthus × giganteus agro-ecosystem. Ecosystems 13:144–156.

Davis, S. C., W. J. Parton, S. J. D. Grosso, C. Keough, E. Marx, P. R. Adler, and E. H. DeLucia. 2012. Impact of second-generation biofuel agriculture on greenhouse-gas emissions in the corn-growing regions of the US. Frontiers in Ecology and the Environment 10:69–74.

DeStefano, S. 2013. Status of exotic grasses and grass-like vegetation and potential impacts on wildlife in New England. Wildlife Society Bulletin 37:486–496.

EDDMapS. 2018. Early detection and distribution mapping system. University of Georgia Center for Invasive Species and Ecosystem Health. http://www.eddmaps.org/. Accessed October 26, 2018.

Energy Information Administration. 2018. U.S. energy consumptions by source, 2017. Monthly energy review, Table 1.3 and 10.1, April 2018, preliminary data. https://www.eia.gov/energyexplained/?page=us_energy_home. Accessed May 16, 2018.

English, B. C., D. G. De La Torre Ugarte, K. Jensen, C. Hellwinckel, J. Menard, B. Wilson, R. Roberts, and M. Walsh. 2006. 25% renewable energy for the United States by 2025: Agricultural and economic impacts. Knoxville: University of Tennessee.

Evans, S. G., L. C. Kelley, and M. D. Potts. 2015. The potential impact of second-generation biofuel landscapes on at-risk species in the US. Global Change Biology Bioenergy 2015:337–348.

Fargione, J., J. Hill, D. Tilman, D. Polasky, and P. Hawthorne. 2008. Land clearing and the biofuel carbon debt. Science 319:1235–1238.

Fargione, J. E., T. R. Cooper, D. J. Flaspohler, J. Hill, C. Lehman, T. McCoy, S. McLeod, E. J. Nelson, K. S. Oberhauser, and D. Tilman. 2009. Bioenergy and wildlife: Threats and opportunities for grassland conservation. BioScience 59:767–777.

Fletcher, R. J. Jr, B. A. Robertson, J. Evans, P. J. Doran, J. R. R. Alavalapati, and D. W. Schemske. 2011. Biodiversity conservation in the era of biofuels: Risks and opportunities. Frontiers in Ecology and the Environment 9:161–168.

Florida Exotic Pest Plant Council. 2017. Florida Exotic Pest Plant Council's 2017 list of invasive plant species. http://bugwoodcloud.org/CDN/fleppc/plantlists/2017/2017FLEPPCLIST-TRIFOLD-FINALAPPROVEDBYKEN-SUBMITTEDTOALTA.pdf.

Gamble, J. D., J. M. Jungers, D. L. Wyse, G. A. Johnson, J. A. Lamb, and C. C. Sheaffer. 2015. Harvest date effects on biomass yield, moisture content, mineral concentration, and mineral export in switchgrass and native polycultures managed for bioenergy. Bioenergy Research. doi:10.1007/s12155-014-9555-0.

Gardiner, M. M., D. A. Landis, C. Gratton, C. D. DiFonzo, M. O'Neal, J. Chacon, M. Wayo, N. Schmidt, E. Mueller, and G. E. Heimpel. 2009. Landscape diversity enhances biological control of an introduced crop pest in the north-central U.S. Ecological Applications 19:143–154.

Gasparatos, A., P. Stromberg, and K. Takeuchi. 2011. Biofuels, ecosystem services and human wellbeing: Putting biofuels in the ecosystem services narrative. Agriculture, Ecosystems and Environment 142:111–128.

Glaser, A., and J. Glick. 2012. Growing risk: Addressing the invasive potential of bioenergy feedstocks. Washington, DC: National Wildlife Federation.

Global Invasive Species Database. 2018. Species profile: *Arundo donax*. http://www.iucngisd.org/gisd/speciesname/Arundo+donax. Accessed May 14, 2018.

Golden, J. S., R. B. Hanfield, J. Daystar, and T. E. McConnell. 2016. An economic impact analysis of the U.S. biobased products industry. US Department of Agriculture. https://www.biopreferred.gov/BPResources/files/BiobasedProductsEconomicAnalysis2016.pdf.

Gordon, D. R., K. J. Tancig, D. A. Onderdonk, and C. A. Gantz. 2011. Assessing the invasive potential of biofuel species proposed for Florida and the United States using the Australian Weed Risk Assessment. Biomass and Bioenergy 35:74–79.

Graham, R. L., W. Liu, and B. C. English. 1995. The environmental benefits of cellulosic energy crops at a landscape scale. Environmental enhancement through agriculture. Conference proceedings, November 15–17. Tufts University, Center for Agriculture, Food and Environment. Boston, MA.

Griffith, A. P., F. M. Epplin, S. D. Fuhlendorf, and R. Gillen. 2011. A comparison of perennial polycultures and monocultures for producing biomass for biorefinery feedstock. Agronomy Journal 103:617–627.

Han, W., Z. Yang, L. Di, and R. Mueller. 2012. CropScape: A Web service–based application for exploring and disseminating US conterminous geospatial cropland data products for decision support. Computers and Electronics in Agriculture 84:111–123.

Harper, C. A., and P. D. Keyser. 2008. Potential impacts on wildlife of switchgrass grown for biofuels. UT Extension Publication SP704A-5M-5/08. Knoxville: University of Tennessee Biofuels Initiative, UT Institute of Agriculture. http://nativegrasses.utk.edu/publications/SP704-A.pdf.

Hartman, J. C., J. B. Nippert, R. A. Orozco, and C. J. Springer. 2011. Potential ecological impacts of switchgrass (*Panicum virgatum* L.) biofuel cultivation in the Central Great Plains, USA. Biomass and Bioenergy 35:3415–3421.

Heaton, E., T. Voigt, and S. P. Long. 2004a. A quantitative review comparing the yields of two candidate C_4 perennial biomass crops in relation to nitrogen, temperature, and water. Biomass Bioenergy 27:21–30.

Heaton, E., S. Long, T. Voigt, M. Jones, and J. Clifton-Brown. 2004b. *Miscanthus* for renewable energy generation: European Union experience and projections for Illinois. Mitigation and Adaptation Strategies for Global Change 9:433–451.

Hoekman, S. K., and A. Broch. 2018. Environmental implications of higher ethanol production and use in the U.S.: A literature review. Part II—Biodiversity, land use change, GHG emissions, and sustainability. Renewable and Sustainable Energy Reviews 81:3159–3177.

Hoekman, S. K., A. Broch, and X. Liu. 2018. Environmental implications of higher ethanol production and use in the U.S.: A literature review. Part I—Impacts on water, soil, and air quality. Renewable and Sustainable Energy Reviews 81:3140–3158.

Holm-Nielsen, J. B., T. Al Seadi, and P. Oleskowicz-Popiel. 2009. The future of anaerobic digestion and biogas utilization. Bioresource Technology 100:5478–5484.

Holzer, K. A., and S. P. Lawler. 2015. Introduced reed canary grass attracts and supports a common native amphibian. Journal of Wildlife Management 79:1081–1090.

Hong, C. O., V. N. Owens, D. K. Lee, and A. Boe. 2013. Switchgrass, big bluestem, and indiangrass monocultures and their two- and three-way mixtures for bioenergy in the Northern Great Plains. Bioenergy Research 6:229–239.

Hribar, C. 2010. Understanding concentrated animal feeding operations and their impact on communities. Bowling Green, OH: National Association of Local Boards of Health. https://www.cdc.gov/nceh/ehs/docs/understanding_cafos_nalboh.pdf.

Huggett, D. A., S. R. Leather, and K. F. Walters. 1999. Suitability of the biomass crop *Miscanthus sinensis* as a host for the aphids *Rhopalosiphum padi* (L.) and *Rhopalosiphum maidis* (f.), and its susceptibility to the plant luteovirus barley yellow dwarf virus. Agricultural and Forest Entomology 1:143–149.

Hull, S., J. Arntzen, C. Bleser, A. Crossley, R. Jackson, E. Lobner, L. Paine, et al. 2011. Wisconsin sustainable planting and harvest guidelines for non-forest biomass. file:///C:/Users/envir/Downloads/WINFBguidelines FinalOct2011.pdf.

Immerzeel, D. J., P. A. Verweij, F. Van Der Hilst, and A. P. C. Faaij. 2014. Biodiversity impacts of bioenergy crop production: A state-of-the-art review. Global Change Biology Bioenergy 6:183–189.

Ingwell, L. L., R. Zemetra, C. Mallory-Smith, and N. A. Bosque-Peréz. 2014. *Arundo donax* infection with barley yellow dwarf virus has implications for biofuel production and non-managed habitats. Biomass and Bioenergy 66:426–433.

IPCC. 2013. Climate change 2013: The physical science basis. Contribution of Working Group I to the fifth assessment report of the Intergovernmental Panel on Climate Change, edited by T. F. Stocker, D. Qin, G.-K. Plattner, M. Tignor, S. K. Allen, J. Boschung, A. Nauels, Y. Xia, V. Bex and P. M. Midgley. Cambridge, UK: Cambridge University Press.

Jarchow, M. E., and M. Liebman. 2010. Incorporating prairies into multifunctional landscapes: Establishing and managing prairies for enhanced environmental quality, livestock grazing and hay production, bioenergy production, and carbon sequestration. PMR 1007. Ames: Iowa State University Extension.

Johnson, D. H. 2005. Grassland bird use of Conservation Reserve Program fields in the Great Plains. In Technical Review 05-2, Fish and Wildlife Benefits of Farm Bill Conservation Programs: 2000–2005 Update, edited by J. B. Haufler, 17–32. Bethesda, MD: Wildlife Society.

Joly, C. A., L. M. Verdade, B. J. Huntley, V. H. Dale, G. Mace, B. Muok, and N. H. Ravindranath. 2015. Biofuel impacts on biodiversity and ecosystem services. Chapter 16 in Bioenergy and Sustainability: Bridging the Gaps, edited by G. M. Souza, R. L., Victoria, C. A. Joly, and M. Verdade, 554–580. Report 72. Paris: Scientific Committee on Problems of the Environment (SCOPE).

Jungers, J. M., T. W. Arnold, and C. Lehman. 2015. Effects of grassland biomass harvest on nesting pheasants and ducks. American Midland Naturalist 173:122–132.

Kang, S., W. M. Post, J. A. Nichols, D. Wang, T. O. West, V. Bandaru, and R. C. Izaurralde. 2013. Marginal lands: Concept, assessment and management. Journal of Agricultural Science 5:129–139.

Kedron, P., and S. Bagchi-Sen. 2016. Limits to policy-led innovation and industry development in US biofuels. Technology Analysis and Strategic Management 29:486–499.

Kibit, L. C., H. Blanco-Canqui, R. B. Mitchell, and W. H. Schacht. 2016. Root biomass and soil carbon response to growing perennial grasses for bioenergy. Energy, Sustainability, and Society 6:1. doi:10.1186/s13705-015-0065-5.

Kim, M., and D. F. Day. 2011. Composition of sugar cane, energy cane, and sweet sorghum suitable for ethanol production at Louisiana sugar mills. Journal of Industrial Microbiology and Biotechnology 38:803–807.

Kirsch, E. M., B. R. Gray, T. J. Fox, and W. E. Thogmartin. 2007. Breeding bird territory placement in riparian wet meadows in relation to invasive reed canary grass, *Phalaris arundinacea*. Wetlands 27:644–655.

Klimstra, R. L., C. E. Moorman, S. J. Converse, J. A. Royle, and C. A. Harper. 2015. Small mammal use of hayed native, warm-season and non-native cool-season forage fields. Wildlife Society Bulletin 39:49–55.

Kulcsar, L. J., T. Selfa, and C. M. Bain. 2016. Privileged access and rural vulnerabilities: Examining social and environmental exploitation in bioenergy development in the American Midwest. Journal of Rural Studies 47:291–299.

Landis, D. A., and B. P. Werling. 2010. Arthropods and biofuel production systems in North America. Insect Science 17:1–17. doi:10.1111/j.1744-7917.2009.01310.x.

Lane, J. 2016. Abengoa's Hugoton cellulosic ethanol project goes on the block. Biofuels Digest, July 18. http://www.biofuelsdigest.com/bdigest/2016/07/18/abengoas-hugoton-cellulosic-ethanol-project-goes-on-the-block/.

Lane, J. 2018. No crop left behind: The Digest's 2018 multi-slide guide to USDA/NIFA's Coordinated Agricultural Projects. Nuu Digital Voice of the Advanced Bioeconomy, April 10. http://biofuelsdigest.com/nuudigest/2018/04/10/no-crop-left-behind-the-digests-2018-multi-slide-guide-to-usda-nifas-coordinated-agricultural-projects/.

Lee, D. K., V. N. Owens, and J. J. Doolittle. 2007. Switchgrass and soil carbon sequestration response to ammonium nitrate, manure, and harvest frequency on Conservation Reserve Program land. Agronomy Journal 99:462–468.

Lee, D. K., E. Aberle, E. K. Anderson, W. Anderson, B. S. Baldwin, D. Baltensperger, M. Barrett, et al. 2018. Biomass production of herbaceous energy crops in the United States: Field trial results and yield potential maps from the multiyear regional feedstock partnership. Global Change Biology Bioenergy. doi:10.1111/gcbb.12493.

Lehrer, N. 2010. (Bio)fueling farm policy: The biofuels boom and the 2008 Farm Bill. Agriculture and Human Values 27:427–444.

Lewandowski, I., J. M. O. Scurlock, E. Lindvall, and M Christou. 2003. The development and current status of perennial rhizomatous grasses as energy crops in the US and Europe. Biomass and Bioenergy 25:335–361.

Li, V. C. F., A. Mulyadi, C. K. Dunn, Y. Deng, and H. J. Qi. 2018. Direct ink write 3D printed cellulose nanofiber aerogel structures. ACS Sustainable Chemistry and Engineering 2018:2011–2022.

Liu, X., S. K. Hoekman, and A. Broch. 2017. Potential water requirements of increased ethanol fuel in the USA. Energy, Sustainability and Society 7:18. doi:10.1186/s13705-017-0121-4.

Low, T., C. Booth, and A. Sheppard. 2011. Weedy biofuels: What can be done? Current Opinion in Environmental Sustainability 3:55–59.

Manning, P., G. Taylor, and M. E. Hanley. 2015. Bioenergy, food production, and biodiversity: An unlikely alliance? Global Change Biology (GCB) Bioenergy 7:570–576.

Marquis, C. 2016. This is advanced energy: Second and third generation biofuels. Advanced Energy Economy: Advanced Energy Perspectives, July 5. https://blog.aee.net/second-and-third-generation-biofuels.

Matsuoka, S., A. J. Kennedy, E. G. D. dos Santos, A. L. Tomazela, and L. C. S. Rubio. 2014. Energy cane: Its concept, development, characteristics, and prospects. Advances in Botany 2014. Article ID587275. http://dx.doi.org/10.1155/2014/597275.

Maves, A. J. 2011. Developing sustainable harvest strategies for cellulose-based biofuels: The effect of intensity of harvest on migratory grassland songbirds. Thesis, South Dakota State University, Brookings.

McCalmont, J. P., A. Hastings, N. P. McNamara, G. M. Richter, P. Robson, I. S. Donnison, and J. Clifton-Brown. 2017. Environmental costs and benefits of growing Miscanthus for bioenergy production in the U.K. Global Change Biology Bioenergy 9:489–507.

McGuire, B. D. 2012. Assessment of bioenergy provisions in the 2008 Farm Bill. Association of Fish and Wildlife Agencies. https://www.fishwildlife.org/ . . . /files/ . . . /08_22_12_bioenergy_report_web_final_1.pdf.

McGuire, B., and S. P. Rupp. 2013. Perennial herbaceous biomass production in the Prairie Pothole Region of the Northern Great Plains: Best management guidelines to achieve sustainability of wildlife resources. Washington DC: National Wildlife Federation.

McLaughlin, S. B., and M. E. Walsh. 1998. Evaluating environmental consequences of producing herbaceous crops for bioenergy. Biomass and Bioenergy 14:317–324.

McLaughlin, S., J. Bouton, D. Bransby, B. Conger, W. Ocumpaugh, D. Parrish, C. Taliaferro, K. Vogel, and S. Wullschleger. 1999. Developing switchgrass as a bioenergy crop. In Perspectives on New Crops and New Uses, edited by Jules Janick, 282–299. Alexandria, VA: American Society for Horticulture Science Press.

Mitchell, R. B., and K. P. Vogel. 2016. Grass invasion into switchgrass managed for biomass energy. BioEnergy Research 9:50–56.

Mitchell, R., K. P. Vogel, and G. Sarath. 2008. Managing and enhancing switchgrass as a bioenergy feedstock. Biofuels, Bioproducts, and Biorefining 2:530–539.

Monge, J. J., L. A. Ribera, J. L. Jifon, J. A. da Silva, and J. W. Richardson. 2014. Economics and uncertainty of lignocellulosic biofuel production from energy cane and sweet sorghum in South Texas. Journal of Agriculture and Applied Economics 46:457–485.

Moorman, C. E., R. L. Klimstra, C. A. Harper, J. F. Marcus, and C. E. Sorenson. 2017. Breeding songbird use of native warm-season and non-native cool-season grass forage fields. Wildlife Society Bulletin 41:42–48.

Mulkey, V., V. Owens, and D. Lee. 2006. Management of switchgrass-dominated Conservation Reserve Program lands for biomass production in South Dakota. Crop Science 46:712–720.

Mulkey, V. R., V. N. Owens, and D. K. Lee. 2008. Management of warm-season grass mixtures for biomass production in South Dakota, USA. Bioresource Technology 99:609–617.

Mundaca, L., and J. L. Richter. 2015. Assessing "green energy economy" stimulus packages: Evidence from the U.S. programs targeting renewable energy. Renewable and Sustainable Energy Reviews 42:1174–1186.

Murray, L. D., and L. B. Best. 2003. Short-term bird response to harvesting switchgrass for biomass in Iowa. Journal of Wildlife Management 67:611–621.

Na, C., L. E. Sollenberger, J. E. Erickson, K. R. Woodard, M. S. Castillo, M. K. Mullenix, J. M. B. Vendramini, and M. L. Silveira. 2015a. Management of perennial warm-season bioenergy grasses, II: Seasonal differences in elephantgrass and energycane morphological characteristics affect responses to harvest frequency and timing. Bioenergy Research 8:618–626.

Na, C., L. E. Sollenberger, J. E. Erickson, K. R. Woodard, J. M. Vendramini, and M. Silveira. 2015b. Management of perennial warm-season bioenergy grasses, I: Biomass harvested, nutrient removal, and persistence responses of elephantgrass and energycane to harvest frequency and timing. Bioenergy Research 8:581–589.

Na, C., J. R. Fedenko, L. E. Sollenberger, and J. E. Erickson. 2016. Harvest management affects biomass composition responses of C4 perennial bioenergy grasses in the humid subtropical USA. Global Change Biology Bioenergy 8:1150–1161.

National Invasive Species Council. 2016. Management plan: 2016–2018. Washington, DC.

Niemuth, N. D., F. R. Quamen, D. E. Naugle, R. E. Reynolds, M. E. Estey, and T. L. Shaffer. 2007. Benefits of the Conservation Reserve Program to grassland bird populations in the Prairie Pothole Region of North Dakota and South Dakota. US Department of Agriculture Farm Service Agency Reimbursable Fund Agreement OS-IA-04000000-N34 Final Report. https://www.fsa .usda.gov/Internet/FSA_File/grassland_birds_fws.pdf.

Pedroli, B., B. Elbersen, P. Frederiksen, U. Grandin, R. Heikkilä, P. H. Krogh, Z. Izakovičová, A. Johansen, L. Mieresonne, and J. Spijker. 2013. Is energy cropping in Europe compatible with biodiversity? Opportunities and threats to biodiversity from land-based production of biomass for bioenergy purposes. Biomass and Bioenergy 55:73–86.

Peplow, M. 2014. Cellulosic ethanol fights for life. Nature 107:153.

Perlack, R. D., L. L. Wright, A. F. Turhollow, R. L. Graham, B. J. Stokes, and D. C. Erbach. 2005. Biomass as feedstock for a bioenergy and bioproducts industry: The technical feasibility of a billion-ton annual supply. US Department of Energy Report DOE/GO-102995-2135.

Picasso, V. D., E. C. Brummer, M. Liebman, P. M. Dixon, and B. J. Wilsey. 2011. Diverse perennial crop mixtures sustain higher productivity over time based on ecological complementarity. Renewable Agriculture and Food Systems 26:317–327.

Pierce, R. A., B. White, T. Reinbott, and R. Wright. 2014. Integrating practices that benefit wildlife with crops grown for biomass in Missouri. Publication g9422. Columbia: University of Missouri Extension.

Quinn, L. D., A. B. Endres, and T. B. Voigt. 2014. Why not harvest existing invaders for bioethanol? Biological Invasions 16:1559–1566.

Rancane, S., A. Karklins, and D. Lazdina. 2012. Perennial grasses for bioenergy production: Characterization of the experimental site. In Research for Rural Development: Proceeding of the Annual 18th International Scientific Conference, 31–37. Latvia University of Agriculture.

Reynolds, R. R. 2005. The conservation reserve program and duck production in the U.S. Prairie Pothole Region. In Technical Review 05-2, Fish and Wildlife Benefits of Farm Bill Conservation Programs: 2000–2005 Update, edited by J. B. Haufler, 33–40. Bethesda, MD: Wildlife Society.

Rieger, J. P., and D. A. Kreager. 1989. Giant reed (*Arundo donax*): A climax community of the riparian zone. USDA Forest Service Gen. Tech. Rep. PSW-UO. https://www.fs .usda.gov/treesearch/pubs/27971.

Rinehart, L. 2006. Switchgrass as a bioenergy crop. ATTRA—National Sustainable Agriculture Information Service. https://attra.ncat.org/attra-pub/viewhtml.php ?id=311. Accessed April 24, 2018.

Rittenhouse, T. A. G. 2011. Anuran larval habitat quality when reed canary grass is present in wetlands. Journal of Herpetology 45:491–496.

Robertson, B. A., P. J. Doran, E. R. Loomis, J. R. Robertson, and D. W. Schemske. 2011. Avian use of perennial biomass feedstocks as post-breeding and migratory stopover habitat. PLOS One 6:e16941.

Robertson, B. A., C. Porter, D. A. Landis, and D. W. Schemske. 2012a. Agroenergy crops influence the diversity, biomass, and guild structure of terrestrial arthropod communities. Bioenergy Research 5:179–188.

Robertson, B. A., R. A. Rice, T. S. Sillett, C. A. Ribic, B. A. Babcock, D. A. Landis, J. R. Herkert, et al. 2012b. Are agrofuels a conservation threat or opportunity for grassland birds in the United States? Condor 114:679–688.

Robertson, B. A., D. A. Landis, T. S. Sillett, E. R. Loomis, and R. A. Rice. 2013. Perennial agroenergy feedstocks as en route habitat for spring migratory birds. Bioenergy Research 6:311–320.

Roeslein Alternative Energy. 2018. How RAE helped Smithfield save the worst hog farm in America. January 4. http://roesleinalternativeenergy.com/how-rae -helped-smithfield-save-the-worst-hog-farm-in-america/.

Roth, A. M., D. W. Sample, C. A. Ribic, L. Paine, D. J. Undersander, and G. A. Bartlet. 2005. Grassland bird response to harvesting switchgrass as a biomass crop. Biomass and Bioenergy 28:490–498.

Rupp, S. P., L. Bies, A. Glaser, C. Kowaleski, T. McCoy, T. Rentz, S. Riffell, J. Sibbing, J. Verschuyl, T. Wigley. 2012. Impacts of bioenergy development on wildlife and wildlife habitat. Technical Review 12-03. Bethesda, MD: Wildlife Society.

Sage, R., M. Cunningham, A. J. Haughton, M. D. Mallot, D. A. Bohan, A. Riche, and A. Karp. 2010. The environmental impacts of biomass crops: Use by birds of *Miscanthus* in summer and winter in southwestern England. Ibis 152:487–499.

Sample, D. W., and M. J. Mossman. 1997. Managing habitat for grassland birds: A guide for Wisconsin. Madison: Wisconsin Department of Natural Resources.

Samson, F. B., and F. L. Knopf. 1994. Prairie conservation in North America. BioScience 44:418–421.

Schill, S. R. 2007. *Miscanthus* versus switchgrass. Ethanol Producer magazine, October 3. http://www.ethanol producer.com/articles/3334/miscanthus-versus-switch grass.

Schulte, L. A., J. Niemi, M. J. Helmers, M. Liebman, J. G. Arbuckle, D. E. James, R. K. Kolka, et al. 2017. Prairie strips improve biodiversity and the delivery of multiple ecosystem services from corn-soybean croplands. Proceedings of the National Academy of Sciences 114:11247–11252.

Semere, T., and F. M. Slater. 2007. Ground flora, small mammal and bird species diversity in miscanthus (*Miscanthus x giganteus*) and reed canary-grass (*Phalaris arundinacea*) fields. Biomass and Bioenergy 31:20–29.

Smith, K. A., and T. D. Searchinger. 2012. Crop-based biofuels and associated environmental concerns. Global Change Biology Bioenergy 4:479–484.

Spyreas, G., B. W. Wilm, A. E. Plocher, D. M. Ketzner, J. W. Matthews, J. L. Ellis, and E. J. Heske. 2010. Biological consequences of invasion by reed canary grass (*Phalaris arundinacea*). Biological Invasions 12:1253–1267.

Stoof, C. R., B. K. Richards, P. B. Woodbury, E. S. Fabio, A. R. Brumbach, J. Cherney, S. Das, et al. 2015. Untapped potential: Opportunities and challenges for sustainable bioenergy production from marginal lands in the northeast USA. Bioenergy Research 8:482–501.

Tarr, N. M., M. J. Rubino, J. K. Costanza, A. J. McKerrow, J. A. Collazo, and R. C. Abt. 2017. Projected gains and losses of wildlife habitat from bioenergy-induced landscape change. Global Change Biology Bioenergy 9:909–923.

Tolbert, V. R., and A. Schiller. 1995. Environmental enhancement using short-rotation woody crops and perennial grasses as alternatives to traditional agricultural crops. Environmental Enhancement through Agriculture. Conference proceedings. Boston, MA.

Torrez, V. C., P. J. Johnson, and A. Boe. 2013. Infestation rates and tiller morphology effects by the switchgrass moth on six cultivars of switchgrass. Bioenergy Research 6:808–812.

Tufekcioglu, A., J. W. Raich, T. M. Isenhart, and R. C. Schultz. 2003. Biomass, carbon, and nitrogen dynamics of multispecies riparian buffers within an agricultural watershed in Iowa, USA. Agroforestry Systems 57:187–198.

Uden, D. R., C. R. Allen, R. B. Mitchell, T. D. McCoy, and Q. Guan. 2015. Predicted avian responses to bioenergy development scenarios in an intensive agricultural landscape. Global Change Biology Bioenergy 7:717–726.

USDA (US Department of Agriculture). 2016. Research education, and economics action plan progress report, 2016. February 2017. https://www.ree.usda.gov/sites /www.ree.usda.gov/files/2017-08/2016USDA_%20 REEProgressReportR2.pdf.

USDA National Institute of Food and Agriculture. 2017. New USDA NIFA AFRI CAP Grant Awardees Combined_Presentations. http://www.caafi.org/resources /pdf/ New_USDA_NIFA_AFRI_CAP_Grant_Awardees _Combined_Presentations.pdf.

USDA National Institute of Food and Agriculture. 2018. AFRI Regional Bioenergy System Coordinated Agricultural Projects. https://nifa.usda.gov/afri-regional -bioenergy-system-coordinated-agricultural-projects. Accessed July 10, 2018.

USDA NRCS (USDA Natural Resources Conservation Service). 2006. Plant fact sheet: Switchgrass (*Panicum virgatum* L.). https://plants.usda.gov/factsheet/pdf/fs _pavi2.pdf.

US Department of Agriculture, US Environmental Protection Agency, US Department of Energy. 2014. Biogas opportunities roadmap: Voluntary actions to reduce methane emissions and increase energy independence. Washington, DC: USDA.

US DOE (US Department of Energy). 2005. Biomass as feedstock for a bioenergy and bioproducts industry: The technical feasibility of a billion-ton annual supply. R. D. Perlack, L. L. Wright, A. F. Turhollow, R. L. Graham, B. J. Stokes, and D. C. Erbach, Leads, ORNL/TM-2005/66. Oak Ridge: Oak Ridge National Laboratory.

US DOE (US Department of Energy). 2011. U.S. billion-ton update: Biomass supply for a bioenergy and bioproducts industry. R. D. Perlack and B. J. Stokes, Leads, ORNL/TM-2011/224. Oak Ridge, TN: Oak Ridge National Laboratory.

US DOE (US Department of Energy). 2017. 2016 billion-ton report: Advancing domestic resources for a thriving bioeconomy, Volume 2, Environmental sustainability effects of select scenarios from Volume 1. R. A. Efroymson, M. H. Langholtz, K. E. Johnson, and B. J. Stokes, Leads, ORNL/TM-2016/727. Oak Ridge, TN: Oak Ridge National Laboratory. doi:10.2172/1338837.

US EPA (US Environmental Protection Agency). 2017a. Final Renewable Fuel Standards for 2017, and the biomass-based diesel volume for 2018 (last updated February 1, 2017). https://www.epa.gov/renewable-fuel-standard-program/final-renewable-fuel-standards-2017-and-biomass-based-diesel-volume.

US EPA (US Environmental Protection Agency). 2017b. Renewable Fuel Standard Program: Standards for 2018 and biomass-based diesel volume for 2019. EPA–HQ–OAR–2017–0091; FRL–9971–73–OAR. Federal Register 82 (237)/Tuesday, December 12, 2017/Rules and Regulations: 58486–58527.

US EPA (US Environmental Protection Agency). 2017c. Ethanol waivers (E15 and E10) (last updated April 14, 2017). https://www.epa.gov/gasoline-standards/ethanol-waivers-e15-and-e10.

US Senate Committee on Agriculture, Nutrition and Forestry. 2018. Agriculture Improvement Act of 2018. https://www.agriculture.senate.gov/2018-farm-bill.

Ventura, S., S. Hull, R. Jackson, G. Radloff, D. Sample, S. Walling, and C. Williams. 2012. Guidelines for sustainable planting and harvest of nonforest biomass in Wisconsin. Journal of Soil and Water Conservation 67:17A–20A.

Viator, H. P., K. J. Han, D. L. Harrell, D. F. Day, M. Salassi, M. W. Alison. 2014. Sweet sorghum production guide. Publication 3357. Baton Rouge: Louisiana State University Sustainable Bioproducts Initiative.

Wiens, J., J. Fargione, and J. Hill. 2011. Biofuels and biodiversity. Ecological Applications 21:1085–1095.

Wilson, M. C., and R. E. Shade. 1966. Survival and development of larvae of the cereal leaf beetle, *Oulema melanopa* (Coleoptera: Chrysomelidae), on various species of Gramineae. Annals of the Entomological Society of America 59:170–173.

Wright, L. L. 2007. Historical perspective on how and why switchgrass was selected as a "model" high-potential energy crop. ORNL/TM-2007/109. Oak Ridge, TN: Oak Ridge National Laboratory.

Yuan, J. S., K. H. Tiller, H. Al-Ahmad, N. R. Stewart, and C. N. Stewart Jr. 2008. Plants to power: Bioenergy to fuel the future. Trends in Plant Science 13:421–429.

Zhuang, Q., Z. Qin, and M. Chen. 2013. Biofuel, land, and water: Maize, switchgrass, or *Miscanthus*? Environmental Research Letters 8:015020.

PART II WIND ENERGY AND WILDLIFE CONSERVATION

5 — Wind Energy Effects on Birds

REGAN DOHM AND
DAVID DRAKE

Introduction

Wind energy as a power source has many environmental benefits compared with fossil fuel consumption and nuclear power generation. Operational wind turbines do not produce air or water pollution or toxic emissions (Saidur et al. 2011, Grodsky et al. 2013). Furthermore, environmental impacts from wind energy tend to be localized, whereas other forms of energy production can have impacts throughout the fuel cycle and more globally (Snyder and Kaiser 2009, Grodsky et al. 2013). However, lingering concerns over potential negative consequences for birds and bats continue to be raised, especially with the proliferation of wind energy facilities (also called wind farms). Research to understand the complete effects of wind development on flying wildlife has not kept pace with the rapid growth of wind power, an industry whose annual energy production has increased by nearly 30% since 2000 (Figure 5.1; Garvin et al. 2011, Leung and Yang 2012).

In this chapter, we focus on the effects of wind development on birds. Chapter 6 (Hein and Hale) addresses wind energy effects on bats, and Chapter 7 (Korfanta and Zero) addresses wind energy effects on other taxa, including prairie grouse, ungulates, and invertebrates. We begin this chapter by exploring a

brief history of wind energy. We then define types of effects birds may experience from both terrestrial and offshore wind turbines and wind energy facilities. We conclude the chapter by discussing wind energy policy relative to birds, strategies to mitigate consequences to birds from wind development, and future directions for research.

History of Wind Energy

Commercial production of wind energy is a relatively recent addition to the global energy portfolio. Industrial-scale wind energy in the United States began shortly after the 1973 oil crisis, prompting the US government to invest in wind energy research and development (Kaldellis and Zafirakis 2011). The first pulse of commercial wind development followed when more than 16,000 turbines were installed in California between 1981 and 1990 (Kaldellis and Zafirakis 2011). In Europe, particularly northern Europe, commercial use of wind energy began in the 1980s and has increased steadily ever since. Wind production in Asia took off in the mid-1990s, with the advent of wind turbines in China. Since then, India also has been a major producer of wind energy in Asia. Although other regions of the world also produce energy

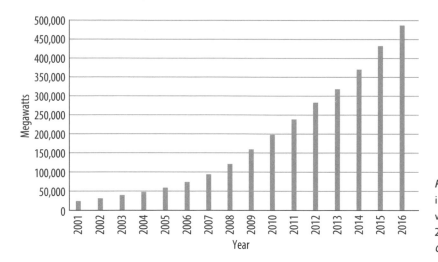

Figure 5.1 Global cumulative installed wind capacity (megawatts) for the years 2001 to 2016. *Source: Global Wind Energy Council 2017a.*

via wind power (e.g., Australia, Brazil, Iran, Mexico, Morocco, Tunisia), the largest producers and consumers of wind energy are China, Europe, and North America (Global Wind Energy Council 2016).

As demand for commercial wind energy has increased, wind turbine design has evolved. Early turbine designs were confined to land-based locations and involved relatively short tower structures, consisting of 3 to 4 legs connected with lattice supporting a single- or double-blade turbine assembly. Rotor diameter of early turbine designs measured about 20 m (Kaldellis and Zafirakis 2011). Modern turbines feature a monopole (i.e., single pole) design, the largest of which measures 140 m tall and supports a 3-blade rotor measuring upward of 128 m in diameter (Figure 5.2). From the ground to a turbine blade's highest reaching point, turbines can stand over 200 m in height, with each blade measuring 62.5 m in length. Currently, the largest turbines are based offshore and can produce 5 megawatts (MW) of electricity each. Typically, 1 MW of wind facility–produced electricity can power 221 houses annually (National Wind Watch 2017).

Types of Effects on Birds

Wind development can affect birds in three ways: (1) directly, through collision mortality; (2) indirectly, through disturbance and displacement; and (3) by physically modifying or destroying habitat (Masden et al. 2009, Schuster et al. 2015). Of these, direct impacts have been the most researched to date. Collision with any part of the turbine structure (e.g., towers, turbine blades, or infrastructure like electrical lines) typically results in avian mortality. Sublethal impacts are caused when a bird collides with some part of the turbine and either persists with its injuries or succumbs to them later; these types of encounters are less studied and still poorly understood.

Construction and operation of a wind facility can indirectly affect birds through disturbance and displacement, in which case birds avoid using space within or around a wind facility. This indirect effect is typically nonlethal to individuals, but may interfere with foraging, reproduction, and travel in flight. If these obstacles manifest as increased energetic demands, birds may face reduced fitness and increased mortality as a result of wind development disturbance and avoidance. Disturbance is not nearly as well understood as collision mortality, in part because the methodology to understand indirect effects can be more involved and more time-intensive to assess (Masden et al. 2009, Garvin et al. 2011).

Finally, wind development can physically destroy or alter avian habitat. As with disturbance, nominal research has been conducted to understand this type of impact. Habitat conversion is most common dur-

Figure 5.2   Example of modern wind turbines at a wind facility in southeastern Wisconsin, US. *Photo: Steven M. Grodsky.*

ing the construction phase for land-based wind energy facilities. Once operational, the footprint per turbine in wind facilities is relatively small. Furthermore, developers usually attempt to mitigate habitat disturbance or loss following construction (Marques et al. 2014). For offshore wind energy, there may be short-term habitat modifications during the construction phase, but avian habitat disturbance or loss occurs mostly during a wind facility's operational phase (Schuster et al. 2015).

Land-Based Wind Energy and Birds
Methods for Studying Effects on Birds

Standard practices to assess collision mortality involve defining search areas centered on a number of randomly selected turbines in a wind energy facility, walking a series of transects within the defined search areas, and collecting and documenting types and numbers of birds killed. Researchers use searcher efficiency and scavenger removal trials to improve accuracy of mortality estimates (Grodsky 2010). For these trials, researchers place a known number of

dead birds within portions of defined search areas and then measure either the ability of searchers to find bird carcasses or the rate at which scavengers remove carcasses. Dead birds used in trials may be specimens previously collected during carcass searches or birds that have been killed as a result of wildlife damage management activities independent of wind facilities (e.g., bird control at dairy farms). Searchers are not made aware of when efficiency trials are being conducted or of where carcasses have been placed (Grodsky et al. 2013).

Avian avoidance of wind facilities cannot be identified without implementing a before-after-control-impact (BACI) experimental design. Under the BACI approach, avian use of the proposed wind development area is evaluated before any construction begins. Preconstruction assessment methodology is then replicated during the post-construction phase, once the wind facility has become operational. Pre- and post-construction assessments are performed both within the wind facility and at reference points located outside the wind facility's footprint. BACI designs typically are not applied to avian impact studies at wind energy facilities, and long-term post-construction datasets are rare. Preconstruction studies are more common, but there is a lack of consistency in application of standardized experimental designs across wind facilities, limiting our ability to explore and understand mechanisms driving avian avoidance of wind facilities (Drewitt and Langston 2006, Garvin et al. 2011).

To evaluate effects on habitat, a pre-, during-, and post-construction evaluation of effects of wind development on surrounding areas may be conducted. A review of rare, threatened, or endangered plants and wildlife may be required as part of the permitting process, although this can be project-, location-, or state-dependent. These biodiversity surveys may entail examining species-account records at the state or federal level to determine whether species of concern have been recorded in the area proposed for siting. An on-site species inventory may also be required (Drewitt and Langston 2006).

Collision Mortality

Collision mortality is the most visible and well-documented effect of wind energy development on birds. Avian collision mortality in North America and Europe generally ranges from 0 to >60 fatalities/turbine/year (Drewitt and Langston 2008). However, per-turbine collision rates can be misleading because collision mortality may be due to one or a few turbines within a wind facility and, therefore, may not be representative of the entire wind facility. Collision mortality typically is reported as collision rate per megawatt of output to more accurately represent the operational wind facility (Drewitt and Langston 2008). Several efforts have been made to estimate avian collision mortality at wind facilities on a national scale, with an estimated 20,000 (Sovacool 2012) to 573,000 bird (Smallwood 2013) deaths at wind facilities each year in the United States alone.

Small passerines (songbirds; order Passeriformes) are among the species most frequently involved in turbine-related collision mortality (Strickland et al. 2011). Between the United States and Canada, Erickson et al. (2014) estimated that 134,000 to 230,000 songbirds fatally collide with wind turbines annually. Songbird migration activity often occurs at night, and these flights commonly take place at low altitudes (Welcker et al. 2017), potentially increasing collision risk for nocturnal migrants. Although songbirds migrating during the night make up the majority of turbine-related fatalities in several studies (Johnson et al. 2002, Grodsky and Drake 2011, Strickland et al. 2011), research suggests that diurnal birds like gulls and raptors may suffer greater losses than nocturnal fliers (Krijgsveld et al. 2009, Welcker et al. 2017). Welcker et al. (2017) combined radar-collected data on bird flight behavior with fatality searches to identify a theoretical collision risk for diurnal migrants that was 19.4 times greater than that of nocturnal migrants.

Critics of mortality estimates argue that they are derived from a limited, nonrepresentative sample of wind facilities (Loss et al. 2013) and that the focus on high-risk wind facilities may result in a "worst-case scenario" estimate of avian mortality (Gove et al. 2013). Regardless, relative to other structures (e.g., communication towers, buildings, power lines), many wind facilities report low collision rates, suggesting that avian collision mortality is lower at wind turbines than at other anthropogenic structures (Erickson et al. 2001, Percival 2005, Drewitt and Langston 2006).

Still, some poorly sited wind facilities have resulted in relatively high mortality rates for birds. Most notably, wind developments in the US Altamont Pass Wind Resource Area (APWRA) of California (Smallwood et al. 2009), Smøla Peninsula of Norway (Bevanger et al. 2010), and Tarifa of Spain (Ferrer et al. 2012) have reputations for their historically high avian mortality rates. Much of the avian mortality reported at facilities in these regions can be attributed to their colocation with areas supporting high raptor activity. Altamont Pass has reported an annual avian mortality rate of 2,710 birds, 1,127 of which are raptors (orders Accipitriformes, Falconiformes, and Strigiformes; Smallwood and Thelander 2008). Similarly, raptor fatalities make up the majority of reported collision mortality at the wind facilities of Tarifa (Barrios and Rodriguez 2004, Fer-

rer et al. 2012). On the Smøla Peninsula, a designated Important Bird Area, an average of 7.8 white-tailed eagles (*Haliaeetus albicilla*) die each year from collisions at wind facilities (Heath and Evans 2000, Bevanger et al. 2010).

Deep Dive: Burrowing Owls at the Altamont Pass Wind Resource Area, California

The western burrowing owl (*Athene cunicularia hypugaea*; Figure 5.3) is a small owl (height ≤24 cm) that is primarily a ground-dweller. Western burrowing owls nest in underground burrows that they excavate themselves or otherwise occupy after burrows have been abandoned by other animals like prairie dogs (*Cynomys* spp.) and California ground squirrels (*Spermophilus beecheyi*). Their burrows occur in open areas (e.g., prairies, farmland, airports) consisting of flat ground and short grass or bare soil. They are opportunistic predators that eat mostly insects and small mammals. Although the owl's range is widespread throughout most of the western United States, populations in south-central Canada and northern Mexico have been declining for years, primarily due to habitat loss but also due to reduced food availability because of rodent-control programs (Sibley 2000, Dunn and Alderfer 2009, Williford et al. 2009).

Collisions with wind turbines have contributed to declines in western burrowing owl abundance in some areas. For example, between 1998 and 2011, Smallwood et al. (2013) estimated that as many as 600 burrowing owls were killed annually at the Altamont Pass Wind Resource Area (APWRA) in California. The APWRA comprises more than 5,400 wind turbines of various designs, with heights

Figure 5.3   The western burrowing owl (*Athene cunicularia hypugaea*) is a small, ground-dwelling raptor; its population is declining across its range because of habitat loss and reduction in its prey base. In some parts of its range, mortality caused by colliding with wind turbines is also a factor. *Photo credit: Robb Hannawacker.*

ranging from 4 m to 43 m above the ground (Smallwood et al. 2007).

Smallwood et al. (2007) identified patterns of owl mortality and behavior at APWRA.

Turbines with blade tips within 15 m of the ground when spinning killed more owls than turbines with blade tips greater than 15 m above the ground, reflecting the burrowing owl's low flight altitude. Burrowing owls tended to avoid areas within APWRA populated with fast-spinning turbine blades, instead opting for sparsely distributed monopole turbines with slower moving blades. Burrowing owls flew through gaps between turbines or rows of turbines and seemed to be killed more often by turbines at the end of a row. Flights by burrowing owls through the rotor-swept zone and collisions with turbines were greater where turbines were within 20 m of grazing cattle, as identified by presence of cattle dung. Because cattle spent more time grazing closer to wind turbines rather than in the rows between turbines, areas of shorter grass occurred closer to turbines. Western burrowing owls were associated with short grass areas, and it was suggested that owls likely were attracted to grazed areas because cattle dung attracted prey items that the owls fed on. Lastly, turbine-caused burrowing owl fatalities were positively correlated with greater densities of ground squirrel burrows (many burrows were created but abandoned by ground squirrels because of a rodent control program) and owl burrows within 90 m of a turbine.

Based on their research, Smallwood et al. (2007) recommended mitigation techniques in an attempt to reduce the burrowing owl's population decline at APWRA. The authors estimated that shutting off turbines during the winter months, when 35% of annual burrowing owl mortalities occurred, would result in only a 14% reduction of electricity generated by the APWRA (Smallwood et al. 2007). Other suggested mitigation strategies included replacing older, shorter turbines with taller turbines with blades that would spin above the flight altitude of burrowing owls and conducting a rodent control program to reduce prey that the burrowing owls were feeding on within APWRA (Smallwood et al. 2007). However, prey reduction of ground squirrels would also potentially reduce availability of underground burrows used by owls for nesting (Smallwood et al. 2007).

Raptors and other large soaring birds (e.g., vultures, order Cathartiformes) seem to be especially vulnerable to turbine-collision mortality (Barrios and Rodriguez 2004, Hoover and Morrison 2005, Gove et al. 2013, Erickson et al. 2015). Although raptors have the highest visual acuity yet identified among birds, their frontal binocular field of vision is fairly narrow (Martin and Katzir 1999, O'Rourke et al. 2010, Martin et al. 2012). In fact, for raptors, it is the lateral fields of view (rather than the frontal field) that possess the highest resolution. Similar to other raptors, vultures in the genus *Gyps* tilt their heads forward during flight to search below for food and conspecifics, thereby acquiring a comprehensive view of the landscape. This positioning creates a blind spot directly in front of these vultures, preventing them from detecting upcoming obstacles and thereby increasing collision risk (Martin and Shaw 2010, Martin et al. 2012).

Morphological traits, particularly relating to body size, combined with site-specific characteristics like topography, influence avian collision risk at wind facilities (Orloff and Flannery 1992, Smallwood and Thelander 2004, Hoover and Morrison 2005, Hötker et al. 2006). Large-bodied, soaring raptors (e.g., vultures) have greater wing loading (i.e., ratio of body weight to wing area) and rely heavily on masses of hot rising wind known as thermals to gain altitude (Dahl et al. 2013, Marques et al. 2014). In the absence of thermals, these large-bodied, thermal soarers gain altitude by riding updrafts created by wind that is deflected upward from the ground as it hits a cliff face or other type of geologic uprising, which creates orographic lift. High wing-loading of these species reduces their in-flight maneuverability and ability to efficiently ride orographic lift. In Tarifa, Barrios and Rodriguez (2004) observed that when constrained to weaker updrafts, vultures struggled to gain altitude when leaving their roosts, resulting in greater turbine-collision mortality. Wind developers often prefer to site turbines along steep ridges or mountains, where wind resources are abundant. This increases the likelihood of collisions for raptor species that often fly low along steep ridges to exploit the orographic lift produced there (Hoover and Morrison 2005, Katzner et al. 2012).

Flight behavior also plays a key role in collision risk. Raptors drawn to high concentrations of prey within wind facilities (Hunt 2002, Smallwood and Thelander 2004, Drewitt and Langston 2006) may become distracted and collide with turbines (Krijsveld et al. 2009, Smallwood et al. 2009). Red-tailed hawks (Buteo jamaicensis) and American kestrels (Falco sparverius) often exhibit kiting or hovering behaviors when hunting, capturing updrafts created by high winds to remain stationary in flight. Static foraging strategies like kiting increase collision risk for these raptors, especially under gusty conditions that may knock hunting raptors off balance (Hoover and Morrison 2005). In the APWRA, Hoover and Morrison (2005) observed a disproportionate number of turbine-related red-tailed hawk mortalities on slopes where kiting behavior was common.

Wind speed and direction appear to be influential collision risk factors for some raptor species. In Tarifa, griffon vultures (Gyps fulvus) often collided with turbines, even during good visibility conditions. Low wind conditions combined with the gentle slopes of the PESUR wind facility failed to produce sufficient updrafts to carry vultures above the spinning turbines (Barrios and Rodriguez 2004). Similarly, Johnston et al. (2014) reported that collision risk for migrating golden eagles (Aquila chrysaetos) decreased with increasing wind speed and increased under headwinds and tailwinds (Johnston et al. 2014). Still, avian collision risk is determined by interactions among several site- and species-specific risk factors. For instance, though vultures at PESUR often struggled to gain altitude in low wind conditions, common buzzards (Buteo buteo) and short-toed eagles (Circaetus gallicus) circling in the same updrafts rarely collided with turbines (Barrios and Rodriguez 2004).

Aside from factors influencing collision risk for raptors and other large soaring birds, a number of factors have been identified that may increase collision risk for non-raptor birds. Behaviors associated with breeding and chick-rearing can influence collision risk. Wind turbines intercepting the main foraging route for a breeding colony of terns (family Sternidae) killed enough birds to increase annual mortality by 0.6%–3.7% (Everaert and Stienen 2006). Everaert and Stienen (2006) noticed a temporal change in turbine collision mortality rates for breeding terns. Although terns avoided turbines early in the breeding season, the imperative to provide for their chicks took priority over time, and terns opted instead to take the most direct route to their feeding grounds, resulting in increased collision mortality (Everaert and Stienen 2006).

Time of year and weather also play an important role in determining avian collision risk, especially during spring and fall migration (Strickland et al. 2011). Migratory birds often rely on landscape features to guide their travels (Gill 2007), and weather conditions that make flying more difficult or reduce visibility (e.g., rain, fog, high winds) may force migrants to fly at lower altitudes (Erickson et al. 2001). In turn, this may increase their likelihood of encountering and colliding with turbines (Langston and Pullan 2003, Langston 2013). Of the 55 avian fatalities observed at Buffalo Ridge wind facility in

Minnesota, between 1996 and 1999, only 3 occurred during calm weather, whereas 31 seemed to be associated with thunderstorms (Erickson et al. 2001).

Few studies have successfully linked turbine-collision mortality to avian population declines. However, even the low mortality rates reported at most wind facilities could prove significant in species characterized by low maturation and reproduction rates, like some raptors species (Drewitt and Langston 2006). Cumulative mortality effects of multiple wind facilities could increase risk of population decline in raptors and rare species through additive mortality (Carrete et al. 2009, Erickson et al. 2015, Beston et al. 2016). In the APWRA, turbine-collision mortality has exceeded reproductive capabilities of the local golden eagle population. Because of the unsustainable mortality rate, the breeding population of golden eagles may now act as a sink, relying on immigration to remain stable (Hunt 2002, Hunt and Hunt 2006).

Disturbance and Avoidance

Wind development can indirectly affect birds through disturbance, a nonlethal by-product of wind facility construction and operation. In response to this disturbance, birds may avoid wind developments. The term avoidance encompasses a spectrum of behaviors that operate at different spatial and temporal scales. Macro-avoidance or displacement occurs when birds avoid a wind facility in its entirety, resulting in functional habitat loss—habitat may still exist, but birds are excluded from it—and a reduction in bird density within the footprint of the wind facility (Gove et al. 2013, May 2015). When responding to individual turbines or turbine clusters in flight, birds engage in localized micro-avoidance (e.g., evasion; May 2015).

Disturbance produced by wind development changes over the course of a wind facility's lifetime. In the construction phase, birds may be disturbed by the visual presence of workers and equipment, construction noise, and dust and pollution produced by construction (BirdLife International 2011). Once a wind facility enters its operational phase, the physical and visual presence of turbines and by-products of turbine operation (e.g., noise, shadow flicker) can disrupt avian communities (BirdLife International 2011). In some species, disturbance during construction elicits a stronger avoidance response than disturbance from an operating wind facility (Pearce-Higgins et al. 2012). For other bird species, the negative effects of wind facility disturbance intensify over time (Stewart et al. 2007, Farfan et al. 2009, Sansom et al. 2016). For example, golden plovers (*Pluvialis apricaria*) showed no signs of decline until a wind facility became operational, at which time the plover breeding population density declined by 79% (Sansom et al. 2016).

Among birds known to avoid wind developments, waterfowl (order Anseriformes) exhibit the largest displacement distances—the distance from turbines at which birds are either absent or occur less frequently than expected (Hötker et al. 2006). Displacement distances for waterfowl have been recorded up to 850 m from a wind development, although distances of 600 m or less are more common (Drewitt and Langston 2006, Gove et al. 2013). Studies of passerine displacement are rare and, in some cases, have not only failed to identify avoidance (Devereux et al. 2008) but have documented a positive association between passerines and wind turbines (Bevanger et al. 2010, Pearce-Higgins et al. 2012). Still, reduced densities of breeding grassland bird species have been observed near wind facilities (Leddy et al. 1999, Pearce-Higgins et al. 2009, Bevanger et al. 2010), indicating that wind development can disturb and displace some passerines, although reasons for this displacement are unknown.

Raptors often avoid turbines and wind facilities. Migrating raptors will adjust their flight trajectories (Cabrera-Cruz et al. 2016) and flight heights (Johnston et al. 2014) to avoid interacting with turbines. Wind developments also can displace resident or breeding raptors, reducing raptor abundance by 40% to 72% (Farfan et al. 2009, Garvin et al. 2011, Campedelli et al. 2014). White-tailed eagles at the Smøla wind facility in Norway, for instance, experienced significantly lower breeding success on territories within 500 m of turbines post-construction compared with the same territories before construction (Dahl et al. 2012). However, it was unclear whether nests were left vacant because of displacement or because of eagle fatalities resulting from collisions with turbines (Dahl et al. 2012).

Even after demonstrating avian avoidance of wind facilities, interpreting the effects of such disturbance is difficult. Long-term datasets implementing BACI designs are still lacking, limiting understanding of the full duration and consequences of disturbance. Disturbance effects may be only temporary if birds habituate to wind facilities. However, not all species are able to habituate and, as the duration of operation increases, greater declines in avian abundance typically are observed (Stewart et al. 2007, Farfan et al. 2009). A handful of studies have documented that long-term response to displacement is taxa specific (Dohm 2017). For example, abundance of both raptors and passerines declined immediately following construction of a wind facility in Spain, but, after 4 to 6 years of operation, raptor abundance improved while the number of passerines continued to decline (Farfan et al. 2017). Similarly, turbine-displacement distances for pink-footed geese (*Anser brachyrhynchus*) declined by up to 67% 8 to 10 years following wind facility construction (Madsen and Boertmann 2008).

Habitat Modification

Wind development can lead to physical destruction and alteration of avian habitat. Habitat loss from wind energy development generally is not viewed as a significant threat to birds. Typically, habitat loss occurs in just 2% to 5% of the total wind facility footprint (Fox et al. 2006). The potential for habitat destruction to affect bird communities depends both on the size of the wind facility and the habitat type or vegetation community at risk (Gove et al. 2013,

Zimmerling et al. 2013). Sensitive or pristine environments may become degraded even after small-scale destruction and alteration via wind development (Gove et al. 2013). For example, increased wind development in forested areas may fragment forest, increasing the amount of edge (Segers and Broders 2014). In turn, forest fragmentation can increase nest predation and parasitism in birds, reducing avian reproduction (Batary and Baldi 2004). However, wind facilities can be built in agricultural landscapes and other marginalized lands where natural areas have already been compromised, thereby minimizing the negative effects of wind development on more intact and less fragmented landscapes (Zimmerling et al. 2013).

Offshore Wind Energy and Birds

Offshore wind energy production has a couple of operational advantages over its land-based counterpart. Relative to land-based locations, winds offshore typically blow harder and more consistently because of less interference, resulting in greater wind energy production (Leung and Yang 2012). Additionally, there are large swaths of unoccupied areas offshore where wind energy facilities can be constructed away from humans to avoid common complaints about turbines (e.g., noise effects, interrupted view sheds; Leung and Yang 2012).

The first offshore wind turbine was installed in 1991 off the coast of Denmark, but global offshore wind energy production didn't increase dramatically until the early twenty-first century (Sun et al. 2012). In 2000, offshore wind facilities produced less than 100 MW of wind energy worldwide (Madariaga et al. 2012). By 2017, over 14,000 MW of wind energy was produced offshore (Global Wind Energy Council 2017b). Currently, 88% of offshore wind energy production is installed off the coasts of 10 European countries, including the United Kingdom (currently supporting the world's largest offshore wind energy market), Germany, Denmark, Holland, and Finland (Global Wind Energy Council 2017b). China established the first offshore wind facility outside of Europe in 2010, and, since then, Japan and South Korea have begun to install wind energy facilities offshore (Leung and Yang 2012). The first offshore facility in North America became operational in 2016 off the coast of Rhode Island (Bureau of Ocean Energy Management 2017).

Offshore wind energy production is projected to continue to increase with research and development of larger turbines, new turbine designs (e.g., vertical-axis turbines rather than horizontal-axis turbines), and advanced methods of supporting turbine infrastructure in deeper water, among other factors (Sun et al. 2012). The cost of constructing, operating, and maintaining offshore turbines is currently the primary barrier to further expanding offshore wind energy. Most offshore wind facilities are in water 30 m deep or less (see Sun et al. 2012) because costs to construct, operate, and maintain turbines, as well as transfer energy from offshore turbines to land, increase with distance from shore (Madariaga et al. 2012).

Methods for Studying Effects on Birds

Most of the research regarding avian impacts from offshore wind facilities has been conducted in Europe. However, relative to the effects from land-based wind energy, much less is known regarding the effects of offshore wind facilities on birds (Schuster et al. 2015). This is partly due to the fact that offshore wind facilities have not been operational as long as have land-based wind facilities. It also is much more difficult to collect avian data offshore than at land-based wind facilities. Carcasses are difficult to locate in the ocean because they sink or are taken away from the area under turbines by wind, currents, or marine scavengers, making it difficult to quantify mortality caused by collisions with offshore wind turbines and associated infrastructure (Schuster et al. 2015). Further, it is more expensive to collect data offshore than on land, regardless of the type of data collected, because of costs to access offshore

wind facilities and survey over water (Snyder and Kaiser 2009).

Human observers employ a variety of observation methods to observe and note numbers and types of birds flying offshore, including observations with the naked eye, binoculars, and spotting scopes (Huppop et al. 2006). Observations typically are conducted in spring and fall because collision risk is thought to increase with migration activity around coastlines (Huppop et al. 2006). Use of ships and aircraft allows observers to document avian species abundance in areas of ocean far from land. Although these surveys can identify birds in the vicinity of proposed or operating offshore wind turbines, they are unlikely to document direct impacts (Goyert et al. 2016). Additionally, human observations are limited to daylight hours with good visibility and, even then, birds must fly relatively close to the observer to be recorded (Fijn et al. 2015). Furthermore, it can be difficult for human observers to accurately estimate flight height and bird density (Fijn et al. 2015). These obstacles restrict the type of avian data that can be collected offshore.

Motion-triggered still cameras, video cameras, and recording microphones can be affixed to offshore wind turbines to collect data in close proximity to turbines and turbine blades (Huppop et al. 2006, Collier et al. 2012). These devices can record flock size, flight behavior, and bird-turbine collisions, as well as facilitate accurate species identification (Huppop et al. 2006, Collier et al. 2012). However, efficacy of camera and recording systems affixed to offshore turbines is limited by weather and light conditions, characteristics of target bird species, angle and trajectory of the bird relative to the camera, and distance of the bird from the camera (Collier et al. 2012).

Radar systems that rotate vertically and horizontally can overcome the shortcomings of human observation by collecting data on density, flight altitude, and flight direction of birds flying offshore during night and day and in a variety of weather conditions (Huppop et al. 2006). Radar systems can be land- or ship-based, allowing them to collect data over large areas in coastal and offshore waters (Huppop et al. 2006). Some of the disadvantages of radar systems are that they typically cannot identify numbers of individual birds and species, detecting single birds >1.5 km from the radar can be difficult, and rough sea conditions can increase radar clutter, thereby decreasing detection probability (Dokter et al. 2013, Fijn et al. 2015).

Attaching very high frequency or global positioning system radio-locators or data loggers to birds flying across water can aid in collecting data on bird location, flight altitude and direction, and time spent in flight, among other variables potentially of interest with respect to the effects of offshore wind generation on birds (Furness et al. 2013, Garthe et al. 2017). These data can help refine calculations of collision risk and macro- and micro-avoidance rates at and near offshore wind facilities (Furness et al. 2013). Lastly, data from birds colliding with other offshore structures such as oil drilling platforms may help predict numbers and types of birds that could be similarly affected by offshore wind turbines (Huppop et al. 2006).

Collision Mortality

Avian mortality from offshore wind facilities is a primary concern, but research documenting birds actually colliding with offshore turbines or turbine infrastructure is rare (Brabant et al. 2015, Fijn et al. 2015, Schuster et al. 2015). Instead, researchers use field observations to model collision risk relative to a variety of factors, including turbine specifications (e.g., number and type of turbines) and observed avian flight heights relative to rotor-swept zones of wind turbines (Furness et al. 2013, Fijn et al. 2015).

Because bird abundances increase during migration events, offshore collision risk may potentially increase during migration (Huppop et al. 2006, Schuster et al. 2015). Although avian migration events can occur year-round, they tend to be seasonal, generally peaking in the fall and spring, and

are species-specific (Huppop et al. 2006, Schuster et al. 2015). It is estimated that as many as 1.3 million seabirds and several hundred million non-seabirds (e.g., passerines) migrate across the North Sea (Brabant et al. 2015). During a 5-year study at a wind facility off the Dutch coast, Krijgsveld et al. (2011) observed 103 species of local and migrating seabirds, including gulls (family Laridae), northern gannets (*Morus bassanus*), divers (family Gaviidae), scoters (*Melanitta* spp.), guillemots (*Cepphus* spp.), and migrating non-seabirds (e.g., passerines; thrushes, *Turdus* spp.; geese, *Anser* and *Branta* spp.). Although inter- and intra-annual variation in abundance and species composition was large, the most commonly observed group of birds flying along the coast included gulls, gannets, cormorants (*Phalacrocorax* spp.), European starlings (*Sturnus vulgaris*), blackbirds (*Turdus* species), and thrushes (Krijgsveld et al. 2011).

Bird occurrence relative to shore also appears to correlate with collision risk. Goyert et al. (2016) modeled and predicted offshore distribution and abundance for 40 marine bird species along the Mid-Atlantic coast of the United States and showed that distance to shore was the most common predictor of bird abundance, whereas avian abundance declined as distance from shore increased. Brabant et al. (2015) documented less than a third of the number of gulls at a North Sea wind facility located 46 km from shore than reported at a wind facility just 10 km off the shore of the Netherlands (Poot et al. 2011). Natural distribution of gull species closer to shore may increase their collision risk at wind facilities, which are concentrated in shallow coastal waters.

Furness et al. (2013) suggested that avian flight altitude relative to the height of the rotor-swept zone is of overwhelming importance compared with other factors (e.g., flight maneuverability, percentage of time flying, amount of nocturnal time flying) in models of avian collision risk at offshore wind facilities. Birds that fly above or below the rotors have little risk of colliding with the turbine blades, whereas birds that fly within the rotor-swept zone are more susceptible to collision. In their review, Furness et al. (2013) concluded that most bird species do not fly at the rotor height of offshore wind turbines and, therefore, are not vulnerable to population-level declines as a result of mortality at offshore wind facilities. Contrary to Furness et al. (2013), Krijgsveld et al. (2011) observed that most of the 103 avian species they recorded flew at altitudes within the collision-risk zone (i.e., rotor height); they suggested that as many as 2 million birds annually may fly through the rotor-swept zone at the Egmond aan Zee wind facility off the Dutch coast.

Certain bird taxa are more vulnerable to colliding with offshore wind turbines because of their flight behavior and characteristics. Passerines appear to be vulnerable because of the relatively low altitude at which they fly and their relatively large flock sizes. Larger seabirds are susceptible because of their limited maneuverability, and ducks may be susceptible because of their nocturnal and high-speed flight patterns (Huppop et al. 2006, Furness et al. 2013, Schuster et al. 2015). Similarly, gull species may be vulnerable to offshore turbine collisions because of their frequent flight at current rotor heights. Brabant et al. (2015) used a collision risk model and visual census data to examine the potential effects on 6 seabird species that flew at rotor height at Belgian offshore wind facilities. Of the 6 seabird species, gulls were predicted to constitute the majority (98%) of the turbine-collision mortality (Brabant et al. 2015).

Passerines may face the greatest collision risk at offshore facilities, with annual fatality rates in the hundreds (Krijgsveld et al. 2011). Although most migrating birds appear to fly above the rotor-swept zone during favorable weather, flight height during migration can be species-specific (Schuster et al. 2015). Up to 17% of migratory songbirds were observed flying offshore at the height of the rotors at a wind facility in Kalmar Sound, Sweden (Pettersson and Fågelvind 2011). Huppop et al. (2006) suggested that 50% of all migrating birds crossing the North Sea may fly within the rotor-swept zone of offshore wind turbines. Hill et al. (2014) suggested that all

avian flight activity across the North Sea over all seasons occurs below 200 m above sea level, thereby commonly intersecting the rotor-swept zone. Using vertical radar, Brabant et al. (2015) observed a particularly intense night of thrush migration over the North Sea. Taking into account model uncertainties and assumptions, extrapolation of the results from this single night of migration under a scenario of 10,000 turbines operating throughout the North Sea would result in a single night mortality event totaling 5,257 thrushes (Brabant et al. 2015).

At current offshore wind energy production levels, mortality from collisions with offshore wind turbines is not likely to result in population declines for many seabird populations (Brabant et al. 2015). Modeling may be used to predict population-level effects of avian turbine-collision mortality. Brabant et al. (2015) suggested that the seabird populations they examined were not susceptible to population declines at current offshore wind production levels, but should the European Union meet its offshore wind energy capacity goal of 43 GW (>14,000 3-MW turbines), lesser black-backed gulls (*Larus fuscus*) and great black-backed gulls (*Larus marinus*), as well as black-legged kittiwakes (*Rissa tridactyla*), could face population declines. Poot et al. (2011) suggested that currently stable or increasing bird populations would not decline regardless of wind facility proximity to shore. Even the 2 species currently in decline—Bewick's swan (*Cygnus columbianus bewickii*) and the Dutch breeding population of herring gull (*Larus argentatus*)—would suffer only a small increase in additive mortality from offshore wind relative to current population trends (Poot et al. 2011).

Disturbance and Avoidance

Direct evidence elucidating trends in avian avoidance of offshore wind facilities is limited. However, it is presumed that presence of large, spinning wind turbines on an otherwise open seascape is the primary cause for avian avoidance (Dierschke et al. 2016). Krijgsveld et al. (2011) observed birds more

Figure 5.4   Service ships routinely visit offshore wind facilities to maintain turbines and can be a reason marine birds avoid areas containing turbines. *Photo credit: Siemens Offshore Wind Facilities.*

often avoiding operational offshore wind turbines than offshore turbines that were not spinning. Avian avoidance of operating offshore wind facilities appears to increase over time, possibly indicating that birds learn to avoid individual turbines or the wind facility as a whole (Plonczkier and Simms 2012). Additional causes for avoidance include regular ship and helicopter traffic inside and around wind facilities that service and maintain wind turbines and a possible decrease in food abundance and availability (Dierschke et al. 2016; Figure 5.4). For example, fishing vessels are prohibited from operating inside boundaries of offshore wind facilities, thus removing resources commonly associated with commercial fishing, including bycatch, discarded fish, and invertebrates, on which some seabirds have grown accustomed to feeding (Dierschke et al. 2016).

Studies at a number of wind facilities in the North Sea estimated that 71% to 99% of migrating birds exhibited macro-avoidance of the wind facilities under observation (Pettersson 2005, Petersen et al. 2006). However, Krijgsveld et al. (2011) estimated macro-avoidance rates between 18% and 34% at the Egmond aan Zee wind facility off the Dutch coast. Dierschke et al. (2016) observed many birds avoiding offshore wind facilities, but avoidance behavior seemed to be species-specific. After examining seabird responses to 20 offshore wind facilities in European waters, Dierschke et al. (2016) determined that divers and northern gannets showed strong avoidance, as did grebes (*Podiceps* spp.) and northern fulmars (*Fulmarus glacialis*), although evidence was more limited for the latter 2 birds. Common guillemots (*Uria aalge*), razorbills (*Alca torda*), long-tailed ducks (*Clangula hyemalis*), common scoters (*Melanitta nigra*), little gulls (*Hydrocoloeus minutus*), and sandwich terns (*Thalasseus sandvicensis*) all demonstrated variable avoidance, and cormorants showed strong attraction to offshore wind facilities because of available perching structures, which seemingly allowed them to forage farther from shore (Dierschke et al. 2016). Krijgsveld et al. (2011) showed that geese and swans avoided the Egmond aan Zee wind facility off the Dutch coast, as did passerines during the night, but not during the day.

Wind facilities can act as a barrier causing birds to divert their flight paths to avoid offshore wind facilities or clusters of offshore wind turbines. By altering their flight paths, birds may take less efficient routes between nesting and feeding grounds, for instance, and thus expend more energy flying. Furthermore, reproductive success for most birds is related to body condition at time of breeding; reduced body mass from increased energy expenditures during migration as a consequence of macro-avoidance of wind facilities may lower reproductive success for some birds (Masden et al. 2009). Birds have been observed altering their flight path 1 km away from offshore wind facilities, and corrections in flight path were observed up to 4 km beyond an offshore wind facility (Pettersson 2005, Krijgsveld et al. 2011).

Avoidance also can occur in closer proximity to offshore wind facilities and turbines. For example, Masden et al. (2009) observed a displacement distance of just 500 m for eiders (*Somateria* spp.), resulting in insignificant energy costs. Their findings support those of Pettersson (2005), who studied eiders at 2 offshore wind facilities near Sweden. Instead of avoiding an entire offshore wind facility, birds may simply respond to individual turbines (e.g., micro-avoidance). Micro-avoidance rate for all avian species observed at the Egmond aan Zee wind facility was 98% (Krijgsveld et al. 2011). Micro-avoidance was greater at night than during the day, and of all observed birds that came within 50 m of an offshore wind turbine, only 7% flew through the rotor-swept zone (Krijgsveld et al. 2011).

Habitat Modification

Construction and operation of an offshore wind facility can change pelagic and benthic environments, although knowledge about the ecological effects over the operational lifespan of offshore wind facilities on sea life and seafloor environments is limited (Miller et al. 2013, Dierschke et al. 2016). Short-term habitat modification occurs when foundations to support offshore wind turbines are drilled or pounded into the seafloor, which disrupts the seabed and resuspends sediments (Miller et al. 2013). Disturbance associated with installing turbine foundations and associated infrastructure (e.g., buried cable) may cause some avian species to leave the area, while attracting others seeking foraging opportunities on injured or dying organisms (Miller et al. 2013). Sessile or relatively immobile marine species may be buried under sediment during drilling, piling, or cable laying and either die or expend additional energy to escape (Miller et al. 2013). Exactly how the food web is affected and what consequential effects trophic interactions have on the avian community are not well understood (Inger et al. 2009).

Once offshore wind turbine foundations are in place, they can have long-term advantageous or disadvantageous effects on birds. Grounded or anchored foundations to support offshore wind turbines create a hard substrate on an otherwise soft sea bottom, thereby providing areas that support benthic invertebrates; in turn, benthic invertebrates may attract fish and other food sources for select avian species (Lindeboom et al. 2011). In a bay in Sweden, epibenthic communities (e.g., barnacles, tube worms) have been shown to cover 100% of hard substrates within a year of installation of submerged steel and concrete pillars (Andersson et al. 2009). However, questions remain as to whether increases in local biomass equate to increases in bioproduction or simply reflect aggregation of species on and around artificial structures like offshore wind turbine foundations (Miller et al. 2013). Additionally, non-native, invasive species may be more likely to outcompete native species and colonize artificial structures (Miller et al. 2013), though what effect such changes might have on birds is not fully known.

Foundations of offshore wind facilities constructed in relatively shallow waters may alter currents in the area of offshore wind facilities and therefore cause sandbanks to shift, disappear, or develop (Dierschke et al. 2016). Ecological consequences of altered sandbanks are not completely understood, but may negatively affect or benefit wading birds if sandbanks disappear or accrete, respectively (Drewitt and Langston 2006). Increased turbidity as a result of currents moving around offshore wind turbine foundations may negatively affect foraging by piscivorous birds because of limited underwater visibility (Drewitt and Langston 2006). Alternatively, offshore wind facilities may serve as a proxy for protected marine areas because activities like commercial fishing are prohibited within offshore wind facility boundaries, thereby enhancing fish populations that provide an available food source for marine birds (Inger et al. 2009, Furness et al. 2013, Dierschke et al. 2016). Offshore turbines may also provide additional roosting sites, which could alter the spatial distributions of some avian populations (Krijgsveld et al. 2011, Dierschke et al. 2016).

Wind Energy Policy and Birds

Governments can encourage or even mandate environmentally responsible siting, construction, and operational practices related to wind energy development through policy. But application of siting regulations, environmental impact assessments/statements, and other policies and procedures to minimize environmental damage varies, sometimes dramatically, among countries and the states, regions, or territories therein. Existing statutes protecting wildlife (e.g., US Endangered Species Act 1973; 16 U.S.C. § 1531–1544) may influence siting and operation of wind facilities, even when such policies were not originally designed to target the wind industry. In this section, we review policy

oversight in the United States and the European Union that guides wind energy development as it relates to management of avian species. Chapter 10 (Arnett) provides comprehensive coverage of renewable energy policy related to wildlife and wind energy development.

United States

There is no standardized or unifying technical procedure for siting wind facilities in the United States. Regulatory requirements for siting new wind projects can vary widely by state, county, site ownership, and project-specific characteristics (Stemler 2007). At the county level, locally elected officials often make siting decisions for wind energy. As of 2012, 26 states used local governments (e.g., county zoning authorities, planning boards) to regulate siting of new wind projects (US Department of Energy 2015). Although this method preserves local autotomy and control over land-use decisions, local commissions may lack the expertise necessary to understand and appropriately review the often complex, technical applications submitted by wind developers (US Department of Energy 2008). In 22 states, primarily state governments authorize new wind projects, usually through a state utility commission (e.g., public utility commission). Some states may even have state-level, environmental guidelines for siting wind projects developed by a state wildlife or natural resource agency, but mandatory environmental regulations are rare (US Department of Energy 2015).

The US Fish and Wildlife Service (USFWS) has released 2 sets of voluntary guidelines designed to lead wind developers through an environmentally responsible siting, construction, and monitoring process. The first is the Land-Based Wind Energy Guidelines (hereafter Wind Energy Guidelines), released in 2012 (USFWS 2012). The Wind Energy Guidelines represent a collaborative effort by industry, the USFWS, state wildlife officials, conservation organizations, science advisers, and tribes to create an extensive set of recommendations for avoiding, minimizing, and mitigating negative effects on wildlife from wind energy development (US Department of Energy 2015). The second is the Eagle Conservation Plan Guidance (hereafter ECP Guidance), released in 2013 (USFWS 2013). Whereas the Wind Energy Guidelines provide broad instructions for safeguarding the environment, the ECP Guidance focuses exclusively on eagle conservation.

Both sets of guidelines follow an iterative, tiered approach to wind energy project planning. This process allows wind developers to identify potential issues and formulate questions to inform future actions at the end of each tier before proceeding to the next tier (USFWS 2012). Wind developers begin by performing preconstruction surveys to identify potential project sites and then describe site-specific characteristics. In the next tier, developers use preconstruction data to predict and assess risk and to develop avoidance, minimization, and, if necessary, compensatory mitigation strategies. Later tiers instruct developers to monitor their projects during the operational phase. Post-construction data is used to evaluate effectiveness of conservation strategies developed in earlier tiers and adapt them as necessary (USFWS 2012, 2013).

Even without a federal mandate to follow these guidelines, wind developers in the United States face legal obstacles. Regardless of state- or local-level regulations, wind developers must comply with 3 federal wildlife protection statutes: (1) the Endangered Species Act 1973 (ESA; 16 U.S.C. §§ 1531–1544); (2) the Migratory Bird Treaty Act 1918 (MBTA; 16 U.S.C. §§ 703–712); and (3) the Bald and Golden Eagle Protection Act (BGEPA; 16 U.S.C. §§ 668–668d). Whereas the ESA protects any avian species listed as "threatened" or "endangered" based on its risk of extinction determined by the USFWS, the MBTA and BGEPA exclusively protect migratory species and eagles, respectively. All 3 statutes protect avian species from "take," which is defined by the USFWS's implementing regulations (C.F.R. § 10.12) as "to pursue, hunt, shoot, wound, kill, trap, capture, or col-

lect." As of 2015, the USFWS was investigating at least 15 wind facilities for illegal wildlife take and has referred several of these cases to the US Department of Justice for possible prosecution (Ahrens and Gabriel 2015).

In addition to take, the ESA extends legal protection of endangered and threatened species to include "harm" defined as "an act which actually kills or injures wildlife . . . [and] may include significant habitat modification or degradation where it actually kills or injures wildlife by significantly impairing essential behavioral patterns, including breeding, feeding, or sheltering" (50 C.F.R. § 17.3). The ESA can potentially delay wind projects or penalize wind developers that threaten to take or harm endangered or threatened species, including birds. To date, only 1 wind developer has faced legal action for a wind project that threatened to take the endangered Indiana bat in West Virginia (*Myotis sodalis*; Animal Welfare Institute 2009).

Rather than risk litigation, wind developers can comply with the ESA by obtaining an incidental take permit. The ESA Section 10(a) authorizes take that occurs incidentally as a result of otherwise lawful activity by issuing an incidental take permit, 16 U.S.C. § 1538(a)(1)(B). To receive an incidental take permit, wind developers must demonstrate that they have reduced take of ESA-protected species to the maximum extent practicable through minimization and mitigation measures; these conservation strategies must be described in a habitat conservation plan. Issuing incidental take permits triggers the National Environmental Protection Act, which requires the USFWS to examine the impact of and alternatives to permit issuance through an environmental impact statement.

Under the MBTA, unlike under the ESA, wind developers cannot receive exemptions by obtaining an incidental take permit. Instead, developers rely on the USFWS's discretion not to prosecute violations under the MBTA (Glen et al. 2013). It appears that, in most cases, the USFWS has declined to prosecute wind facilities under the MBTA when avian mortality rates are relatively low (US General Administration Office 2005). To date, the USFWS has reached plea agreements with just 2 wind developers found violating the MBTA. In the first criminal enforcement of the MBTA in 2013, Duke Energy Renewables faced $1 million in fines, restitution, and community service after 2 wind facilities killed 14 golden eagles and 149 individual birds that were protected migratory species (US Department of Justice 2013). One year later, PacifiCorp Energy was found guilty of killing 38 golden eagles and 336 individual birds of protected migratory species and agreed to a $2.5 million settlement (US Department of Justice 2014). In both cases, Duke and PacifiCorp were accused of failing to make all reasonable efforts to avoid avian turbine-collision mortality despite warnings from the USFWS (US Department of Justice 2013, 2014).

Finally, the BGEPA prohibits the take of the United States' 2 eagle species—bald eagle (*Haliaeetus leucocephalus*) and golden eagle. Between 1997 and 2012, 32 wind facilities killed at minimum 85 eagles, 93% of which were golden eagles (Pagel et al. 2013). Actual eagle mortality rates may be even greater; the aforementioned estimates relied on incidental mortality observations and excluded eagle mortality at the APWRA, where an estimated 67 golden eagles perished annually (Smallwood and Thelander 2008). As under the ESA, wind developers may apply for an incidental eagle take permit to avoid prosecution under the BGEPA (50 C.F.R. § 22.26). The USFWS can issue incidental take permits only if the proposed eagle take (1) is compatible with eagle conservation; (2) is necessary to protect the economic interests of a locality; (3) is associated with but not the purpose of the activity; and (4) has been reduced to the maximum extent practicable (50 C.F.R. § 22.26).

The USFWS first promulgated the incidental permitting process in 2009 but finalized significant changes to the rule in 2016, after performing a new programmatic environmental impact statement. Based on this new report, the USFWS redefined the geographic limits of their eagle management units

and their corresponding eagle take limits (Eagle Permits 2016). But the 2016 rule change also introduced new flexibility and policy security for wind developers by building more leniency into meeting issuance criteria and by extending the maximum permit duration from 5 to 30 years (Eagle Permits 2016). The 30-year permit terms better align with the operational lifespan of wind facilities while also allowing the USFWS to implement an adaptive management strategy to avoid and minimize eagle take. The incidental eagle take permitting process is time-consuming and costly both for the wind developer and for the USFWS, which is charged with consulting on and reviewing permit applications and their corresponding eagle conservation plans. To offset these high labor costs, the USFWS requires that applicants pay a $36,000 processing fee (Eagle Permits 2016).

European Union

Contrasting the United States' voluntary guidelines, the European Union regulates wind development using mandatory directives. Although directives set a mandatory goal or objective, individual member countries choose how to implement them within their own national legal system (Hansen 2011). The Directive on Environmental Impact Assessment (EIA; Directive 85/337/EEC) and the Directive on Strategic Environmental Assessment (SEA; Directive 2001/42/EC) are the 2 most relevant legal obstacles facing wind developers in the European Union. While EIAs influence individual projects, SEAs are used to assess public plans or programs (Hansen 2011). The EIAs also produce new data through site-specific monitoring and modeling, while SEAs rely on existing available information (Hansen 2011). Neither of these directives targets the wind industry specifically, but together they effectively require that all large wind facilities perform a systematic assessment of their environmental impact (Hansen 2011, Ferrer et al. 2012).

Because of the Directive on EIA, use of EIAs is widespread in Europe (Hansen 2011, Ferrer et al.

2012). Many European Union member countries not only mandate EIAs, but also have published guidelines for performing them (Hansen 2011). Specifically, an EIA is a systematic process to identify, predict, and evaluate environmental effects of proposed actions and projects (Hansen 2011), culminating in a "declaration on the environmental impact"; this declaration states the significance and acceptability of the project's expected effects (Ferrer et al. 2012). EIAs not only identify the negative environmental impacts of a project, but they also consider availability of less damaging project alternatives, of productive opportunities to support biodiversity, and of avoidance, minimization, and mitigation techniques (Gove et al. 2013).

Best practice dictates that EIAs begin by screening or scoping the proposed project for any potential environmental impacts. Environmental sensitivity, the presence of any vulnerable species (avian and otherwise), and the extent of the wind facility's impact as an individual project and cumulatively across all wind facilities in the area are considered at this phase (Gove et al. 2013). Next, alternatives to the proposed location, project design, and operational characteristics are considered, followed by a description of the baseline environmental conditions of the selected site (Hansen 2011). Because of variability between wind project sites, baseline (e.g., preconstruction) data collection methods and requirements must be adjusted on a project-to-project basis (Ferrer et al. 2012). In the case of birds, it is recommended that EIA baseline data be collected for at least 12 months to describe complete annual avian activity within a site across all seasons (Langston and Pullan 2003). If a project threatens a protected species, preconstruction monitoring may be extended beyond 12 months (Ferrer et al. 2012).

The comprehensive nature of EIAs makes them time-intensive and costly. Unfortunately, perhaps in an effort to expedite the process, not all EIAs follow the best standard practices. EIAs often fail to assess project alternatives to identify the most environmentally responsible development option, implement

inadequate pre- and post-construction monitoring, and neglect public participation (Gove et al. 2013). Perhaps more concerning is the possibility that EIAs don't accurately estimate avian mortality risk. A review of 20 wind facilities in Spain showed no clear relationship between predicted risk identified during EIAs and actual avian mortality reported (Ferrer et al. 2012). It is possible that the criteria currently used to predict avian mortality risk are insufficient, leading to the construction of wind facilities in sensitive avian habitat (Ferrer et al. 2012).

Mitigation

Mitigation refers to managing entire wind facilities or individual turbines to reduce or eliminate deleterious effects on birds created by construction and operation of wind turbines. Mitigation practices have been applied to land-based wind facilities more commonly than to offshore wind facilities, yet there is a dearth of empirical evidence supporting the efficacy of most proposed mitigation measures for wind energy on land or at sea (Langston 2013). A mitigation hierarchy has been proposed in which avoiding negative impacts to wildlife by careful siting of wind facilities and individual turbines is the first priority, followed by minimizing and reducing impacts once a wind facility is operational, and lastly, compensation that is supplemented with monitoring and reactive management (Koppel et al. 2014).

Avoidance through Careful Siting

Proper wind facility siting has been one of the most commonly suggested and implemented mitigation practices to minimize negative avian consequences (Drewitt and Langston 2006, Langston 2013). The requirement for careful siting may be codified in policy at different levels of government and implemented at the planning and permitting stages (Koppel et al. 2014). Careful siting typically involves avoiding areas of conservation importance, including feeding, nesting, or display areas, as well as along migration and other important travel routes (Langston 2013).

In addition to careful siting of entire wind facilities, informed micro-siting of individual turbines has been used to avoid potential negative effects on avian communities. Grouping individual turbines in close proximity to one another can minimize the overall size of a wind facility's footprint, thereby reducing the general area birds might avoid. Rows of turbines can be aligned to avoid running clusters of wind turbines perpendicular to avian flight paths, potentially reducing barrier effects and avian collision risk (Drewitt and Langston 2006). Alternatively, providing flight corridors aligned with flight paths between clusters of turbines in larger wind facilities has been suggested (Drewitt and Langston 2006). Krijgsveld et al. (2011) observed more frequent flight activity by birds within an offshore wind facility where spacing between turbines was greater than areas of the same wind facility where turbines were spaced more closely together. Krijgsveld et al. (2011) also observed birds flying past a single line of offshore turbines more often than passing the main body of the wind facility, leading them to conclude that avoidance behavior associated with a particular part of an entire wind facility may be a result of overall wind facility design.

Minimizing and Compensating for Negative Effects on Birds

Several mitigation strategies have been attempted or proposed to minimize or reduce the effects of operating wind facilities on avian species. Curtailment involves completely shutting down individual wind turbines or, in extreme cases, all wind turbines within a wind facility at times when high avian mortality is expected—typically during peak avian migration periods. The amount of time curtailment occurs can be variable based on the likelihood of avian mortality at wind facilities, but typically does not total more than a few days out of an entire year. Curtailment can reduce avian turbine-collision mortality while only minimally interfering with wind energy production. For example, de Lucas et al. (2012) conducted a curtailment study to examine the effect that ceasing turbine operations would have on griffon vultures at wind facilities in Spain. Selective curtailment of 10 turbines across 6 wind facilities that were evaluated to be most hazardous to griffon vultures resulted in a 50% reduction in vulture mortality and only 0.07% in average turbine stoppage time (de Lucas et al. 2012).

Rather than completely stopping turbines, "feathering" turbine blades may minimize avian-collision mortality. Feathering turns each blade on a turbine parallel to the wind at or below the cut-in speed—the minimum wind speed at which turbines begin to produce electricity. Many turbines spin at wind speeds below their cut-in speed but produce no electricity. By feathering turbine blades at or below cut-in speeds, the profile of each blade is reduced, thereby reducing the opportunity for birds to collide with spinning blades (Langston 2013, Arnett and May 2016).

Repowering wind facilities is a retroactive approach to minimize effects on birds where older and smaller turbines are replaced with fewer but larger turbines that produce more electricity because of their greater overall height and larger diameter of the rotor-swept zone (Koppel et al. 2014). Research results are mixed on the effectiveness of repowering as a mitigation measure for birds. In the United States, raptor mortality was 54% less and overall avian mortality was reduced by 64% at a repowered wind facility in California, suggesting that repowering could reduce avian mortality while simultaneously increasing power generation efficiency (Smallwood and Karas 2009). Dahl et al. (2015) modeled avian turbine-collision risk under 2 different repowering scenarios at a wind facility in Norway and predicted a 32% to 71% reduction in turbine-collision risk, depending on the number and size of new turbines that would replace the existing ones. However, Krijgsveld et al. (2009) reported that avian turbine-collision

Figure 5.5   Painting turbine blades to increase visibility (*left*) is a proposed mitigation technique to reduce or eliminate bird collisions with wind turbines. Thus far, empirical evidence is lacking to support this as an effective mitigation technique. *Photo credit: lmwindpower.com.*

risk at smaller, earlier-generation turbines was similar to newer, larger turbines. De Lucas et al. (2008) determined raptor collision probability increased with turbine height at wind facilities in Spain; this finding is similar to Loss et al. (2013), who examined a broader range of avian species in the United States relative to turbine height. Loss et al. (2013) suggested that installation of larger turbines could result in increased avian mortality.

An assortment of noise-based and visual mitigation strategies have been proposed or tried to reduce bird collisions with wind turbines. Approaches using high-intensity or bioacoustic sounds like distress calls to deter birds from wind turbines have been suggested, but have not been evaluated using scientific trials (Marques et al. 2014). Visual approaches have primarily taken the form of painting 1 or more turbine blades to increase visibility to birds (Marques et al. 2014; Figure 5.5). Lights affixed to offshore turbines for marine and aerial navigation may attract flying birds and increase the chance for turbine-collision risk, although the complete effect lights have on avian-collision risk with vertical structures is not well understood (Drewitt and Langston 2006, Furness et al. 2013). Adjusting the color, wavelength, and frequency of lights flashing off and on or not flashing at all have been suggested as ways to reduce bird collisions with wind turbines (Arnett and May 2016).

The idea of compensatory mitigation is to offset any negative consequences of wind energy development by attracting birds to an area away from wind turbine(s) or a wind facility (Koppel et al. 2014, Arnett and May 2016). This is not a common approach, but the idea has been attempted on- and off-site and during construction and operational phases of wind facilities to reduce avian presence and turbine-collision risk (Koppel et al. 2014, Arnett and May 2016). Enhancing or creating habitat outside of wind facilities for avian breeding, roosting, and wintering purposes are some approaches to compensatory mitigation. Other mitigation suggestions have included supplementary feeding or prey fostering outside of wind facilities and decreasing prey abundance/availability and predator control within wind facilities (Koppel et al. 2014, Marques et al. 2014, Arnett and May 2016). Generally, empirical evidence is lacking to support or refute attempts at compensatory mitigation related to wind facilities and birds (Arnett and May 2016).

Future Research Directions

It is inevitable that wind energy will continue to expand to help meet global energy demands. Looking ahead, wind energy is expected to expand into areas not previously developed (e.g., further offshore), and turbine designs will incorporate taller towers with larger rotor-swept areas to produce more energy than do currently operating wind turbines. Although a considerable amount of research has been conducted to understand the effects of wind energy facilities on avian species, scant information is available in some areas regarding the interactions between birds and these facilities and still other areas where data are available but data gaps exist.

Expansion to New Areas and Use of Different Turbine Technology

As production of wind energy expands, new areas for wind development on land and offshore will need to be explored. Although a substantial amount of research has been conducted to understand turbine-collision impacts on birds from land-based wind facilities, data gaps remain. For example, comparisons of results from studies of 1 wind facility to other studies using similar methodologies are generally missing from the literature. Information about effects on avian populations caused by avoidance of wind facilities or individual wind turbines and effects as a result of habitat loss or modification from construction and operation of land-based wind facilities is gener-

ally unavailable. An understanding of the full range of consequences to birds from the production of offshore wind energy is woefully lacking. As wind energy expands into new areas both onshore and offshore, information about impacts to avian species in those new areas will need to be collected and analyzed.

In addition to wind energy expanding into new areas, wind turbine designs are advancing rapidly. Consensus regarding the effects of wind turbine technology on avian collision-mortality risk has not yet been reached. Lattice-tower turbines are often associated with increased turbine-collision risk because the lattice provides birds with roosts and perches just below the spinning turbine blades (Orloff and Flannery 1992, Hunt 2002). Although monopole towers have fewer perching locations and therefore are regarded as a safer design for birds, they often support larger rotors, resulting in a larger rotor-swept zone and therefore larger area of turbine-collision risk (Thelander and Rugge 2000, Marques et al. 2014). Larger turbines that can harness more wind to produce more power are being developed and constructed, especially for use offshore. As the mixed research results from repowering wind facilities demonstrates, fewer, but larger, turbines does not necessarily translate into reduced negative effects for birds. Completely new turbine designs also are being proposed and tested (Tong 2010). For example, turbines with blades that spin on a vertical axis, rather than a horizontal axis, have been developed and installed. Other designs include only 2 turbine blades arranged 1 behind the other at the front of the turbine, or 1 turbine blade at the front and the other at the rear of the unit (e.g., coaxial contrarotating wind turbine, intelligent wind power unit). Research is needed to identify the potential impacts these emerging wind turbine designs pose to bird populations. Regardless of the specific type of turbine design, the most critical factor relative to avian turbine-collision mortality likely is overall turbine height. Based on studies that have examined avian mortality from communication towers, estimated tower mortality increases exponentially with tower height (Longcore et al. 2012).

Long-Term Monitoring Studies across Multiple Wind Facilities

Most studies examining the effects on birds from wind energy production focus on a single wind facility for a relatively short period of time (~2 years) post-construction. Few studies apply proper experimental design to allow conclusions to be drawn from preconstruction data to isolate impacts following the construction and operation of a wind facility. To determine how generalizable results are across wind facilities, it is essential that replicated studies using a control be conducted. BACI study designs should be implemented at wind facilities whenever possible to capture the range of avian impacts from pre- to post-construction. Furthermore, longer-term studies beyond 2 years post-construction are necessary to understand whether the magnitude of effects on avian species changes over time.

Understanding Population-Level Effects

Although our general understanding of avian turbine-collision mortality (e.g., mortality rates, vulnerable species) is improving, research on the effects of disturbance and habitat modification on birds is mostly lacking. To assemble a complete understanding of wind development's impact on bird populations, assumptions about demographic attributes, additive versus compensatory mortality, and mechanisms driving displacement, among other factors, need to be made. By testing these and other assumptions, we can then extrapolate results of studies and gain an industry-wide estimate of numbers and types of birds affected by currently operating and future wind facilities. The ultimate goal is to understand population-level effects of wind energy on avian species, especially those that are endangered or threatened, and whether the mortality experienced by avian species from wind energy is compensatory or additive.

Empirical Evidence to Support Mitigation Efficacy

Many different types of mitigation practices dealing with the effects of wind facilities on avian species have been suggested, but there is little empirical evidence to suggest which are effective. Careful siting of wind facilities is often touted as a primary method to proactively avoid known negative wind energy-avian interactions, but data are lacking that demonstrate that the models and other decision tools are accurate when choosing among multiple possible sites. For other proposed mitigation practices like navigation lights on turbines, compensatory mitigation, and various acoustic and visual mitigation efforts, data are sorely lacking. Of great interest to wind energy companies with any mitigation approach is the cost-effectiveness of the practice, so data demonstrating the trade-off between reducing or eliminating negative consequences for avian species and the cost of purchasing and implementing mitigation practices will be necessary moving forward.

LITERATURE CITED

Ahrens, M. H., and M. Gabriel. 2015. Wind, birds, and bats: Recent legal migrations. North American Windpower. https://www.milbank.com/images/content/2/1/v5/21717/NAW-9-15-15-Milbank.pdf.

Andersson, M., M. Berggren, D. Wilhelmsson, and M. C. Ohman. 2009. Epibenthic colonization of concrete and steel pilings in a cold-temperate embayment: A field experiment. Helgoland Marine Research 63:249–260.

Animal Welfare Institute v. Beech Ridge Energy, LLC. 8:09-cv-01519. 2009. June 10. https://www.plainsite.org/dockets/krz4q7wy/maryland-district-court/animal-welfare-institute-et-al-v-beech-ridge-energy-llc-et-al/.

Arnett, E. B., and R. F. May. 2016. Mitigating wind energy impacts on wildlife: Approaches for multiple taxa. Human-Wildlife Interactions 10:28–41.

Barrios, L., and A. Rodriguez. 2004. Behavioural and environmental correlates of soaring-bird mortality at on-shore wind turbines. Journal of Applied Ecology 41:72–81.

Batary, P., and A. Baldi. 2004. Evidence of an edge effect on avian nest success. Conservation Biology 18:389–400.

Beston, J. A., J. E. Diffendorfer, S. R. Loss, and D. H. Johnson. 2016. Prioritizing avian species for their risk of population-level consequences from wind energy development. PLOS One 3:e0150813. doi:10.1371/journal.pone.0150813.

Bevanger, K., F. Berntsen, S. Clausen, E. L. Dahl, Ø. Flagstad, A. Follestad, D. Halley, et al. 2010. Pre- and post-construction studies of conflicts between birds and wind turbines in coastal Norway (Bird-Wind). Report on findings 2007–2010. Trondheim: Norwegian Institute for Nature Research.

BirdLife International. 2011. Meeting Europe's renewable energy targets in harmony with nature. Edited by I. Scrase and B. Gove. Sandy, UK: BirdLife Europe, Royal Society for the Protection of Birds.

Brabant, R., N. Vanermen, E. W. M. Stienen, and S. Degraer. 2015. Towards a cumulative collision risk assessment of local and migrating birds in North Sea offshore wind facilities. Hydrobiologia 756:63–74.

Bureau of Ocean Energy Management. 2017. Offshore wind energy. https://www.boem.gov/Offshore-Wind-Energy/. Accessed July 12, 2017.

Cabrera-Cruz, S. A., and R. Villegas-Patraca. 2016. Response of migrating raptors to an increasing number of wind facilities. Journal of Applied Ecology 53:1667–1675.

Campedelli, T., G. Londi, S. Cutini, A. Sorace, and G. T. Florenzano. 2014. Raptor displacement due to the construction of a wind facility: Preliminary results after the first 2 years since the construction. Ethology, Ecology, and Evolution 26:376–391.

Carrete, M., J. A. Sanchez-Zapata, J. R. Benitez, M. Lobon, and J. A. Donazar. 2009. Large scale risk-assessment of wind-facilities on population viability of a globally endangered, long-lived raptor. Biological Conservation 142:2954–2961.

Collier, M. P., S. Dirksen, and K. L. Krijgsveld. 2012. A review of methods to monitor collisions or micro-avoidance of birds with offshore wind turbines. Part 2: Feasibility study of systems to monitor collisions. Strategic Ornithological Support Services Project SOSS-03A. Culemborg, Netherlands: Bureau Waardenburg.

Dahl, E. L., K. Bevanger, T. Nygard, E. Roskaft, and B. G. Stokke. 2012. Reduced breeding success in white-tailed eagles at Smola windfarm, western Norway, is caused by mortality and displacement. Biological Conservation 145:79–85.

Dahl, E. L., R. May, P. L. Hoel, K. Bevanger, H. C. Pedersen, E. Røskaft, and B. G. Stokke. 2013. White-tailed eagles (*Haliaeetus albicilla*) at the Smøla wind-power plant, central Norway, lack behavioral flight responses to wind turbines. Wildlife Society Bulletin 37:66–74.

Dahl, E. L., R. May, T. Nygard, J. Astrom, and O. H. Diserud. 2015. Repowering Smola wind power plant: An assessment of avian conflicts. Trondheim: Norwegian Institute for Nature Research.

de Lucas, M., M. Ferrer, M. J. Becahrd, and A. R. Munoz. 2012. Griffon vulture mortality at wind projects in southern Spain: Distribution of fatalities and active mitigation measures. Biological Conservation 147:184–189.

Devereux, C. L., M. J. H. Denny, and M. J. Whittingham. 2008. Minimal effects of wind turbines on the distribution of wintering farmland birds. Journal of Applied Ecology 45:1689–1694.

Dierschke, V., R. W. Furness, and S. Garthe. 2016. Seabirds and offshore wind facilities in European waters: Avoidance and attraction. Biological Conservation 202:59–68.

Dohm, R. L. 2017. Sharing the wind: The interactions and management of raptors and wind power development. Thesis, University of Wisconsin-Madison.

Dokter, A. M., M. J. Baptist, B. J. Ens, K. L. Krijgsvild, and E. Emiel van Loon. 2013. Bird radar validation in the field by time-referencing line-transect surveys. PLOS One. https://doi.org/10.1371/journal.pone.0074129.

Drewitt, A. L., and R. H. W. Langston. 2006. Assessing the impacts of wind facilities on birds. Ibis 148:29–42.

Drewitt, A. L., and R. H. W. Langston. 2008. Collision effects of wind-power generators and other obstacles on birds. Annals of the New York Academy of Sciences 1134:233–266.

Dunn, J. L., and J. Alderfer. 2009. National Geographic illustrated birds of North America. Washington DC: National Geographic Society.

Eagle Permits; Revisions to Regulations for Eagle Incidental take and take of Eagle Nests; Proposed Rule. 2016. May 6. (to be codified at 50 CFR pts. 13 and 22).

Erickson, W. P., G. Johnson, D. Strickland, D. Young, K. J. Sernka, and R. Good. 2001. Avian collisions with wind turbines: A summary of existing studies and comparisons to other sources of avian collision mortality in the United States. Western Resource Technologies, Inc. Report. Washington, DC: National Wind Coordinating Committee.

Erickson, W. P., M. M. Wolfe, K. J. Bay, D. H. Johnson, and J. L. Gehring. 2014. A comprehensive analysis of small-passerine fatalities from collision with turbines at wind energy facilities. PLOS One 9:e107491.

Erickson, R. A., E. A. Eager, J. C. Stanton, J. A. Beston, J. E. Diffendorfer, and W. E. Thogmartin. 2015. Assessing local population vulnerability with branching process models: An application to wind energy development. Ecosphere 6:254. http://dx.doi.org/10.1890/ES15-00103.1.

Everaert, J., and E. W. M. Stienen. 2006. Impact of wind turbines on birds in Zeebrugge (Belgium): Significant effect on breeding tern colony due to collisions. Biodiversity and Conservation. doi:10.1007/s10531-006-9082-1.

Farfan, M. A., J. M. Vargas, J. Duarate, and R. Real. 2009. What is the impact of wind facilities on birds? A case study in southern Spain. Biodiversity and Conservation 18:3743–3758.

Farfan, M. A., J. Duarte, R. Real, A. R. Munoz, J. E. Fa, and J. M. Vargas. 2017. Differential recovery of habitat use by birds after wind facility installation: A multi-year comparison. Environmental Impact Assessment Review 64:8–15.

Ferrer, M., M. de Lucas, G. F. E., Janss, E. Casado, A. R. Muñoz, M. J. Bechard, and C. P. Calabuig. 2012. Weak relationship between risk assessment studies and recorded mortality in wind facilities. Journal of Applied Ecology 49:38–46.

Fijn, R. C., K. L. Krijgsveld, M. J. M. Poot, and S. Dirksen. 2015. Bird movements at rotor heights measured continuously with vertical radar at a Dutch offshore wind facility. Ibis 157:558–566.

Fox, A. D., M. Desholm, J. Kahlert, T. K. Christensen, and I. K. Petersen. 2006. Information needs to support environmental impact assessment of the effects of European marine offshore wind facilities on birds. Ibis 148:129–144.

Furness, R. W., H. M. Wade, and E. A. Masden. 2013. Assessing vulnerability of marine bird populations to offshore wind facilities. Journal of Environmental Management 119:56–66.

Garthe, S. N. Markones, and A. Corman. 2017. Possible impacts of offshore wind facilities on seabirds: A pilot study in Northern Gannets in the southern North Sea. Journal of Ornithology 158:345–349.

Garvin, J. C., C. S. Jennelle, D. Drake, and S. M. Grodsky. 2011. Response of raptors to a wind facility. Journal of Applied Ecology 48:199–209.

Gill, F. 2007. Ornithology. 3rd ed. New York: W. H. Freeman.

Glen, A. M., R. D. Barho, and L. M. Evans. 2013. How federal wildlife laws impact the development of wind energy. April 8. https://www.americanbar.org/publications/trends/2012_13/march_april/how_federal_wildlife_laws_impact_development_of_wind_energy.html.

Global Wind Energy Council. 2016. Global wind report 2016. http://gwec.net/publications/global-wind-report-2/global-wind-report-2016/.

Global Wind Energy Council. 2017a. Global wind statistics 2016. http://www.gwec.net/wp-content/uploads/vip/GWEC_PRstats2016_EN_WEB.pdf.

Global Wind Energy Council. 2017b. Offshore wind power. http://www.gwec.net/global-figures/global-offshore/. Accessed July 12, 2017.

Gove, B., R. H. W., Langston, A. McCluskie, J. D. Pullan, and I. Scrase. 2013. Wind farms and birds: An updated analysis of the effects of wind farms on birds, and best practice guidance on integrated planning and impact assessment. Report T-PVS/Inf 15. Birdlife International. https://rm.coe.int/1680746245.

Goyert, H. F., B. Gardner, R. Sollmann, R. R. Veit, A. T. Gilbert, E. E. Connelly, and K. A. Williams. 2016. Predicting the offshore distribution and abundance of marine birds with a hierarchal distance sampling model. Ecological Applications 26:1797–1815.

Grodsky, S. M. 2010. Aspects of bird and bat mortality at a wind energy facility in southeastern Wisconsin: Impacts, relationships, and cause of death. Thesis, University of Wisconsin-Madison.

Grodsky, S. M., and D. Drake. 2011. Assessing bird and bat mortality at the Forward Energy Center. Final Report filed with the Wisconsin Public Service Commission. http://psc.wi.gov/apps35/ERF_view/viewdoc.aspx?docid=152052.

Grodsky, S. M., C. S. Jennelle, and D. Drake. 2013. Bird mortality at a wind-energy facility near a wetland of international importance. Condor: Ornithological Applications 115:700–711.

Hansen, H. S. 2011. Obstacles for wind energy development due to EU legislation. Copenhagen: South Baltic Programme.

Heath, M. F., and M. I. Evans. 2000. Important bird areas in Europe: Priority sites for conservation, volume 1, Northern Europe. Birdlife Conservation Series. Cambridge, UK: BirdLife International.

Hill, R., K. Hill, R. Aumuller, A. Schulz, T. Dittmann, C. Kulemeyer, and T. Coppack. 2014. Of birds, blades and barriers: Detecting and analyzing mass migration events at Alpha Ventus. In Ecological Research at the Offshore Wind facility Alpha Ventus, edited by Federal Maritime and Hydrographic Agency, Federal Ministry for the Environment, Nature Conservation and Nuclear Safety, 111–131. Wiesbaden: Springer Fachmedien.

Hoover, S. L., and M. L. Morrison. 2005. Behavior of red-tailed hawks in a wind turbine development. Journal of Wildlife Management 69:150–159.

Hötker, H., K. M. Thomsen, and H. Jeromin. 2006. Impacts on biodiversity of exploitation of renewable energy sources: The example of birds and bats—facts, gaps in knowledge, demands for further research, and ornithological guidelines for the development of renewable energy exploitation. Bergenhusen, Germany: Michael-Otto-Institut im NABU http://www.proj6.turbo.pl /upload/file/389.

Hunt, W. G. 2002. Golden eagles in a perilous landscape: Predicting the effects of mitigation for wind turbine blade-strike mortality. California Energy Commission Report P500-02-043F. University of California, Santa Cruz. https://tethys.pnnl.gov/sites/default/files /publications/Hunt-2002.pdf.

Hunt, W. G., and T. Hunt. 2006. The trend of golden eagle territory occupancy in the vicinity of the Altamont Pass Wind Resource Area: 2005 survey. California Energy Commission, PIER Energy-Related Environmental Research. CEC-500-2006-056. https://listserver.energy .ca.gov/2006publications/CEC-500-2006-056/CEC-500 -2006-056.PDF.

Huppop, O., J. Dierschke, K. M. Exo, E. Fredrich, and R. Hill. 2006. Bird migration studies and potential collision risk with offshore wind turbines. Ibis 148:90–109.

Inger, R. M. J. Attrill, S. Bearhop. A. C. Broderick, W. J. Grecian, D. J. Hodgson, C. Mills, et al. 2009. Marine renewable energy: Potential benefits to biodiversity? An urgent call for research. Journal of Applied Ecology 46:1145–1153.

Johnson, G. D., W. P. Erickson, M. D. Strickland, M. F. Shepherd, D. A. Shepherd, and S. A. Sarappo. 2002. Collision mortality of local and migrant birds at a large-scale windpower development on Buffalo Ridge, Minnesota. Wildlife Society Bulletin 30:879–887.

Johnston N. N., J. E. Bradley, and K. A. Otter. 2014. Increased flight altitudes among migrating golden eagles suggest turbine avoidance at a Rocky Mountain wind installation. PLOS One 3:e93030. doi:10.1371/journal. pone.0093030.

Kaldellis, J. K., and D. Zafirakis. 2011. The wind energy (r) evolution: A short review of a long history. Renewable Energy 36:1887–1901.

Katzner, T. E., D. Brandes, T. Miller, M. Lanzone, C. Maisonneuve, J. A. Tremblay, R. Mulvihill, and G. T. Merovich Jr. 2012. Topography drives migratory flight altitude of golden eagles: Implications for on-shore wind energy development. Journal of Applied Ecology 49: 1178–1186.

Koppel, J., M. Dahmen, J. Helfrich, E. Schuster, and L. Bulling. 2014. Cautious but committed: Moving toward adaptive planning and operation strategies for renewable energy's wildlife implications. Environmental Management 54:744–755.

Krijgsveld, K. L., K. Akershoek, F. Schenk, F. Dijk, and S. Dirksen. 2009. Collision risk of birds with modern large wind turbines. Ardea 97:357–366.

Krijgsveld, K. L., R. C. Fijn, M. Japink, P. W. van Horssen, C. Heunks, M. P. Collier, M. J. M. Poot, D. Beuker, and S. Dirksen. 2011. Effect studies offshore wind facility Egmond aan Zee: Final report on fluxes, flight altitudes, and behaviour of flying birds. Bureau Waardenburd Report 10-219. https://www.buwa.nl/fileadmin/buwa _upload/Bureau_Waardenburg_rapporten/06-467 _effectstudies_offshore_windpark_Egmond_02.pdf.

Langston, R. H. W. 2013. Birds and wind projects across the pond: A UK perspective. Wildlife Society Bulletin 37:5–18.

Langston, R. H. W., and J. D. Pullan. 2003. Windfarms and birds: An analysis of the effects of windfarms on birds, and guidance on environmental assessment criteria and site selection issues. Report T-PVS/Inf 12. Birdlife International. https://tethys.pnnl.gov/sites/default/files /publications/Langston%20and%20Pullan%202003.pdf.

Leddy, K. L., K. F. Higgins, and D. E. Naugle. 1999. Effects of wind turbines on upland nesting birds in Conservation Reserve Program grasslands. Wilson Bulletin 111:100–104.

Leung, D. Y. C., and Y. Yang. 2012. Wind energy development and its environmental impact: A review. Renewable and Sustainable Energy Reviews 16:1031–1039.

Lindeboom, H. J., H. J. Kouwenhoven, M. J. N. Bergman, S. Bouma, S. Brasseur, R. Daan, R. C. Fijn, D. de Haan, S. Dirksen, and R. van Hall. 2011. Short-term ecological effects of an offshore wind facility in the Dutch coastal zone; A compilation. Environmental Research Letters 6:1–13.

Longcore, T., C. Rich, P. Mineau, B. MacDonald, D. G. Bert, L. M. Sullivan, E. Mutrie, et al. 2012. An estimate of avian mortality at communication towers in the United States and Canada. PLOS One 7:e34025.

Loss, S. R., T. Will, and P. P. Marra. 2013. Estimates of bird collision mortality at wind facilities in the contiguous United States. Biological Conservation 168:201–209.

Madariaga, A., I. Martinez de Alegria, J. L. Martin, P. Eguia, and S. Ceballos. 2012. Current facts about offshore wind facilities. Renewable and Sustainable Energy Reviews 16:3105–3116.

Madsen, J. 1995. Impacts of disturbance on migratory waterfowl. Ibis 137:S67–S74.

Madsen, J., and D. Boertmann. 2008. Animal behavioral adaptation to changing landscapes: Spring-staging geese habituate to wind facilities. Landscape Ecology 23: 1007–1011.

Marques, A. T., H. Batalha, S. Rodrigues, H. Costa, M. J. R. Pereira, C. Fonseca, M. Mascarenhas, and J. Bernardino. 2014. Understanding bird collisions at wind facilities: An updated review on the causes and possible mitigation strategies. Biological Conservation 179:40–52.

Martin, G. R., and G. Katzir. 1999. Visual field in short-toed eagles *Circaetus gallicus* and the function of binocularity in birds. Brain Behavioral Evolution 53:55–66.

Martin, G. R., and J. M. Shaw. 2010. Bird collisions with power lines: Failing to see the way ahead? Biological Conservation 143:2695–2702.

Martin, G. R., S. J. Portugal, and C. P. Murn. 2012. Visual fields, foraging, and collision vulnerability in *Gyps* vultures. Ibis 154:626–631.

Masden, E. A., D. T. Hayden, A. D. Fox, R. W. Furness, R. Bullman, and M. Desholm. 2009. Barriers to movement: Impacts of wind facilities on migrating birds. ICES Journal of Marine Science 66:746–753.

May, R. F. 2015. A unifying framework for the underlying mechanisms of avian avoidance of wind turbines. Biological Conservation 190:179–187.

Miller, R. G., Z. L. Hutchinson, A. K. Macleod, M. T. Burrows, E. J. Cook, K. S. Last, and B. Wilson. 2013. Marine renewable energy development: Assessing the benthic footprint at multiple scales. Frontiers in Ecology and the Environment 11:433–440.

National Wind Watch. 2017. Presenting the facts about industrial wind power. https://www.wind-watch.org/faq-output.php. Accessed December 13, 2017.

Orloff, S., and A. Flannery. 1992. Wind turbine effects on avian activity, habitat use, and mortality in Altamont Pass and Solano County Wind Resource Areas, 1989–1991. California Energy Commission Report P700-92-001. Biosystems Analysis, Inc. Tiburon, CA. https://nationalwind.org/wp-content/uploads/2013/05/Orloff-and-Flannery-1992.pdf.

O'Rourke, C., M. Hall, T. Pitlik, and E. Fernandez-Juricic. 2010. Hawk eyes. I: Diurnal raptors differ in visual fields and degree of eye movement. PLOS One 5:e12802.

Pagel, J. E., K. J. Kritz, B. A. Millsap, R. K. Murphy, E. L. Kershner, and S. Covington. 2013. Bald eagle and golden eagle mortalities at wind energy facilities in the contiguous United States. Journal of Raptor Research 47:311–315.

Pearce-Higgins, J. W., L. Stephen, R. H. W. Langston, I. P. Bainbridge, and R. Bullman. 2009. The distribution of breeding birds around upland wind farms. Journal of Applied Ecology 46:1323–1331.

Pearce-Higgins, J. W., L. Stephen, A. Douse, and R. H. W. Langston. 2012. Greater impacts of wind facilities on bird populations during construction than subsequent operation: Results of a multi-site and multi-species analysis. Journal of Applied Ecology 49:386–394.

Percival, S. 2005. Birds and wind facilities: What are the real issues? British Birds 98:194–204.

Petersen, I. K., T. K. Christensen, J. Kahlert, M. Desholm, and A. D. Fox. 2006. Final results of bird studies at the offshore wind facilities at Nysted and Horns Rev, Denmark: Report request. Commissioned by DONG energy and Vattenfall A/S. National Environmental Research Institute. https://tethys.pnnl.gov/sites/default/files/publications/NERI_Bird_Studies.pdf.

Pettersson, J. 2005. The impact of offshore wind facilities on bird life in Southern Kalmar Sound, Sweden. A final report based on studies 1999–2003. https://tethys.pnnl.gov/sites/default/files/publications/The_Impact_of_Offshore_Wind_Farms_on_Bird_Life.pdf. Accessed August 17, 2017.

Pettersson, J., and J. P. Fågelvind. 2011. Night migration of songbirds and waterfowl at the Utgrunden off-shore wind facility. A radar-assisted study in the southern Kalmar Sound, Sweden. Report 6438. Stockholm, Sweden. http://swedishepa.se/Documents/publikationer 6400/978-91-620-6438-9.pdf.

Plonczkier, P., and I. C. Simms. 2012. Radar monitoring of migrating pinkfooted geese: Behavioural responses to offshore wind facility development. Journal of Applied Ecology 49:1187–1194.

Poot, M. J. M., P. W. van Horssen, M. P. Collier, R. Lensink, and S. Dirksen. 2011. Effect studies offshore Wind Egmond aan Zee: Cumulative effects on seabirds. A modeling approach to estimate effects on population levels in seabirds. Bureau Waardenburg Report 06-466. https://www.buwa.nl/fileadmin/buwa_upload/Bureau_Waardenburg_rapporten/06-466_BW_research_OWEZ_cumulative_effects-web.pdf.

Saidur, R., N. A. Rahim, M. R. Islam, and K. H. Solangi. 2011. Environmental impact of wind energy. Renewable and Sustainable Energy Reviews 15:2423–2430.

Sansom A., J. W. Pearce-Higgins, and D. J. T. Douglas. 2016. Negative impact of wind energy development on a breeding shorebird assessed with a BACI study design. Ibis 158:541–555.

Schuster, E., L. Bulling, and J. Koppel. 2015. Consolidating the state of knowledge: A synoptical review of wind energy's wildlife effects. Environmental Management 56:300–331.

Segers, J. L., and H. G. Broders. 2014. Interspecific effects of forest fragmentation on bats. Canadian Journal of Zoology 92:665–673.

Sibley, D. A. 2000. The Sibley guide to birds. New York: Alfred A. Knopf.

Smallwood, K. S. 2013. Comparing bird and bat fatality-rate estimates among North American wind energy projects. Wildlife Society Bulletin 37:19–33.

Smallwood, K. S., and B. Karas. 2009. Avian and bat fatality rates at old-generation and repowered wind turbines in California. Journal of Wildlife Management 73:1062–1071.

Smallwood, K. S., and C. G. Thelander. 2004. Developing methods to reduce mortality in the Altamont Pass Wind Resource Area. Final Report by BioResource Consultants to the California Energy Commission, Public Interest Energy Research 500-04-052. https://www.energy.ca.gov /reports/500-04-052/500-04-052_00_EXEC_SUM.PDF.

Smallwood, K. S., C. G. Thelander, M. L. Morrison, and L. M. Rugge. 2007. Burrowing owl mortality in the Altamont Pass Wind Resource Area. Journal of Wildlife Management 71:1513–1524.

Smallwood, K. S., and C. G. Thelander. 2008. Bird mortality in the Altamont Pass Wind Resource Area, California. Journal of Wildlife Management 72:215–223.

Smallwood, K. S., L. Rugge, and M. L. Morrison. 2009. Influence of behavior on bird mortality in wind energy developments. Journal of Wildlife Management 73:1082–1098.

Smallwood, K. S., L. Neher, J. Mount, and R. C. E. Culver. 2013. Nesting burrowing owl density and abundance in the Altamont Pass Wind Resource Area, California. Wildlife Society Bulletin 37:787–795.

Snyder, B., and M. J. Kaiser. 2009. Ecological and economic cost-benefit analysis of offshore wind energy. Renewable Energy 34:1567–1578.

Sovacool, B. K. 2012. The avian benefits of wind energy: A 2009 update. Renewable Energy 49:19–24.

Stemler, J. 2007. Wind power siting regulations and wildlife guidelines in the United States. Jodi Stemler Report. Denver, CO: Association of Fish and Wildlife Agencies and US Fish and Wildlife Service.

Stewart, G. B., A. S. Pullin, and C. F. Coles. 2007. Poor evidence-base for assessment of wind facility impacts on birds. Environmental Conservation 34:1–11.

Strickland, M. D., E. B. Arnett, W. P. Erickson, D. H. Johnson, G. D. Johnson, M. L., Morrison, J. A. Shaffer, and W. Warren-Hicks. 2011. Comprehensive guide to studying wind energy/wildlife interactions. Prepared for the National Wind Coordinating Collaborative, Washington, DC. http://www.nationalwind.org/wp -content/uploads/2018/04/Comprehensive_Guide_to _Studying_Wind_Energy_Wildlife_Interactions_2011 _Updated.pdf.

Sun, X., D. Huang, and G. Wu. 2012. The current state of offshore wind energy technology development. Energy 41:298–312.

Thelander, C. G., and L. Rugge. 2000. Avian risk behavior and fatalities at the Altamont Wind Resource Area. Golden, CO: National Renewable Energy Laboratory.

Tong, W. 2010. Wind power generation and wind turbine design. Southampton, UK: WIT Press.

US Department of Energy. 2008. 20% wind energy by 2030: Increasing wind energy's contribution to the U.S. electrical supply. http://www.nrel.gov/docs/fy08osti /41869.pdf.

US Department of Energy. 2015. Wind vision: A new era for wind power in the United States. https://www .energy.gov/sites/prod/files/WindVision_Report_final .pdf.

US Department of Justice. 2013. Utility company sentenced in Wyoming for killing protected birds at wind projects. Press release, November 22. https://www.justice.gov/opa /pr/utility-company-sentenced-wyoming-killing-protected -birds-wind-projects.

USFWS (US Fish and Wildlife Service). 2012. U.S. Fish and Wildlife Service land-based wind energy guidelines. https://www.fws.gov/ecological-services/es-library/pdfs /WEG_final.pdf.

USFWS (US Fish and Wildlife Service). 2013. Eagle conservation plan guidance module 1: Land-based wind energy, version 2. https://www.fws.gov/migratory birds/pdf/management/eagleconservationplanguidance .pdf.

US General Administration Office. 2005. Wind power: Impacts on wildlife and government responsibilities for regulating development and protecting wildlife. http:// www.gao.gov/assets/250/247787.pdf.

Welcker, J., M. Liesenjohann, J. Blew, G. Nehls, and T. Grünkorn. 2017. Nocturnal migrants do not incur higher collision risk at wind turbines than diurnally active species. Ibis 159:366–373.

Williford, D. L., M. C. Woodin, M. K. Skoruppa, and G. C. Hickman. 2009. Rodents new to the diet of the western burrowing owl (Athene cunicularia hypugaea). South-western Naturalist 54:87–90.

Zimmerling, J. R., A. C. Pomeroy, M. V. d'Entremont, and C. M. Francis. 2013. Canadian estimate of bird mortality due to collisions and direct habitat loss associated with wind turbine developments. Avian Conservation and Ecology 8:10.

6 — Wind Energy Effects on Bats

CRIS D. HEIN AND
AMANDA M. HALE

Utility-scale wind energy facilities require no fuel, consume no water, and produce no greenhouse gas emissions or other pollutants during energy production. Wind power currently supplies nearly 6.4% of the electricity consumed in the United States, with continued growth expected in the coming years. Although further expansion of wind power is anticipated to provide environmental and economic benefits, there are increasing concerns about bat fatalities occurring at wind energy facilities across North America. Recent estimates place the number of bat fatalities in the several hundreds of thousands annually, and that number is projected to rise. As a result, population-level consequences of wind turbine-caused mortality are of increasing concern. For example, modeling efforts for a widespread species, the hoary bat (*Lasiurus cinereus*), indicate that the population could decline by 90% within 50 years, assuming no growth in installed wind power capacity and no significant implementation of conservation measures. Given the increasing demand for wind energy and increasing evidence that bats are attracted to wind turbines, the need to develop cost-effective and practical impact-minimization strategies is a high priority. Current strategies include siting restrictions and operational minimization, both of which

limit wind power generation. Potential solutions that do not limit power generation include broadcasting ultrasound from wind turbines or using ultraviolet (UV) light to deter bats from approaching and entering the rotor-swept area where fatalities occur. Another possible solution is the development of turbine-surface materials that reduce the relative attractiveness of wind turbine towers to bats. In this chapter, we provide a succinct summary of the known effects of wind energy generation on bats, present current and future research priorities, and describe the challenges and opportunities associated with developing and implementing effective solutions to minimize wind turbine–caused bat fatality.

Introduction

Wind power currently supplies nearly 6.4% of the electricity consumed in the United States and may grow to supply 20% by 2030 (US Department of Energy 2015, Oteri et al. 2018; Figure 6.1). Given the current and projected development in installed capacity, there are increasing concerns about the unprecedented number of bat fatalities occurring at wind energy facilities. In 2004, an estimated 1,398 to 4,031 bats were killed at a wind energy facility in

Figure 6.1 Sunset at a wind energy facility in the southern Great Plains, US. *Photo credit: Cole Lindsey.*

West Virginia (Kerns and Kerlinger 2004), demonstrating the adverse effect of wind turbines at a single facility. At the time, the installed wind power capacity in the United States and Canada was 6,544 MW (CanWEA 2016, AWEA 2017). By 2012, installed capacity in the 2 countries had grown to 52,195 MW (CanWEA 2016, AWEA 2017), and the estimated cumulative wind turbine–caused fatality of bats was 650,104 to 1,308,378 individuals (Arnett and Baerwald 2013). Cumulative fatality estimates since 2012 at wind energy facilities are lacking, but this mortality rate appears unsustainable for at least one widespread species, the hoary bat (*Lasiurus cinereus*), and may be an additive source of mortality for species already decimated by white-nose syndrome (WNS; Grodsky et al. 2012, Erickson et al. 2016, Frick et al. 2017).

Despite economic and political uncertainties, wind energy will continue to grow in the United States, with estimates for installed capacity nearing 224 GW by 2030 (US Department of Energy 2015). Significant expansions of wind energy development, reductions in turbine manufacturing costs, and increases in energy production and reliability are required to achieve this outcome (Baranowski et al. 2017). Simultaneously, public and private sector investments that lower manufacturing costs, increases in turbine height and blade length, and enhancements in wind resource characterization have allowed wind development to rapidly expand into once unfavorable regions onshore and previously unattainable regions offshore (US Department of Energy 2015).

Given these advancements and the potential for growth in wind energy capacity, it has become a high priority for conservationists and the wind industry to develop and implement cost-effective, scientifically proven measures that reduce wind turbine–caused fatalities of bats. Strategies that reduce fatalities include siting facilities away from areas where bats congregate (e.g., migratory pathways, maternity colonies, hibernacula), slowing blade rotation during periods when high numbers of fatalities have been reported, and developing technologies that discourage bat activity near wind turbines. However, the data to make informed decisions regarding mitigation are scarce. Thus, the need to move beyond basic pre- and post-construction monitoring to address crucial research questions is essential for resolving this complex issue.

Several recent publications have summarized the current understanding of bat–wind turbine interactions and the existing methods used to collect and analyze data from wind energy facilities (Strickland et al. 2011, Arnett and Baerwald 2013, Hein and Schirmacher 2016, Hein 2017). In this chapter, we review the gaps in our understanding of how bats interact with wind turbines, summarize current and future research priorities, and highlight challenges and opportunities associated with addressing the remaining issues.

Siting Wind Energy Facilities and Wind Turbines

Many factors are considered when siting wind energy facilities or individual wind turbines, including suitability of the project area (e.g., wind speeds, access, connectivity to the electrical grid) and potential environmental impacts (e.g., wind turbine–related bat fatalities). Biological and technological implications of siting should be considered early in the planning phase of the wind project. Siting of wind energy facilities and individual turbines may influence bat activity, behavior, and mortality, yet effects on bats remain difficult to quantify because we do not fully understand why or how bats respond to siting at different scales.

Siting wind energy facilities or wind turbines near large bat colonies (e.g., maternity roosts or hibernacula) may increase risk to bats. To provide the appropriate siting guidance away from a colony, an understanding of the resource requirements, such as swarming areas and surrounding hibernacula, is necessary. For example, the US Fish and Wildlife Service (USFWS) compiled radio-tracking data to assess the commuting distance of Indiana bats (*Myotis sodalis*) near priority hibernacula (as defined by population size) and suggested siting wind energy facilities at least 32.2 and 16.1 km from Priority 1 or 2 and Priority 3 or 4 hibernacula, respectively (USFWS 2011). However, these recommendations are an exception, as there is no guidance for other species, including other federally listed species such as the northern long-eared bat (*Myotis septentrionalis*) and non-protected species such as the Brazilian free-tailed bat (*Tadarida brasiliensis*).

Risk may increase with development near important bat resources, including migratory corridors, roosting sites, and foraging areas. Baerwald and Barclay (2009) documented greater fatality rates at wind energy facilities near the foothills of the US Rocky Mountains compared with those in adjacent grasslands. They suggested that mountains may influence bat migration by serving as a navigational reference or by providing roosting locations. Altering surrounding habitat during construction, such as clearing forests for roads and turbine pads, can increase bat activity by creating features (e.g., forest gaps and edge) that are used by bats as foraging and commuting corridors (Limpens and Kapteyn 1991, Verboom and Huitema 1997). Confirmation of a relationship between resource use and risk is lacking because few studies have explored the subject, and those that have are inconclusive (Piorkowski and O'Connell 2010).

Given the paucity of data on the movement ecology of bats, resource use of bats, and how siting may attract bats, it is difficult to provide the level of guidance necessary to site wind energy facilities or wind turbines to reduce risk. Historically, wildlife managers relied on pre-construction acoustic surveys and post-construction fatality monitoring to provide data on seasonal activity patterns and periods of risk, respectively. Yet, attempts to relate activity and fatality to predict risk have not been successful (Hein et al. 2013, Lintott et al. 2016). More recently, species distribution models have been used to assess potential risk to bats in a region or vegetation community (Santos et al. 2013, Hayes et al. 2015), but developing these models is only the first step. To be truly effective, models must be validated by comparing the predicted risk to actual fatality data from post-construction monitoring. This is likely a long-term endeavor, but the results of such a process could be valuable. Use of global positioning system (GPS) transmitters and methodology for long-term attachment of transmitters to bats are providing insights into the long-distance movement patterns of bats (Castle et al. 2015, Weller et al. 2016). Currently, use of this technology is limited because GPS transmitters are still too heavy to secure to most bats residing in the United States, and bats must be recaptured to download the data. An alternative tracking system, MOTUS (www.motus.org), uses smaller radio tags and a series of stationary towers to record observations of wildlife as they pass nearby. Krauel et al. (2017) used this system to track Indiana bats during both their spring and autumn migrations. This technology may provide a more effective and efficient means of acquiring data on regional movement patterns of bats compared with traditional tracking efforts.

Wind turbine technology plays a large role in siting decisions. The type of turbine constructed and its height, rotor-swept area, cut-in speed, and supervisory control and data acquisition system may influence interactions between bats and wind turbines. The foreseeable trend is for wind turbines to increase in height and blade length, which allows them to take advantage of more stable wind conditions at higher altitudes (Zayas et al. 2015). This allows for greater energy capture and makes generating electricity at lower wind speeds more economical (US Department of Energy 2015). The improved performance and economics also may increase the potential for geographic expansion of wind energy development, such as in the southeastern United States (Baranowski et al. 2017), and increases the possibility for wind development to interact with different species of bats. Moreover, the larger rotor-swept area may present greater risk to bats. Barclay et al. (2007) suggested that bat fatalities increase with increasing size of wind turbines, but additional data are needed to confirm this relationship. Although it may be economically feasible to site wind energy facilities in new areas, the costs of monitoring will likely increase. Larger wind turbines may require larger search plots during post-construction monitoring because dead or injured bats may fall farther from the base of the wind turbine. Larger turbines will also result in greater landowner payments to compensate for larger areas of lost crops, increased mowing and plot maintenance, and greater monitoring efforts for wind turbines co-sited with agricultural crops.

Offshore wind energy development represents an even greater challenge with respect to understanding bat activity and behavior, monitoring, and making informed siting decisions to reduce risks to bats. Wind projects off the eastern coast of the United States are already under way, and rapid offshore expansion is expected (Gilman et al. 2016). With few exceptions, our understanding of offshore bat activity comes from anecdotal observations. Peterson et al. (2014) summarized several accounts of bats flying near or roosting on ships in the Atlantic Ocean at distances ranging from 8 to 800 km from the US coastline. Results from their acoustic monitoring study across several islands matched these observations and showed the regular presence of several bat species at sea between July 15 and October 15, 2009–2011. Cryan and Brown (2007) summarized 38 years (1968–2006) of observations of hoary bats recorded between August and October on the Farallon Islands, which are situated 32 km due south of Point Reyes, California. McGuire et al. (2012) conducted a radio-telemetry study documenting movement of silver-haired bats (*Lasionycteris noctivagans*) along and across the shoreline of Lake Erie in the United States. Their findings indicated patterns of bat activity offshore that correspond with data collected at land-based wind energy facilities (Arnett and Baerwald 2013). If bats perceive turbines as a valuable resource (e.g., roosting, mating, foraging opportunity) on land, where similar resources are relatively available, then attraction may be amplified at offshore wind energy facilities in an environment devoid of such resources. Potential effects of offshore wind energy facilities on bats may depend on the activity levels of bats at various distances from the coastline and where offshore facilities are sited. Additional studies assessing the patterns of bat activity relative to the shorelines are necessary to determine whether risk increases or decreases with the distance from the shoreline.

Challenges and Opportunities Related to Siting

To our knowledge, there has been no quantitative assessment of reductions in bat mortality related to siting restrictions, and it is unlikely that we will be able to quantify this outcome anytime soon. It would require accumulating fatality data from across the

United States, developing predictive models of high- and low-risk areas for bats, and validating those models by comparing fatality data from wind energy facilities that develop in the different areas. Access to credible data to generate predictive models and data to validate those models are limited. Nonetheless, permitting agencies and wind energy developers should continue to work together to avoid siting wind energy facilities where large concentrations of bats occur, such as migratory pathways, hibernacula, and maternity colonies.

Despite more than a decade of studies and advancement of monitoring technology (e.g., GPS transmitters, acoustic detectors), we still lack basic information about bat ecology and behavior, as well as how they are affected by wind energy development. The problem stems, in part, from how difficult it is to study small, cryptic, nocturnal, and volant species like bats, but also from a lack of publicly available data on bat activity and fatality at wind energy facilities.

Opportunities still exist to increase our knowledge of habitat use and migratory patterns for bats to inform sustainable siting of future wind energy facilities. New transmitter technologies offer exciting possibilities to track bats over great distances and expanses of time. Current advances in molecular and biochemical techniques and analysis methods provide novel insights into the patterns and processes contributing to bat fatalities at wind energy facilities. The use of biochemical signatures such as stable isotopes provide much-needed information on the geographic origins of bats killed at wind energy facilities (Voigt et al. 2016, Pylant et al. 2016), thereby elucidating heretofore unknown migratory routes and population connectivity. Similarly, DNA barcoding techniques can provide irrefutable sex and species identification, even for carcasses found in an advanced state of decay or carcasses that have been scavenged (Korstian et al. 2013, Korstian et al. 2016). As new data become available, it will be important to update guidelines and decision makers on best practices for siting wind energy facilities.

Pre- and Post-Construction Monitoring

Our understanding of the effects of wind energy development on bats comes from nearly 20 years of pre- and post-construction studies. For bats, pre-construction studies involve placing acoustic detectors across the project site and recording the number of bat passes at each location. Total bat passes, or bat passes of a given species from an individual location, can then be compared with other locations within the project area. Post-construction monitoring studies estimate fatality based on carcasses recovered from systematic searches. One of the early research priorities for bats and wind energy was to develop a predictive relationship between pre-construction bat activity and post-construction fatality. Hein et al. (2013) reviewed 94 and 75 pre- and post-construction studies, respectively. Of these, only 12 wind energy facilities had paired data (i.e., the same site had paired activity and fatality data). Although a slight positive relationship was observed between pre-construction activity and post-construction fatality, the low adjusted R^2 value and wide prediction intervals, which included zero, indicated a poor relationship for the available data. Using data from the United Kingdom, Lintott et al. (2016) reported similar findings. They found that Ecological Impact Assessments, the primary tool used around the world to assess the effects of renewable energy, are poor predictors of risk. There are several reasons for the lack of a relationship between pre-construction bat activity and post-construction fatality, including the ways in which data were collected, access to data, changes to habitat during construction, and the apparent attraction of bats to wind turbines (Hein et al. 2013). Yet even if improvements to the methodology are made, such as using only data collected from detectors positioned in rotor-swept areas (Collins and Jones 2009, Stanton and Poulton 2012), the variability in detector types, weatherproofing strategies, and analysis software make it difficult to standardize acoustic surveys.

Although pre-construction bat activity has not been shown to be predictive of fatality, the patterns between the two are similar. For example, acoustic surveys show seasonal peaks in bat activity levels that tend to match those of fatality rates (i.e., late summer and autumn). Acoustic studies can also provide data on species presence, relationships between bat activity and weather variables, and relative activity patterns associated with habitat features. However, in regions where activity and fatality levels are well established, implementing additional pre-construction acoustic surveys may not be the best use of limited resources. In 2013, the Pennsylvania Game Commission removed their requirements for acoustic surveys in their updated version of the Wind Energy Voluntary Cooperation Agreement Amendment I (Pennsylvania Game Commission 2013). Conversely, in understudied regions (e.g., the southeastern United States and offshore), pre-construction acoustic surveys may be necessary to understand the activity patterns and presence of understudied species.

For post-construction monitoring, the issue is not whether it should be done, but how it should be conducted (Figure 6.2). Considerations for methods include how many turbines to monitor, how often to monitor (e.g., daily, every 3 days, etc.) for how many years, how large should the plots be, and how fatality should be estimated. Answers to these questions depend on the research or monitoring question(s) and conditions at the site. Although there is flexibility in how to design a post-construction monitoring study, it is important to capture the spatiotemporal variation of fatality at each site. It also is important to understand and articulate the biases associated with the data and to use an appropriate fatality estimator that accounts for those biases.

Historically, post-construction monitoring studies have been designed and implemented for each facility, with little consideration of how results from one facility relate to others. Some state agencies provide

Figure 6.2 Post-construction fatality monitoring at a wind energy facility in the southern Great Plains, US. Inset: Brazilian free-tailed bat, *Tadarida brasiliensis*, found during a turbine search. *Photo credit: Amanda Hale.*

guidelines for studies, which help standardize statewide monitoring methodologies (e.g., Ohio Department of Natural Resources 2009, Pennsylvania Game Commission 2013). However, if the permitting or regulatory process requires a specific estimator that is inappropriate (i.e., because the data do not meet the assumptions of the estimator) or requires 2 or more different estimators, it prevents comparability and causes confusion (see Deep Dive). Lack of comparability among studies makes it difficult to examine larger-scale patterns in bat fatality or estimate cumulative bat fatality. Recently, 3 separate efforts estimated cumulative bat fatality at wind energy facilities (Arnett and Baerwald 2013, Hayes 2013, Smallwood 2013), each indicating that hundreds of thousands of bats are killed each year in the United States. Huso and Dalthorp (2014), while recognizing the importance of understanding the cumulative effect on bats, discussed how these fatality estimates are inaccurate given the shortcomings of the available data. As they described it, the problem is twofold. First, data used were not representative, meaning they were based on publicly available data and not a random sample across all wind energy facilities. Second, individual site estimates were derived from such different methodologies that they were not comparable. Although there is a need for greater dissemination of study results, in this case, it would only partly resolve our ability to compare fatality among wind energy facilities and develop cumulative estimates of bat fatality. The larger challenge relates to statistical analyses and points to the need for a generalized fatality estimator that accounts for the various biases encountered in studies, provides an accurate and precise estimate of fatality, and is comparable among studies. Yet, a reliable cumulative fatality estimate only provides half of what is needed to understand the effects of wind turbines on bats. Population estimates also are required to put fatality into context (see Population Analyses section below).

Data on how and when bats use the offshore environment are lacking, and the tools and current land-based methodologies to collect pre- and post-construction data are insufficient for offshore facilities. For example, there are limited locations to install acoustic detectors and standard carcass searches are not applicable. Therefore, alternative strategies for studies of bat fatalities at offshore wind energy facilities must be developed. Suryan et al. (2016) tested a synchronized array of sensors that continuously monitored wildlife interactions with wind turbines, including impacts, using 3 sensor nodes: (1) vibration (accelerometers and contact microphones; (2) optical (visual and infrared spectrum cameras); and (3) bioacoustics (acoustic and ultrasonic microphones). Yet, this and other similar systems are still in the proof-of-concept phase, and direct comparison between a new system and standard post-construction monitoring practices is necessary before its wide-scale use. Until these technologies are developed, we will not be able to accurately assess the effects of offshore wind energy development on bats.

Challenges and Opportunities Related to Monitoring

Our understanding of wind turbine–caused bat fatalities is limited because wind development has yet to occur in some regions of North America, data typically are not publicly available, and data that are available may have been improperly collected and analyzed. If study designs are not scientifically credible and are not made public, the expense and effort to collect the data provide little to no value. Moreover, improperly collected and analyzed data can cause more harm than no data at all by confounding our understanding of this issue. Consistency across jurisdictional boundaries and comparability among studies are critical to assess regional patterns and estimate the cumulative effects of bat fatalities at wind energy facilities.

Pre- and post-construction monitoring studies are expensive, but as our understanding increases and new methodologies become available, studies can be more robust and cost-effective. Monitoring may not always be necessary, but it remains essential in regions for which little to no data exists, where different species are present, and with advances in turbine technology. Significant changes to how we conduct and analyze post-construction fatality data are forthcoming. Several groups are independently investigating the efficacy of searching only roads and turbine pads to estimate bat fatality. This methodology could reduce the costs of monitoring and those associated with clearing and maintaining large search plots. In addition, development of a generalized fatality estimator will offer comparability among results and reduce bias and confusion related to the myriad of fatality estimators. Validating these and other new tools is an important step prior to their use, but if proven credible, broad implementation of new, standardized methods should begin immediately.

Strategies for Reducing Bat Fatalities

With concerns about bat population declines fueled by reports of continuing high numbers of fatalities at wind energy facilities, site managers and wind developers find themselves subject to increasing pressure to reduce the negative effects of wind energy production on bats and other wildlife (Kunz et al. 2007, Northrup and Wittemyer 2013, Arnett 2017). Although the obvious first step in the mitigation hierarchy is avoidance of adverse effects, perhaps through improved siting of wind energy facilities, a paucity of data are available to test the efficacy of avoidance. Moreover, this problem is further compounded by our as yet limited ability to successfully predict collision risk for bats, either through acoustic surveys or traditional impact assessment approaches, prior to construction (Hein et al. 2013, Lintott et al. 2017).

The second step in the mitigation hierarchy is minimization of adverse effects, which could be implemented when bats are known to be in the area and therefore potentially at risk of collision with turbines. For example, early monitoring efforts revealed that bat activity was greater during periods of relatively low wind speeds, and lower during periods of rain, low temperatures, and high wind speeds (Reynolds 2006, Horn et al. 2008). Subsequent studies showed a positive relationship between acoustic bat activity, bat fatalities, and low wind speeds at operational wind energy facilities (Baerwald and Barclay 2009, Arnett et al. 2011). As a result, it was proposed that raising the manufacturer's cut-in speed (the minimum wind speed at which electricity is generated—typically between 2.5 and 3.5 m/s) would effectively reduce the number of bat–wind turbine collisions by slowing or stopping blade rotation at low wind speeds when bats are active (e.g., Baerwald et al. 2009, Arnett et al. 2011). Where tested in the United States and Canada, minimizing wind turbine operations by feathering blades (changing the angle of the blades to slow rotation to 1–3 rpm) and raising the cut-in speed has proven to be an effective means of reducing bat fatality by 50% or more (Arnett et al. 2013a, Martin et al. 2017). For example, Baerwald et al. (2009) showed that raising the cut-in speed from 4.0 to 5.5 m/s reduced bat fatality by

up to 60%. A similar study by Arnett et al. (2011) demonstrated that raising cut-in speed from 3.5 to 5.0 m/s and to 6.5 m/s reduced bat fatality by 82% (95% CI: 52–93) in 2008 and 72% (95% CI: 44–86) in 2009. Including additional factors (e.g., temperature) may further improve minimization strategies. Martin et al. (2017) demonstrated a 62% (95% CI: 34–78) reduction in bat fatality by raising cut-in speed from 4.0 to 6.0 m/s when temperatures were >9.5°C.

Although such operational minimization strategies have general application, have proven effective at reducing bat fatalities, and seem to have reasonable economic costs based on the few studies that have made those data publicly available (Baerwald et al. 2009, Arnett et al. 2011, Martin et al. 2017), gaining further insight into when and where limiting wind turbine operations will be most effective at reducing bat fatality has important ecological and economic benefits. One such strategy to reduce bat fatalities with minimal power loss involves feathering wind turbine blades below the manufacturer's normal operational cut-in speed at night and during periods of high risk (World Bank Group 2015). This "best management practice" has been adopted by the American Wind Energy Association as a voluntary bat conservation program. The concept was recommended because turbine blades can rotate at speeds that are lethal to bats even when electricity is not being generated (Arnett et al. 2013a). Feathering, or pitching the turbine blades parallel to the wind so that the tip speed is < 50 mph (~1–3 rpm), below the manufacturer's cut-in speed is considered a mutually beneficial strategy in which bat collision risk is reduced with little effect on power production. Nonetheless, several questions remain unanswered regarding the effectiveness and financial costs of feathering below cut-in speed as a minimization strategy. For example, what level of fatality reduction can be expected by feathering turbine blades below cut-in speed? To our knowledge, only one publicly available study has demonstrated a significant reduction in bat fatalities using this minimization strategy. Good et al. (2012) reported a 36.3% (90% CI: 12.4–53.8) reduction in bat fatalities by feathering below the manufacturer's cut-in speed of 3.5 m/s. However, species-specific differences were observed, with eastern red bats (*Lasiurus borealis*) showing a nearly 50% reduction in carcasses recovered compared with the rate for hoary bats, which had similar fatality between control and treatment groups. Although there are clear benefits to reducing bat fatalities when electricity is not being produced, broader applicability of the results in other regions and under future economic contexts is unknown.

Because of concerns about power loss and uncertainty about the viability of operational minimization in the future, there is substantial interest in exploring technologies that reduce bat fatalities while allowing turbines to operate at full capacity. One such possible technological solution is based on deploying ultrasonic acoustic deterrents on wind turbines to create a disorienting airspace for echolocating bats. The ultrasound produced by such deterrents overlaps with the intensity and frequency range of bat echolocation calls, and thus may discourage bats from approaching and entering the rotor-swept area of wind turbines, thereby reducing fatalities. An early field study demonstrated the potential for deterrents to reduce bat activity at ponds (Szewczak and Arnett 2017), and a subsequent study at an operational wind energy facility suggested that deterrents could reduce bat fatalities at wind turbines by 18% to 62% compared with control turbines (Arnett et al. 2013b). Given the promise shown in these early studies, additional research is under way to improve this deterrent technology, optimize the deterrent signal, and understand how species may vary in their responses to deterrents over space and time. For example, a recent ground-based deterrent study by Lindsey (2017) showed a reduction in effectiveness with distance at approximately 30 m from the source signal, but no difference in bat activity among the continuous and pulsed ultrasonic signals tested. The deterrent also influenced bat flight behavior, with significantly fewer complex foraging flight paths and signifi-

cantly more straight-line flight paths observed during deterrent tests compared with control periods (Lindsey 2017). The results of ground-based studies such as these are informing ongoing tests of deterrent technology on operational wind turbines. Of interest is determining the optimal placement and orientation of the sound sources along the turbine tower, nacelle (housing for the generator, gear box, drivetrain, and brake assembly), and blades. Because of the rapid attenuation of ultrasound, deterrent effectiveness likely diminishes with distance, and multiple sound sources will be required per turbine to provide adequate coverage of the rotor-swept zone.

In addition to ultrasonic acoustic deterrents, research is under way investigating dim UV light as a possible bat deterrent (Gorresen et al. 2015). Results from Gorresen et al. (2015) indicated a reduction in bat activity in trees illuminated with dim flickering UV light, despite a concomitant increase in insect numbers. If the observed reduction in activity at trees translates into a reduction in fatality at operational wind turbines, then dim UV light illumination may prove to be an effective minimization technology.

Studies with live-captured bats have shown that, in captivity, bats make fewer contacts and close passes with texture-treated surfaces compared with smooth surfaces (Bienz 2016). Additional research is under way to investigate the behavioral response of free-flying bats to texture-treated turbine tower monopoles at an operational wind energy facility. If the texture coating leads to a reduction in the number of bats making close investigative approaches at the turbine tower surfaces and the amount of time spent close to the rotor-swept zone, then texture coating for the nonmoving turbine parts may also prove to be an effective mitigation technology to reduce bat fatalities at wind turbines.

The third and final pillar of the mitigation hierarchy, compensatory mitigation, can be a useful tool in wildlife conservation; however, with respect to wind turbine-related bat fatalities, compensatory mitigation is an unproven conservation strategy. Creating new habitat or protecting existing habitat is local and often focused on a single hibernaculum or forest patch, for example. Given the broad seasonal movement patterns of bats, there is no assurance of survival or increased fitness outside of the mitigated area. Moreover, with the current and projected growth of wind energy development, the likelihood of a bat encountering wind turbines near the mitigated area or elsewhere along its migration route is high. Furthermore, no data yet exist to support the idea that compensatory mitigation ensures greater reproductive output by bats using the area, and even if it did, we have no way of measuring it. As a more cost-effective approach, funding for compensatory mitigation should be diverted, at least in part, to developing and testing strategies to reduce the adverse effects of wind turbines on bats. These strategies have a greater chance of providing real conservation value by protecting existing bats, and their results are measurable. Alternatively, funding research to better understand resource use by bats may help in determining the proper habitat area and quality necessary for bat conservation.

Challenges and Opportunities Related to Reducing Bat Fatalities

Under the current regulatory framework, it is unlikely that operational minimization requiring higher cut-in speeds will be broadly implemented because of the concomitant loss in power generation. This is unfortunate, as the best estimates suggest that wind turbine–associated tree bat mortality (the 3 *Lasiurus* species with high reported rates of bat fatality at wind turbines across the United States and Canada) could be reduced by 50% or greater by raising cut-in speed to 5.0 m/s from July to September. Given the lack of reliable population estimates for migratory tree bats, however, it is not possible to put this mortality reduction into context or enact policy to protect these bat species that have no special conservation status, at least in the United States.

Implementing and evaluating the effectiveness of minimization strategies is more expensive than typical post-construction fatality studies and requires large amounts of data collected over several field seasons. Moreover, the cost of technology research and development is expensive and, understandably, wind energy companies are hesitant to fully bear the costs of these types of endeavors. In addition, given the permitting environment, it may be a challenge to obtain a permit allowing a subset of wind turbines to operate under "normal" conditions, which is required to evaluate the effectiveness of the strategy in question. Another challenge is to understand how much uncertainty is acceptable when testing and evaluating these strategies. For example, how many studies are necessary before a strategy is either accepted or rejected?

Advances in technology may affect minimization strategies. Increases in blade lengths will likely decrease the effective range for nacelle-mounted ultrasonic acoustic deterrents because high frequency sound attenuates rapidly in the atmosphere (Arnett et al. 2013b). In addition, operational minimization or feathering turbine blades during periods of high risk for bats may no longer be an option because of changes in blade length. Arnett et al. (2008) summarized findings from several studies and determined that bat fatalities were greater when wind speeds were relatively low. Moreover, feathering turbine blades, even below the manufacturer's cut-in speed, may become obsolete if it is not financially viable to feather blades at sites where only low wind speed conditions exist.

One clear research opportunity is to develop more robust predictive models of bat activity and fatality at wind turbines that incorporate a broad spectrum of environmental conditions. For example, there may be environmental variables that can potentially predict bat activity or fatality other than just wind speed and temperature (Weller and Baldwin 2012, Martin et al. 2017). Identifying these variables and incorporating them into operating procedures may lead to more ecologically informed and more cost-effective operational minimization.

We fully anticipate that advances in technological solutions to characterize and reduce bat–wind turbine collisions are forthcoming. Significant progress is being made in the areas of acoustic deterrent technology, blade-strike sensors, and real-time acoustic or video monitoring within the rotor-swept area of wind turbines that may inform and improve strategies to reduce bat fatalities at wind turbines. If regulatory agencies, nongovernmental organizations, and wind energy developers can collectively support and implement "research as mitigation," this would greatly accelerate the pace at which new technological solutions are developed and tested at operational wind energy facilities, as the risk would be spread among all involved stakeholders. It also is likely that in the future, we will see 2 or more minimization strategies used in concert to achieve desirable levels of bat fatality reductions for the smallest possible loss in power generation.

Population Analyses

Although not all bat species are affected equally by wind energy development, the combination of high wind turbine–related fatality rates for some species and other sources of bat mortality, such as WNS, place the population status of many North American bat species in question (Thogmartin et al. 2013, Rodhouse et al. 2015, Erickson et al. 2016, Frick et al. 2017). For most bat species, including those affected by wind energy production, our understanding of their population sizes, demographic trends, or the extent of population structure is limited (Kunz et al. 2007), making it exceedingly difficult to put mortality into context and devise scientifically based recommendations for the long-term conservation of bats. Nonetheless, it is anticipated that mortality from wind energy development will be additive rather than compensatory for bat populations that are already thought to be in decline (Winhold et al. 2008, Ford et al. 2011, Francl et al. 2012). Furthermore, as

bats have low reproductive rates (Barclay and Harder 2003) and population dynamics are sensitive to variation in adult survival (e.g., Schorcht et al. 2009, Thogmartin et al. 2013), recovery from population declines is likely to be slow for bats. For these reasons, researchers and wildlife managers are concerned that fatalities from wind turbines are likely contributing to bat population declines (Barclay and Harder 2003, Hein and Schirmacher 2016, Frick et al. 2017).

Traditional survey techniques and approaches used to estimate population sizes and trends in other wildlife taxa have limitations when applied to bats, because they are nocturnal, cryptic, and difficult to follow over time because of extensive seasonal movements between summer breeding areas and over-wintering sites (Cryan 2003). This is especially true for migratory tree bats, given their solitary nature. Increasingly, researchers are turning to intrinsic markers such as genetically based biomarkers and chemical signals to measure demographic trends, estimate population parameters, and identify migratory pathways (e.g., Lehnert et al. 2014, Korstian et al. 2015, Pylant et al. 2016).

An improved understanding of the geographic origins of bats killed at wind energy facilities can provide important insights into the diversity and population structures of affected species, as well as identify the scales at which these species are affected by wind energy development (Figure 6.3). Stable hydrogen isotope ratios (δ^2H) have proven to be an excellent tool for inferring the geographic origins of migratory wildlife (Rubenstein and Hobson 2004, Hobson 2008), including bat species killed at wind energy facilities (e.g., Cryan et al. 2014a). Using this approach, it was revealed that noctule bat (*Nyctalus noctula*) fatalities at wind energy facilities in eastern Germany comprised both resident and long-distance migrant bats (Lehnert et al. 2014). This migratory species is the most frequently detected bat species in fatality searches in this part of Germany, and research indicated that individual losses could affect both local and distant populations (Lehnert et al.

Figure 6.3   Researcher setting up a mist-net to catch and collect stable isotope data from live bats. Inset: Eastern red bat, *Lasiurus borealis*, captured in a mist net. *Photo credit: Texas Christian University.*

2014). Of particular concern was the large proportion of dead adult females, because recruitment and population stability of noctule bats are especially sensitive to changes in adult female survival (Petit and Mayer 2000, Jones et al. 2003). In the United States, Pylant et al. (2016) investigated the origins of eastern red bats and hoary bats killed at wind energy facilities in the central Appalachian Mountains. For eastern red bats, 57% of fatalities were from nonlocal sources, whereas only 1% of hoary bat fatalities were from nonlocal sources. This study showed that, even among the migratory tree bats in North America, different species may be affected differently by wind power development, and baseline monitoring efforts such as these may help us track these effects over time.

Similarly, genetic diversity and population structure of affected bat species likely influences resiliency to sustained wind turbine mortality and the likelihood of population persistence. For example, species with small effective population sizes (N_e) contain low levels of genetic diversity that may limit their evolutionary potential to respond to selection and avoid inbreeding (Frankham et al. 2009). Thus, these species may be affected by wind turbine mortality to a greater extent than species with large N_e. Population genetic analyses can also provide insights into historical population sizes, including both range expansions and population bottlenecks, identify evolutionarily unique subpopulations, and be used to monitor changes in population sizes over time (Frankham et al. 2009). Several recent studies have used genetic approaches to address these questions for the 3 species of bats that appear to be most affected by wind energy development in North America (e.g. Korstian et al. 2015, Vonhof and Russell

2015, Sovic et al. 2016). For the eastern red bat, several studies estimating N_e have come to similar conclusions that the effective population size is in the hundreds of thousands to millions of individuals, with no evidence of barriers to gene flow among groups of samples (Korstian et al. 2015, Vonhof and Russell 2015, Pylant et al. 2016, Sovic et al. 2016). Similarly, there is no evidence of population genetic structure in the hoary bat (Korstian et al. 2015, Pylant et al. 2016, Sovic et al. 2016), although the estimates of N_e are smaller for this species compared with the eastern red bat. Sovic et al. (2016) also analysed genetic samples from silver-haired bats and came to similar conclusions: there is no evidence for population structure in this species, although it had the smallest estimated N_e of these 3 species. In summary, high levels of gene flow and connectivity across the population ranges of three North American tree bat species indicates that monitoring and management efforts must integrate information from across their entire ranges as potential impacts of mortality in any given region may have far-reaching implications.

Given the challenges of obtaining empirical demographic data to predict population- and species-level effects from wind turbine mortality, several different modeling approaches are also being implemented to provide quantitative assessments of the potential effects of wind energy development on bats. For example, Erickson et al. (2016) created a full-annual-cycle model to simultaneously explore the potential impacts of wind energy development and WNS across the range of the federally endangered Indiana bat. This spatially explicit modeling approach incorporated mortality and reproduction for all seasons of this species' life cycle. Based on simulated mortality rates, Erickson et al. (2016) concluded that the combination of mortality from wind turbines and WNS appeared to be additive, with wind turbine mortality likely leading to extirpation of small hibernacula and WNS mortality likely leading to extirpation of large hibernacula. This study was the first to investigate the potential impacts of

these 2 stressors together and showed a synergistic effect. The potential effect of both stressors together was greater than when considering either stressor alone, further highlighting the need for a better understanding of Indiana bat–wind turbine collision risk.

As another example, Frick et al. (2017) used a population projection modeling approach to explore whether fatalities from wind turbines threaten the population viability of the hoary bat, the species most frequently detected during carcass searches at wind turbines across North America (Arnett and Baerwald 2013). Given the lack of available empirical data for hoary bats, this modeling approach relied on expert elicitation (Martin et al. 2012), bat population growth-rate data from other species, and estimates of mortality from wind turbines across North America to assess the likelihood that mortality from wind turbines (at the installed capacity in 2014; AWEA 2016, CanWEA 2016) poses a species-level threat to hoary bats in North America. The results of the simulations showed that current mortality from wind turbines could result in rapid and severe declines in hoary bat populations within 50 years and increased risk of extinction within 100 years. Moreover, the approach taken by Frick et al. (2017) is conservative in that they used the lowest published estimates of bat fatality rates and held installed wind energy capacity constant at 2014 levels. Installed wind energy capacity is projected to increase; therefore, it is likely that species-level impacts will also increase, unless mitigation strategies are broadly implemented across North America.

Challenges and Opportunities Related to Population Analyses

Without a better understanding of population size, demographics, and the effects of fatalities on the viability of bat populations, it is not possible to fully quantify effects of wind turbines on bat populations, nor the effectiveness of mitigation strategies at wind energy facilities. These knowledge gaps also present significant challenges for nongovernmental organizations, policy makers, and regulatory agencies. Thus, some questions remain largely unanswered: How large do fatality reductions need to be to ensure some level of assurance regarding bat population persistence? How do we know whether fatality thresholds established at a single site or region are biologically meaningful? How are these site-specific thresholds accounting for mortality occurring beyond their jurisdiction? (Arnett et al. 2013c).

Given the challenges associated with gathering empirical demographic and population data for the North American migratory tree bats, we may need to rely on elicitation of expert opinion to guide conservation decision making for the foreseeable future (Martin et al. 2012) and inferences from intrinsic markers to monitor and evaluate the potential impacts of wind turbine–associated mortality on bats. To date, there is no strong genetic evidence of population declines in migratory tree bat species; however, data from other sources, such as capture rates and acoustic detections, suggest their populations may be at risk. Regional declines have been reported in eastern red bats based on decreases in capture rates in mist-netting surveys and the numbers of bats submitted for rabies testing (Winhold et al. 2008), whereas declines in hoary bats are based on decreases in acoustic activity patterns (Ford et al. 2011) and capture rates in mist-netting surveys (Francl et al. 2012). Nonetheless, Korstian et al. (2015) cautioned that genetic monitoring of migratory tree bats specifically for detecting population declines caused by wind turbines may be impractical because of the large effective population sizes and high levels of gene flow in these species. Instead, they suggested that future efforts should focus on developing genomic resources for these species, obtaining better estimates of mutation rates, and conducting rangewide population genetic studies to better estimate historical and current population sizes. Population sizes of threatened and endangered bat species are already small, which suggests that levels of fatality at wind energy facilities will also be small. Yet, rare

mortality events are exceedingly difficult to accurately estimate, so it may not be possible to know what the effects of wind turbine fatality will be for these bats.

New advances in genetic approaches and stable isotopes have provided key insights into the geographic origins of bats killed at wind energy facilities, in both Europe and North America, and it is becoming clear that these facilities may affect both local and distant bat populations. By combining an analysis of patterns of mortality across existing wind energy facilities with information gained from intrinsic markers, it may be possible to identify bat migration routes and flyways in North America. An improved understanding of bat migration could also be used to increase the effectiveness of operational minimization strategies and inform siting of future wind energy development.

Future research efforts should focus on understanding the population-level consequences of wind turbine–caused bat mortality. Being able to place fatality estimates into a population-level context will help shape the conversation about the sustainability of wind power generation and could lead to actions to protect migratory tree bat populations while they still are intact. In particular, the modeling approach employed by Frick et al. (2017) using expert elicitation could be used to inform relative risk to other North American migratory bat species. At a minimum, these efforts would provide a baseline against which future data sets could be compared.

Reducing the take of bats via operational minimization could ensure the persistence of many bat species and even aid in the potential recovery of those species critically affected by WNS. Thus, until we know why bats are active near wind turbines and, in turn, devise ways to prevent such interactions, curtailing wind turbines is currently the most effective solution to reducing bat fatalities. The challenge is to improve the efficiency of this minimization strategy to facilitate population-level responses in bat species, while simultaneously reducing costs to wind energy operators.

Given that fatality monitoring is taking place at wind energy facilities across North America, it's not too far-fetched to envision developing a coordinated system or repository that could enable a holistic look at which species are killed when and provide tracking for how population numbers may change over time. With coordinated efforts, collecting hair and tissue samples for stable isotope and genetic analyses could become routine, and collectively, these data could allow assessment of the cumulative demographic and genetic effects of wind turbine-caused bat mortality.

Bat Behavior at Wind Turbines

Although the proximate causes of bat fatality at wind turbines are relatively well understood (i.e., bats may die from barotrauma, Baerwald et al. 2008, but see Rollins et al. 2012; collision with the rotating blades, Horn et al. 2008; or a combination of the two, Grodsky et al. 2011), the ultimate causes are not as clear (Kunz et al. 2007, Arnett et al. 2008, Cryan and Barclay 2009). Increasingly, however, several lines of evidence suggest that bats may be attracted to wind turbines, and several specific hypotheses have been proposed to explain this phenomenon (e.g., Cryan and Barclay 2009). One possibility is that bats find something about the turbines themselves to be interesting (Cryan and Barclay 2009). For example, red aviation lights on top of turbine towers have been considered a possible attractant to bats; however, studies have shown that bat mortality at towers with aviation lights is similar to or even lower than mortality at towers without aviation lights (Arnett et al. 2008, Baerwald 2008, Bennett and Hale 2014). Another possibility is that bats are attracted to noises generated from the blades or nacelle, blade movement, or Doppler effects resulting from blade movement (Long et al. 2009, 2010).

Alternatively, another attraction hypothesis is that bats may misperceive wind turbines as potential resources. As fatalities are concentrated in late summer and early fall, coinciding with both the migratory and mating seasons of some of the species most frequently found during carcass searches (Barclay et al. 2017), it has been hypothesized that bats are mistaking wind turbines for trees that could serve as congregation points and display sites for attracting and finding mates (Cryan and Brown 2007, Cryan 2008). A subsequent study using thermal imagery to observe bats interacting with turbines supported this hypothesis; bats were seen approaching turbines most often from the leeward side in the same manner as they would approach trees for roosting, mating, and social opportunities (Cryan et al. 2014b). Another study hypothesized that bats may misperceive wind turbine towers as water (McAlexander 2013), as previous research demonstrated that echolocating bats misidentify artificial smooth surfaces to be water (Greif and Siemers 2010, Russo et al. 2012). Another possible explanation is that wind turbines may provide bats with resources like water (e.g., condensation on the tower), roosting sites, and foraging opportunities. For example, Long et al. (2011) reported that insects are drawn to light-colored turbines in particular; because turbines are commonly painted light colors, bats may be attracted to wind energy facilities as a result of insect aggregations on and around the turbine towers.

Many attraction hypotheses have not yet been rigorously tested because bats are difficult to observe interacting with turbine blades, nacelles, and towers high above the ground at night. Nonetheless, with advances in technology, we can see and hear bats interacting with wind turbines in various ways, and these studies are providing much-needed insights to address the central question: Why are bats being killed at wind turbines? With the use of infrared videography, bats can be seen exploring and contacting turbine monopoles, nacelles, and blades, especially during low wind speed nights and often from the leeward side of turbines (e.g., Ahlén et al. 2007, Horn et al. 2008, McAlexander 2013, Cryan et al. 2014b). Moreover, flight patterns of these bats are suggestive of aerial pursuits and gleaning of prey items, roost investigation, mating behavior, and even drinking behavior at wind turbines, thereby providing support for several of the resource-based attraction hypotheses.

When acoustic detectors are deployed in or near the rotor-swept zone of operational wind turbines, they can provide additional information about specific types of bat activities observed in the immediate vicinity of turbines. For example, in the midwestern United States, social calls of hoary bats were recorded on the leeward side of turbines, a finding consistent with the hypothesis that bats were seeking social opportunities (Cryan et al. 2014b). In another study, a range of echolocation call types, including foraging and approach phase calls, in addition to terminal buzzes (Altringham 2011), were recorded for 6 species of bats at wind turbine towers in the US southern Great Plains (Foo et al. 2017). Included in this data set were recordings of terminal buzzes from within the rotor-swept zone, which provides compelling evidence that bats are attempting to capture prey items at heights at which they are susceptible to collision with rotating blades. It is important to note, however, that terminal buzzes also may be indicative of bats locating landing sites (e.g., Melcón et al. 2007) and drinking water (Griffiths 2013). To date, no published study has demonstrated that wind turbines provide bats with water, but a recent study by Bennett et al. (2017) shows that bats will roost on turbines. Thus, more research is needed to fully understand the relative frequency of these activities and how they may contribute to overall wind turbine–mortality risk in bats.

As video and audio technologies have advanced our understanding of bat behavior at wind turbines, so too have studies of bat carcasses (and their stomach contents) recovered from post-construction fatality searches (Valdez and Cryan 2013, Rydell et al. 2016, Foo et al. 2017). These studies provide important insights into what bats are doing just prior to death, including foraging for insect prey items that are often abundant around turbine towers. These findings indicate that foraging ecology in a broad context may be contributing to bat fatalities at wind turbines, especially as insect aggregations are known to occur on and around the turbine towers (Long et al. 2011). Although there have not been direct observations of bats capturing prey items from the surfaces of turbine towers, diurnal flies and flightless insect taxa were identified in the stomachs of bats killed in Sweden (Rydell et al. 2016), and field crickets, which are primarily terrestrial, were frequently found in the stomachs of hoary bats and eastern red bats killed at wind turbines in the state of Texas (Foo et al. 2017). The findings of these recent diet analyses, in conjunction with the numerous published observations of bats making close "investigative" approaches at turbine towers (e.g., Horn et al. 2008, Cryan et al. 2014b), suggest that at least some aerial-hawking bat species may be able to capture insects that rest on wind turbine tower surfaces. Additional research is needed to more fully understand the extent to which aerial-hawking bats may be able to switch foraging strategies and whether their foraging success at wind turbines could be enhanced by the acoustic mirror effect, which renders prey objects on or near smooth surfaces more conspicuous to echolocating bats (Siemers et al. 2005).

Challenges and Opportunities Related to Bat Behavior

Recent and ongoing research indicates that at least some bat species are attracted to wind turbines, and that attraction, in a broad sense, is likely contributing to bat fatalities at wind turbines worldwide. Moreover, the various attractors may not be mutually exclusive, and their relative importance likely varies by species, sex, age, time of year, and geographic location. Thus, it may not be possible to identify a single ultimate explanation for bat fatality or a single solution that would work to minimize risk to all bat species across all wind energy facilities. Bat attraction to wind turbines may help explain the difficulty to date in predicting risk to bats based on pre-construction activity surveys or other measures of habitat suitability. For example, if turbines reliably attract insects, which, in turn, attract bats, then pre-construction bat activity surveys at potential wind energy facilities could drastically underestimate post-construction bat fatality rates. If reliable and abundant foraging opportunities continue to attract migrating bats to wind turbines, thereby increasing collision risk, then future mitigation efforts should focus on technological innovations (e.g., acoustic deterrents) and operational changes (e.g., raising the cut-in speed on low wind speed nights) to reduce bat fatalities at wind turbines.

Undoubtedly advancements will continue in technologies that allow us to see, hear, and track bats in the dark (e.g., Roeleke et al. 2016), giving us the opportunity to gain new insights into their behavior. Although the problem is multifaceted and complex, the more we understand why bats are encountering wind turbines, the better equipped we will be to devise and implement effective solutions to inform sustainable wind facility siting, wind turbine operations, and the development of new technologies to reduce bat fatalities.

Deep Dive: A Generalized Estimator for Quantifying Bat Fatality at Wind Energy Facilities

Estimating fatality rates is fundamental to understanding the effects of wind turbines on bats. Multiple methods for estimating fatalities have been developed in the last 20 years and several are still in use, including those of Shoenfeld (2004), Huso et al. (2012), Korner-Nievergelt et al. (2013), and Wolpert (2013). These methods adjust raw survey data to account for imperfect detections (e.g., carcasses missed by searchers or removed by scavengers); however, the published estimators differ in their underlying assumptions and complexity and can produce statistically significant differences in estimates of fatality from the same data when model assumptions are not met. Moreover, the current use of multiple estimators is problematic for comparing results among studies, developing cumulative estimates, and implementing operational minimization strategies based on some threshold of fatality.

Notably, the problem with interpreting results from 2 different fatality estimators was encountered at a wind energy facility in the state of Pennsylvania. Fatality data were collected in accordance with the Pennsylvania Game Commission's Wind Energy Cooperation Agreement (Pennsylvania Game Commission 2013). As per the agreement, 2 different estimates were required: (1) the Shoenfeld (2004) estimator, also referred to as the Erickson estimator; and (2) the Huso et al. (2012) estimator. In addition, the agreement established a bat fatality threshold equal to 30 bats/turbine/year, but did not specify which fatality estimator should be used in determining whether the threshold was exceeded, nor did it specify whether the threshold had to be above the point estimate of fatality or the upper $(1-\alpha)$ confidence limit.

The difference between the 2 fatality estimators was statistically significant, and the mean of the Shoenfeld and Huso estimators were below and above the threshold, respectively. Causes of the discrepancy are related to the assumptions of the estimators with respect to searcher efficiency and carcass persistence. For example, the Shoenfeld estimator assumes the probability of finding a carcass does not change with each search. The consequence of this assumption, when persistence times are long, is that after multiple searches, the probability that a carcass will eventually be found approaches

100%, which is unlikely (Korner-Nievergelt et al. 2011). Wolpert et al. (2012) suggest that searcher efficiency likely diminishes over time. Assuming constant search efficiency will lead to overestimation of searcher efficiency and underestimation of fatality. In contrast, the Huso estimator assumes the probability of finding a carcass after the first search is 0%, which also is unlikely. However, the user of the Huso estimator satisfies this assumption by using only "fresh" carcasses or those determined to have died since the previous search in the observed carcass counts.

There also are different assumptions in the estimators related to carcass persistence. The Shoenfeld estimator assumes carcass removal occurs at a constant rate (i.e., an exponential distribution), which is rare in field trials (Bispo et al. 2013). The Huso estimator allows for model selection of the most appropriate carcass persistence distribution (i.e., exponential, Weibull, log-logistic, log-normal) based on available data. In the Pennsylvania case, the best model for carcass persistence was the log-logistic distribution with initially rapid removal followed by high persistence. Assuming the exponential distribution rather than the best model based on the data resulted in an overestimation of carcass persistence, which led to an underestimation of fatality.

Inappropriate use of an estimator or using one that is biased because it fails to appropriately account for all sources of imperfect detection results in inaccurate estimates and can lead to poor management decisions. Moreover, these estimates cannot meaningfully be compared or used in broader analyses, such as those designed to estimate regional or national impacts (Huso and Dalthorp 2014). The example presented here demonstrates the confusion inherent in using multiple fatality estimators. In this case, the Shoenfeld estimator is not the most appropriate one to use, but knowing which estimator to base decisions on is beyond the capacity of most natural resource managers and wind energy developers.

Careful analysis of the myriad available estimators indicates that each estimator can potentially be viewed as a special case of a generalized estimator. A working group comprising several statisticians (many of whom created the estimators in current use), together with nongovernmental organizations and the wind industry, has been tasked with developing a generalized estimator (aka GenEst). These collaborators have recognized the commonalities among estimators and will formalize the theoretical bases to unify, under a single theoretical umbrella, several seemingly disparate approaches to estimating bat fatality at wind facilities. This generalized fatality estimator will allow end users to test assumptions regarding input parameters and select the approach that best reflects their situation and data. Thus, results of this effort are expected to reduce confusion, provide comparability among studies, and offer a more accurate estimate of bat fatalities.

Conclusions

Although our understanding of the effects of wind turbines on bats has improved in recent years, we have only scratched the surface of understanding bat–wind turbine interactions. Our ability to answer basic questions is hindered by a lack of robust data, and we are in danger of repeating past mistakes as wind energy development expands into new regions. To resolve these issues, a new paradigm is required— one of comparability and transparency. We must develop and use consistent methods for estimating effects that take into consideration the localized effects of an individual facility as well as the cumulative impact on a region or population. Concurrently, the peer-review process and dissemination of data is paramount. Collecting data in a manner suitable for publication and sharing results with the wind and wildlife communities should become the norm so we can fill in existing knowledge gaps and address broader-scale questions.

Advances in technology have improved our ability to observe bats interacting with wind turbines and track bat movement patterns over longer periods of time and greater distances. As new tools become available, it will be important to understand the applicability, biases, and limitations of these technologies, as well as how they can be paired with other techniques to maximize effectiveness. Technology is also paving the way to reduce bat fatalities in a cost-effective manner. Yet, to ensure the sustainability of bat populations, we must keep in mind the projected future growth of wind energy development and advances in wind turbine technology when implementing policies or designing strategies to minimize bat fatalities at wind facilities. As promising as smart curtailment strategies and deterrent devices appear to be, it is imperative that their effectiveness be thoroughly tested across a range of bat species and wind energy facilities. We caution against accepting new programs or technologies that lack a substantiated body of work or peer-reviewed results.

Finally, collaborations are essential to addressing complex conservation issues, such as those presented in this chapter. The wind energy and wildlife communities must work together to tackle unresolved research questions associated with comparable monitoring protocols and analytical tools, cumulative assessments of the effects of wind turbines on bats, population estimates for numerous species, timing and conditions under which bats are interacting with wind turbines, practicable mitigation strategies, and monitoring protocols and mitigation strategies for offshore wind turbines. Only by combining the expertise and resources from academia, nongovernmental organizations, industry, and government agencies can we develop and implement mutually beneficial guidelines, methods, and mitigation strategies that minimize negative consequences for bats while maximizing renewable energy production.

LITERATURE CITED

Ahlén, I., L. Bach, H. J. Baagøe, and J. Pettersson. 2007. Bats and offshore wind turbines studied in southern Scandinavia. Report 5571:1–35. Stockholm: Swedish Environmental Protection Agency.

Altringham, J. D. 2011. Bats: From evolution to conservation. 2nd ed. Oxford: Oxford University Press.

Arnett, E. B. 2017. Mitigating bat collision. In Wildlife and Wind Farms, Conflicts and Solutions, Volume 2, Onshore: Monitoring and Mitigation, edited by M. Perrow, 167–184. Exeter, UK: Pelagic Publishing.

Arnett, E. B., and E. F. Baerwald. 2013. Impacts of wind energy development on bats: Implications for conservation. In Bat Evolution, Ecology and Conservation, edited by R. A. Adams and S. C. Pedersen, 435–456. New York: Springer.

Arnett, E. B., K. Brown, W. P. Erickson, J. K. Fiedler, B. L. Hamilton, T. H. Henry, A. Jain, et al. 2008. Patterns of bat fatalities at wind energy facilities in North America. Journal of Wildlife Management 72:61–78.

Arnett, E. B., M. M. P. Huso, M. R. Schirmacher, and J. P. Hayes. 2011. Altering wind turbine speed reduces bat mortality at wind-energy facilities. Frontiers in Ecology and the Environment 9:209–214.

Arnett, E. B., G. D. Johnson, W. P. Erickson, and C. D. Hein. 2013a. A synthesis of operational mitigation studies to reduce bat fatalities at wind energy facilities in North America. Unpublished report submitted to the National Renewable Energy Laboratory, Austin, Texas, by Bat Conservation International. http://www.batsandwind .org/pdf/Pre-Post-construction-Synthesis_FINAL REPORT.pdf.

Arnett, E. B., C. D. Hein, M. R. Schirmacher, M. M. P. Huso, and J. M. Szewczak. 2013b. Evaluating the effectiveness of an ultrasonic acoustic deterrent for reducing bat fatalities at wind turbines. PLOS One 6:e65794. https://doi.org/10.1371/journal.pone.0065794.

Arnett, E. B., R. M. R. Barclay, and C. D. Hein. 2013c. Thresholds for bats killed by wind turbines. Frontiers in Ecology and the Environment 11:171.

AWEA (American Wind Energy Association). 2017. U.S. wind industry first quarter 2018 market report: Public version. http://www.awea.org/1q2017. Accessed February 1, 2018.

Baerwald, E. F. 2008. Variation in the activity and fatality of migratory bats at wind energy facilities in southern Alberta: Causes and consequences. Thesis, University of Calgary, Alberta, Canada.

Baerwald, E. F., and R. M. R. Barclay. 2009. Geographic variation in activity and fatality of migratory bats at wind energy facilities. Journal of Mammalogy 90:1341–1349.

Baerwald, E. F., G. H. D'Amours, B. J. Klug, and R. M. R. Barclay. 2008. Barotrauma is a significant cause of bat fatalities at wind turbines. Current Biology 18:R695–R696.

Baerwald, E. F., J. Edworthy, M. Holder M, and R. M. R. Barclay. 2009. A large-scale mitigation experiment to reduce bat fatalities at wind energy facilities. Journal of Wildlife Management 73:1077–1081.

Baranowski, R., F. Oteri, I. Baring-Gould, and S. Tegen. 2017. 2016 state of wind development in the United States by region. Technical Report NREL/TP-5000-67624. Golden, CO: National Renewable Energy Laboratory.

Barclay, R. M. R., and L. D. Harder. 2003. Life histories of bats: Life in the slow lane. In Bat Ecology, edited by T. H. Kunz and M. B. Fenton, 209–256. Chicago: University of Chicago Press.

Barclay, R. M. R., E. F. Baerwald, and J. C. Gruver. 2007. Variation in bat and bird fatalities at wind energy facilities: Assessing the effects of rotor size and tower height. Canadian Journal of Zoology 85:381–387.

Barclay, R. M. R., E. F. Baerwald, and J. Rydell. 2017. Bats. In Wildlife and Wind Farms, Conflicts and Solutions, Volume 1, Onshore: Potential Effects, edited by M. Perrow, 191–221. Exeter, UK: Pelagic Publishing.

Bennett, V. J., and A. M. Hale. 2014. Red aviation lights on wind turbines do not increase bat-turbine collisions. Animal Conservation 17:354–358.

Bennett, V. J., A. M. Hale, and D. A. Williams. 2017. When the excrement hits the fan: Fecal surveys reveal species-specific bat activity at wind turbines. Mammalian Biology 87:125–129.

Bienz, C. R. 2016. Surface texture discrimination by wild-caught bats: Implications for reducing mortality at wind turbines. Thesis, Texas Christian University, Fort Worth.

Bispo, R., J. Bernardino, T. Marques, and D. Pestana. 2013. Modeling carcass removal time for avian mortality assessments in wind farms using survival analysis. Environmental and Ecological Statistics 20:147–165.

CanWEA (Canadian Wind Energy Association). 2016. Powering Canada's future. http://canwea.ca/wind-energy/installed-capacity/. Accessed May 10, 2017.

Castle, K. T., T. J. Weller, P. M. Cryan, C. D. Hein, and M. R. Schirmacher. 2015. Using sutures to attach miniature tracking tags to small bats for multimonth movement and behavioral studies. Ecology and Evolution 5:2980–2989.

Collins, J., and G. Jones. 2009. Differences in bat activity in relation to bat detector height: Implications for bat surveys at proposed wind farm sites. Acta Chiropterologica 11:343–350.

Cryan, P. M. 2003. Seasonal distribution of migratory tree bats (Lasiurus and Lasionycteris) in North America. Journal of Mammalogy 84:579–593.

Cryan, P. M. 2008. Mating behavior as a possible cause of bat fatalities at wind turbines. Journal of Wildlife Management 72:845–849.

Cryan, P. M., and A. C. Brown. 2007. Migration of bats past a remote island offers clues toward the problem of bat fatalities at wind turbines. Biological Conservation 139:1–11.

Cryan, P. M., and R. M. R Barclay. 2009. Causes of bat fatalities at wind turbines: Hypotheses and predictions. Journal of Mammalogy 90:1330–1340.

Cryan, P. M., C. A. Stricker, and M. B. Wunder. 2014a. Continental-scale, seasonal movements of a heterothermic migratory tree bat. Ecological Applications 24:602–616.

Cryan, P. M., P. M. Gorresen, C. D. Hein, M. R. Schirmacher, R. H. Diehl, M. M. Huso, D. T. S. Hayman, et al. 2014b. Behavior of bats at wind turbines. Proceedings of the National Academy of Sciences USA 111:15126–15131.

Erickson, R. A., W. E. Thogmartin, J. E. Diffendorfer, R. E. Russell, and J. A. Szymanski. 2016. Effects of wind energy generation and white-nose syndrome on the viability of the Indiana bat. PeerJ 4:e2830. https://doi.org/10.7717/peerj.2830.

Foo, C. F., V. J. Bennett, A. M. Hale, J. M. Korstian, A. J. Schildt, and D. A. Williams. 2017. Increasing evidence that bats actively forage at wind turbines. PeerJ 5:e3985. http://doi.org/10.7717/peerj.3985.

Ford, W. M., E. R. Britzke, C. A. Dobony, J. L. Rodrigue, and J. B. Johnson. 2011. Patterns of acoustical activity of bats prior to and following white-nose syndrome occurrence. Journal of Fish and Wildlife Management 2:125–134.

Francl, K. E., W. M. Ford, D. W. Sparks, and V. Brack Jr. 2012. Capture and reproductive trends in summer bat communities in West Virginia: Assessing the impact of white-nose syndrome. Journal of Fish and Wildlife Management 3:33–42.

Frankham, R., J. D. Ballou, and D. A. Briscoe. 2009. Introduction to conservation genetics. 2nd ed. New York: Cambridge University Press.

Frick, W. F., E. F. Baerwald, J. F. Pollock, R. M. R. Barclay, J. A. Szymanski, T. J. Weller, A. L. Russell, S. C. Loeb, R. A. Medellin, and L. P. McGuire. 2017. Fatalities at wind turbines may threaten population viability of a migratory bat. Biological Conservation 209:172–177.

Good, R. E., A. Merrill, S. Simon, K. Murray, and K. Bay. 2012. Bat monitoring at the Fowler Ridge Wind Farm, Benton County, Indiana, April 1–October 31, 2011. Unpublished report prepared for the Fowler Ridge Wind Farm. Bloomington, IN: Western EcoSystems Technology.

Gilman, P., B. Maurer, L. Feinberg, A. Duerr, L. Peterson, W. Musial, P. Beiter, et al. 2016. National offshore wind strategy: Facilitating the development of the offshore wind industry in the United States. Technical Report DOE/GO-102016-4866. https://energy.gov/sites/prod /files/2016/09/f33/National-Offshore-Wind-Strategy -report-09082016.pdf.

Gorresen, P. M., P. M. Cryan, D. C. Dalton, S. Wolf, J. A. Johnson, C. M. Todd, and F. J. Bonaccorso. 2015. Dim ultraviolet light as a means of deterring activity by the Hawaiian hoary bat Lasiurus cinereus semotus. Endangered Species Research 28:249–257.

Greif, S., and B. M. Siemers. 2010. Innate recognition of water bodies in echolocating bats. Nature Communications 1:107. doi:10.1038/ncomms1110.

Griffiths, S. R. 2013. Echolocating bats emit terminal phase buzz calls when drinking on the wing. Behavioural Processes 98:58–60.

Grodsky, S. M., M. J. Behr, A. Gendler, D. Drake, B. D. Dieterle, R. J. Rudd, and N. L. Walrath. 2011. Investigating the causes of death for wind turbine-associated bat fatalities. Journal of Mammalogy 92:917–925.

Grodsky, S. M., C. S. Jennelle, D. Drake, and T. Virzi. 2012. Bat mortality at a wind-energy facility in southeastern Wisconsin. Wildlife Society Bulletin 36:773–783.

Hayes, M. A. 2013. Bats killed in large numbers at United States wind energy facilities. BioScience 63:975–979.

Hayes, M. A., P. M. Cryan, and M. B. Wunder. 2015. Seasonally-dynamic presence-only species distribution models for a cryptic migratory bat impacted by wind energy development. PLOS One 7:e0132599. https://doi .org/10.1371/journal.pone.0132599.

Hein, C. D. 2017. Monitoring bats. In Wildlife and Wind Farms, Conflicts and Solutions, Volume 2, Onshore: Monitoring and Mitigation, edited by M. Perrow, 31–57. Exeter, UK: Pelagic Publishing.

Hein, C. D., and M. R. Schirmacher. 2016. Impact of wind energy on bats: A summary of our current knowledge. Human-Wildlife Interactions 10:19–27.

Hein, C. D., J. Gruver, and E. B. Arnett. 2013. Relating pre-construction bat activity and post-construction bat

fatality to predict risk at wind energy facilities: A synthesis. Unpublished report submitted to the National Renewable Energy Laboratory, Austin, Texas, by Bat Conservation International. http://www.batsandwind .org/pdf/Pre-Post-construction-Synthesis_FINAL REPORT.pdf.

Hobson, K. A. 2008. Applying isotopic methods to tracking animal movements. In Tracking Animal Migration with Stable Isotopes, edited by K. A. Hobson and L. I. Wassenaar, 45–78. Boston, MA: Elsevier.

Horn, J. W., E. B. Arnett, and T. H. Kunz. 2008. Behavioral responses of bats to operating wind turbines. Journal of Wildlife Management 72:123–132.

Huso, M. M. P., and D. Dalthorp. 2014. A comment on "bats killed in large numbers at United States wind energy facilities." BioScience 64:546–547.

Huso, M. M. P., N. Som, and L. Ladd. 2012. Fatality estimator user's guide. Version 1.1, December 2015. U.S. Geological Survey Data Series 729. http://pubs.usgs.gov /ds/729/. Accessed December 15, 2017.

Jones, K. E., A. Purvis, and J. L. Gittleman. 2003. Biological correlates of extinction risk in bats. American Naturalist 161:601–614.

Kerns, J., and P. Kerlinger. 2004. A study of bird and bat collisions fatalities at the Mountaineer Wind Energy Center, Tucker County, West Virginia. Annual report for 2003. Unpublished report submitted to FPL Energy and Mountaineer Wind Energy Center Technical Review Committee. McLean, VA: Curry and Kerlinger. http:// www.batsandwind.org/pdf/BWEC%20BIBLIOGRAPHY _updated%202016.pdf.

Korner-Nievergelt, F., P. Korner-Nievergelt, O. Behr, I. Niermann, R. Brinkmann, and B. Hellriegel. 2011. A new method to determine bird and bat fatality at wind energy turbines from carcass searches. Wildlife Biology 17:350–363.

Korner-Nievergelt, F., R. Brinkmann, I. Niermann, and O. Behr. 2013. Estimating bat and bird mortality occurring at wind energy turbines from covariates and carcass searches using mixture models. PLOS One 7:e67997. https://doi.org/10.1371/journal.pone .0067997.

Korstian, J. M., A. M. Hale, V. J. Bennett, and D. A. Williams. 2013. Advances in sex determination in bats and its utility in wind-wildlife studies. Molecular Ecology Resources 13:776–780.

Korstian, J. M., A. M. Hale, and D. A. Williams. 2015. High genetic diversity, large historic population size, and lack of population structure in two North American tree bats. Journal of Mammalogy 96:972–980.

Korstian, J. M., A. M. Hale, V. J. Bennett, and D. A. Williams. 2016. Using DNA barcoding to improve bat carcass identification at wind farms in the United States. Conservation Genetics Resources 8:27–34.

Krauel, J. J., L. P. McGuire, and J. G. Boyles. 2017. Testing traditional assumptions about regional migration in bats. Mammal Research. https://doi.org/10.1007/s13364-017-0346-9.

Kunz, T. H., E. B. Arnett, W. P. Erickson, A. R. Hoar, G. D. Johnson, R. P. Larkin, M. D. Strickland, R. W. Thresher, and M. D. Tuttle. 2007. Ecological impacts of wind energy development on bats: Questions, research needs, and hypotheses. Frontiers in Ecology and the Environment 5:315–324.

Lehnert, L. S., S. Kramer-Schadt, S. Schönborn, O. Lindecke, I. Niermann, and C. C. Voigt. 2014. Wind farm facilities in Germany kill noctule bats from near and far. PLOS One 8:e103106. https://doi.org/10.1371/journal.pone.0103106.

Limpens, H. J. G. A., and K. Kapteyn. 1991. Bats, their behaviour and linear landscape elements. Myotis 29:39–48.

Lindsey, C. T. 2017. Assessing changes in bat activity in response to an acoustic deterrent: Implications for decreasing bat fatalities at wind facilities. Thesis, Texas Christian University, Fort Worth.

Lintott, P. R., S. M. Richardson, D. J. Hosken, S. A. Fensome, and F. Mathews. 2016. Ecological impact assessments fail to reduce risk of bat casualties at wind farms. Current Biology 26: R1135–R1136.

Long, C. V., J. A. Flint, P. A. Lepper, and S. A. Dible. 2009. Wind turbines and bat mortality: Interactions of bat echolocation pulses with moving turbine rotor blades. Proceedings of the Institute of Acoustics 31:185–192.

Long, C. V., J. A. Flint, and P. A. Lepper. 2010. Wind turbines and bat mortality: Doppler shift profiles and ultrasonic bat-like pulse reflection from moving turbine blades. Journal of the Acoustical Society of America 128:2238–2245.

Long, C. V., J. A. Flint, and P. A. Lepper. 2011. Insect attraction to wind turbines: Does colour play a role? European Journal of Wildlife Research 57:323–331.

Martin, T. G., M. A. Burgman, F. Fidler, P. M. Kuhnert, S. Low-Choy, M. McBride, and K. Mengersen. 2012. Eliciting expert knowledge in conservation science. Conservation Biology 26:29–38.

Martin, C. M., E. B. Arnett, R. D. Stevens, and M. C. Wallace. 2017. Reducing bat fatalities at wind turbines while improving the economic efficiency of operational mitigation. Journal of Mammalogy 98:378–385.

McAlexander, A. M. 2013. Evidence that bats perceive wind turbine surfaces to be water. Thesis, Texas Christian University, Fort Worth.

McGuire, L. P., C. G. Guglielmo, S. A. Mackenzie, and P. D. Taylor. 2012. Migratory stopover in the long-distance migrant silver-haired bat, Lasionycteris noctivagans. Journal of Animal Ecology 81:377–385.

Melcón, M. L., A. Denzinger, and H.-U. Schnitzler. 2007. Aerial hawking and landing: Approach behaviour in Natterer's bats, Myotis nattereri (Kuhl 1818). Journal of Experimental Biology 210:4457–4464.

Northrup, J. M., and G. Wittemyer. 2013. Characterising the impacts of emerging energy development on wildlife, with an eye towards mitigation. Ecological Letters 16:112–125.

Ohio Department of Natural Resources. 2009. On-shore bird and bat pre- and post-construction monitoring protocol for commercial wind energy facilities in Ohio. http://wildlife.ohiodnr.gov/portals/wildlife/pdfs/species%20and%20habitats/windwildlifemonitoringprotocol.pdf.

Oteri, F., R. Baranowski, I. Baring-Gould, and S. Tegen. 2018. 2017 state of wind development in the United States by region. NREL/TP-5000-70738. Golden, CO: National Renewable Energy Laboratory. https://www.nrel.gov/docs/fy180sti/70738.pdf.

Pennsylvania Game Commission. 2013. Pennsylvania Game Commission Wind Energy Voluntary Cooperation Agreement Amendment I. http://www.pgc.pa.gov/InformationResources/AgencyBusinessCenter/WindEnergy/Documents/Amended%20Cooperative%20Agreement%20and%20Exhibits%20-%202013.pdf.

Peterson, T. S., S. K. Pelletier, S. A. Boyden, and K. S. Watrous. 2014. Offshore acoustic monitoring of bats in the Gulf of Maine. Northeastern Naturalist 21:86–107.

Petit, E., and F. Mayer. 2000. A population genetic analysis of migration: The case of the noctule bat (Nyctalus noctula). Molecular Ecology 9:683–690.

Piorkowski, M. D., and T. J. O'Connell. 2010. Spatial pattern of summer bat mortality from collisions with wind turbines in mixed-grass prairie. American Midland Naturalist 164:260–269.

Pylant, C. L., D. M. Nelson, M. C. Fitzpatrick, J. E. Gates, and S. R. Keller. 2016. Geographic origins and population genetics of bats killed at wind-energy facilities. Ecological Applications 26:1381–1395.

Reynolds, D. S. 2006. Monitoring the potential impact of a wind development site on bats in the northeast. Journal of Wildlife Management 70:1219–1227.

Rodhouse, T. J., P. C. Ormsbee, K. M. Irvine, L. A. Vierling, J. M. Szewczak, and K. T. Vierling. 2015. Establishing conservation baselines with dynamic distribution models for bat populations facing imminent decline. Diversity and Distributions 21:1401–1413.

Roeleke, M., T. Blohm, S. Kramer-Schadt, Y. Yovel, and C. C. Voigt. 2016. Habitat use of bats in relation to wind turbines revealed by GPS tracking. Scientific Reports 6:28961. https://doi:10.1038/srep28961.

Rollins, K. E., D. K. Meyerholz, G. D. Johnson, A. P. Capparella, and S. S. Loew. 2012. A forensic investigation into the etiology of bat mortality at a wind farm: Barotrauma or traumatic injury? Veterinary Pathology 49:362–371.

Rubenstein, D. R., and K. A. Hobson. 2004. From birds to butterflies: Animal movement patterns and stable isotopes. Trends in Ecology and Evolution 19:256–263.

Russo, D., L. Cistrone, and G. Jones. 2012. Sensory ecology of water detection by bats: A field experiment. PLOS One 10:e48144. https://doi.org/10.1371/journal.pone.0048144.

Rydell, J., W. Bodganowicz, A. Boonman, S. Pettersson, E. Suchecka, and J. J. Pomorski. 2016. Bats may eat diurnal flies that rest on wind turbines. Mammalian Biology 81:331–339.

Santos. H., L. Rodrigues, G. Jones, and H. Rebelo. 2013. Using species distribution modelling to predict bat fatality risk at wind farms. Biological Conservation 157:178–186.

Schorcht, W., F. Bontadina, and M. Schaub. 2009. Variation of adult survival drives population dynamics in a migrating forest bat. Journal of Animal Ecology 78:1182–1190.

Shoenfeld, P. 2004. Suggestions regarding avian mortality extrapolation. Unpublished report for the West Virginia Highlands Conservancy. https://www.nationalwind.org/wp-content/uploads/2013/05/Shoenfeld-2004-Suggestions-Regarding-Avian-Mortality-Extrapolation.pdf.

Siemers, B. M., E. Baur, and H-U. Schnitzler. 2005. Acoustic mirror effect increases prey detection distance in trawling bats. Naturwissenschaften 92:272–276. https://doi.org/10.1007/s00114-005-0622-4.

Smallwood, K. S. 2013. Comparing bird and bat fatality-rate estimates among North American wind-energy projects. Wildlife Society Bulletin 37:19–33.

Sovic, M. G., B. C. Carstens, and H. L. Gibbs. 2016. Genetic diversity in migratory bats: Results from RADseq data for three tree bat species at an Ohio windfarm. PeerJ 4:e1647. https://doi.org/10.7717/peerj.1647.

Stanton, T., and S. Poulton. 2012. Seasonal variation in bat activity in relation to detector height: A case study. Acta Chiropterologica 14:401–408.

Strickland, M. D., E. B. Arnett, W. P. Erickson, D. H. Johnson, G. D. Johnson, M. L. Morrison, J. A. Shaffer, and W. Warren-Hicks. 2011. Comprehensive guide to studying wind energy/wildlife interactions. Prepared for the National Wind Coordinating Collaborative, Washington, DC. https://www.nationalwind.org/wp-content/uploads/assets/publications/Comprehensive_Guide_to_Studying_Wind_Energy_Wildlife_Interactions_2011_Updated.pdf.

Suryan, R., R. Albertani, and B. Polagye. 2016. A synchronized sensor array for remote monitoring of avian and bat interactions with offshore renewable energy facilities. Report by Oregon State University and University of Washington. https://tethys.pnnl.gov/sites/default/files/publications/Suryan-et-al-2016.pdf.

Szewczak, J., and E. Arnett. 2007. Field test results of a potential acoustic deterrent to reduce bat mortality from wind turbines. Report submitted to the Bats and Wind Energy Cooperative and Bat Conservation International, Austin, Texas. https://tethys.pnnl.gov/publications/field-test-results-potential-acoustic-deterrent-reduce-bat-mortality-wind-turbines.

Thogmartin, W. E., C. A. Sanders-Reed, J. A. Szymanski, P. C. McKann, L. Pruitt, R. A. King, M. C. Runge, and R. E. Russell. 2013. White-nose syndrome is likely to extirpate the endangered Indiana bat over larger parts of its range. Biological Conservation 160:162–172.

US Department of Energy. 2015. Wind vision: A new era for wind power in the United States. USDOE Office of Scientific and Technical Information Report DOE/GO-102015-4557. Oak Ridge, TN.

USFWS (US Fish and Wildlife Service). 2011. Indian bat Section 7 and Section 10 guidance for wind energy projects, revised October 26, 2011. http://www.fws.gov/midwest/endangered/mammals/inba/pdf/inbaS7and10WindGuidanceFinal26Oct2011.pdf.

Valdez, E. W., and P. M. Cryan. 2013. Insect prey eaten by hoary bats (Lasiurus cinereus) prior to fatal collisions with wind turbines. Western North American Naturalist 73:516–524.

Verboom, B., and H. Huitema. 1997. The importance of linear landscape elements for the pipistrelle Pipistrellus pipistrellus and the serotine bat Eptesicus serotinus. Landscape Ecology 12:117–125.

Voigt, C. C., O. Lindecke, S. Schönborn, S. Kramer-Schadt, and D. Lehmann. 2016. Habitat use of migratory bats killed during autumn at wind turbines. Ecological Applications 26:771–783.

Vonhof, M. J., and A. L. Russell. 2015. Genetic approaches to the conservation of migratory bats: A study of the eastern red bat (Lasiurus borealis). PeerJ 3:e983. http://doi.org/10.7717/peerj.983.

Weller, T. J., and J. A. Baldwin. 2012. Using echolocation monitoring to model bat occupancy and inform mitigations at wind energy facilities. Journal of Wildlife Management 76:619–631.

Weller, T. J., K. T. Castle, F. Liechti, C. D. Hein, M. R. Schirmacher, and P. M. Cryan. 2016. First direct evidence of long-distance seasonal movements and hibernation in a migratory bat. Scientific Reports 6:34585. https://doi.org/10.1038/srep34585.

Winhold, L., A. Kurta, and R. Foster. 2008. Long-term change in an assemblage of North American bats: Are eastern red bats declining? Acta Chiropterologica 10:359–366.

Wolpert, R. 2013. A partially periodic equation for estimating avian mortality rates. In Appendix B in Improving Methods for Estimating Fatality of Birds and Bats at Wind Energy Facilities, edited by W. Warren-Hicks. Berkeley: California Wind Energy Association.

Wolpert, R., W. Warren-Hicks, B. Karas, L. Tran, and J. Newman. 2012. Improving methods for estimating fatality of birds and bats at wind energy facilities: Evaluation of accuracy of existing equations, including assumptions and statistical bias. National Wind Coordinating Collaborative Wind and Wildlife Research Meeting 9, Broomfield, CO.

World Bank Group. 2015. Environmental, health, and safety guidelines for wind energy (English). Washington, DC. http://documents.worldbank.org/curated/en/498831479463882556/Environmental-health-and-safety-guidelines-for-wind-energy.

Zayas, J., M. Derby, P. Gilman, S. Ananthan, E. Lantz, J. Cotrell, and R. Tusin. 2015. Enabling wind power nationwide. DOE/EE-1218. Washington, DC: US Department of Energy. http://www.energy.gov/sites/prod/files/2015/05/f22/Enabling-Wind-Power-Nationwide_18MAY2015_FINAL.pdf.

7 — Effects of Wind Energy on Wildlife

NICOLE M. KORFANTA AND
VICTORIA H. ZERO

Emerging Issues and Underrepresented Taxa

Introduction

Bird and bat carcasses beneath wind turbines were clear and early indications of impacts of wind energy production on wildlife, generating substantial public concern and scientific interest in volant species. For birds in particular, understanding effects of wind energy on wildlife and how to mitigate them was further spurred by key policies, including the Endangered Species Act of 1973 (16 US Code Chapter 35), the Bald and Golden Eagle Protection Act of 1940 (16 US Code 668-668c), and the Migratory Bird Treaty Act of 1918 (16 US Code 703-712). Together, legal protections for birds and growing conservation concern for bats have helped motivate a relatively robust and growing literature on wind energy impacts. Chapter 5 (Dohm and Drake) and Chapter 6 (Hein and Hale) provide comprehensive coverage of wind energy effects on birds and bats, respectively.

For other taxonomic groups, research on wind energy effects is in its infancy. For example, many ungulate species are of cultural and economic importance to human communities, yet we know comparatively little about how these animals respond to wind energy development relative to what we know about some bird and bat species. Although raptors and other birds of prey have received the most

attention, birds that do not fly into turbine blades may nonetheless be affected by towering infrastructure, roads, and increased human activity associated with wind development. Many prairie-chicken (*Tympanuchus* spp.) and sage-grouse (*Centrocercus* spp.) populations (hereafter, collectively, "prairie grouse") inhabit the same windy plains where wind turbines are popping up; yet, studies of wind energy impacts on declining prairie grouse species are few. And insects, generally of little interest to the public and wildlife scientists alike (e.g., Grodsky et al. 2015), surely interact with wind facilities in interesting ways, yet our knowledge about wind energy–insect interactions is minimal. A study showed insect carcasses that build up on wind turbine blades can reduce wind power production by 50% (Corten and Veldkamp 2001). That is a problem for wind industry efficiency, but is it a problem for insect populations? Research to answer that question is just beginning.

Various taxonomic groups—as disparate as elk (*Cervus elaphus*) and bumble bees (*Bombus* spp.)— are emerging as contemporary subjects of regulatory and policy concern, stimulating new research on species not typically associated with wind energy impacts. For example, a growing appreciation for the migration routes that connect summer and winter ranges of mule deer (*Odocoileus hemionus*), pronghorn (*Antilocapra americana*), and other ungulate species has led several states to implement migration corridor policies. These new policies may help shape siting and mitigation strategies in places where wind development overlaps with critical ungulate habitat, including migration routes. Recent petitions for endangered species protection of insects (e.g., rusty patched bumble bee, *Bombus affinis*) also raised concerns regarding siting of wind facilities in imperiled insect habitat. And of course, the specter of endangered species listing for the greater sage-grouse (*Centrocercus urophasianus*) continues to shape industrial wind development in the western United States. Some states already have created "core-area" policies that limit wind and other energy development in key sage-grouse habitat. Each of these policy priorities demands that we better understand how these various taxonomic groups respond to wind energy development.

This chapter explores the early science of the effects of terrestrial wind development on species for which there are emerging policy and regulatory drivers of conservation interest. We build on existing summaries of wind energy effects on wildlife (e.g., Arnett et al. 2007, National Research Council 2007, Northrup and Wittemyer 2013, Schuster et al. 2015) and update the state of knowledge of the effects of wind development on previously underrepresented wildlife. Based on limited peer-reviewed literature, and while acknowledging scientific uncertainty, we explore wind energy–wildlife interactions and mitigation options for ungulates, prairie grouse, insects, and other little-studied taxonomic groups. We conclude with ideas about the broader conservation context by weighing local effects of wind energy on wildlife against the benefits of renewable energy for mitigating climate change impacts to wildlife.

Ungulates
Emerging Issues

The evolving policy and regulatory landscape of the western United States is prompting new interest in wind energy impacts on ungulates, even in the absence of endangered species considerations that have compelled interest in other taxonomic groups (Arnett et al. 2007). Early wind development in the western United States occurred mostly on private lands with fewer regulatory hurdles than would be associated with development on public lands (AWEA 2017). However, to meet the ambitious renewable energy goals of many western states, wind development is quickly expanding into the public lands of the Interior West (Adams 2006); these lands support some of the largest remaining ungulate populations on Earth. Concurrently, regional declines in mule deer and local declines in other ungulate populations (summarized by Johnson et al. 2016) have spurred more interest in how wind energy development contributes to landscape-scale habitat changes and pressures experienced by ungulate species (Nellemann et al. 2003, Leu et al. 2008, Johnson et al. 2016). Additionally, a growing appreciation for migration and its role in maintaining robust ungulate populations has led states to expand "crucial range" definitions beyond winter and parturition areas to include migration corridors (e.g., Wyoming's Ungulate Migration Corridor Strategy; Wyoming Executive Order 2015-4). Each of these factors—more wind development on federal lands, declining ungulate populations, and greater attention to ungulate migrations—requires developers and managers to better understand and anticipate the effects of wind energy across a suite of ungulate species with year-round habitat needs.

Potential Impacts
Habitat Loss

Direct habitat loss from wind energy development occurs when land and vegetation cover are replaced by infrastructure, including pads, roads, transmission lines, and associated buildings. Future wind energy development, estimated from existing wind energy leases and the Western Governors' Association's renewable energy zones, is predicted to affect shrublands that provide habitat for elk, mule deer, and pronghorn more than any other ecosystem type in the western United States (Pocewicz et al. 2011). Shrub communities are also notoriously difficult to recover in semiarid and arid environments, making wind development impacts on vegetation long lasting (Mason et al. 2011, Korfanta et al. 2015, Minnick and Alward 2015).

Indirect habitat loss—also known as displacement—can happen when animals avoid or spend less time in habitat adjacent to wind energy infrastructure or activity. Indirect habitat loss through behavioral avoidance can be larger than direct habitat loss and, together, they can reduce carrying capacity for ungulate populations unless impacts are mitigated (Sawyer et al. 2006, 2009, Northrup et al. 2015). For instance, over 50% of mule deer winter range in the Piceance Basin of Colorado was lost to displacement and direct habitat loss related to oil and gas development (Northrup et al. 2015). Yet, avoidance behavior in response to energy infrastructure and disturbance can be highly variable across species. For example, mule deer have shown greater avoidance (Sawyer et al. 2006) than pronghorn (Beckmann et al. 2012) to oil and gas development. Moreover, responses of individuals within a population vary from year to year (Sawyer et al. 2017) and even within a day (Northrup et al. 2015). Topographic and vegetative diversity provide refugia and can ameliorate ungulate response to disturbance from energy development (Lendrum et al. 2012, Buchanan et al. 2014, Northrup et al. 2015). In general, ungulate avoidance levels are greater when (1) human activity or disturbance is elevated, for example, during the construction phase (Sawyer et al. 2009) or during daytime operations (Dzialak et al. 2011, Lendrum et al. 2012, Northrup et al. 2015); (2) individuals are in good body condition, for example, during mild winters (e.g., Sawyer et al. 2017); and (3) individuals occupy open areas with little topographic relief (e.g., Northrup et al. 2015, Sawyer et al. 2017). Conversely, avoidance levels tend to be attenuated at lower levels of activity, during harsh winters (Saw-

yer et al. 2017), and in regions with rugged topography and more forest cover (Northrup et al. 2015). Although it is commonly assumed that ungulates habituate to development, a 17-year study of mule deer and gas development revealed animals did not habituate and avoidance impacts were long term (Sawyer et al. 2017).

Migratory Connectivity

Most ungulates migrate, sometimes great distances, between seasonal ranges, which necessitates analysis of wind energy impacts across the annual movement cycle of ungulate species. Winter ranges often are spatially limited by geographic features or forage availability. As a result, habitat loss in winter ranges can reduce carrying capacity for ungulates, and disturbance can exacerbate already energetically stressful conditions. Often ignored, summer range is important to the year-round nutritional condition of ungulates and can be more important than winter range in predicting overwinter survival (Cook et al. 2004, 2016, Tollefson et al. 2010, Monteith et al. 2013).

Only recently have researchers begun to unravel the biological and nutritional importance of migration routes that connect seasonal ranges. Migratory animals far outnumber nonmigratory residents (Fryxell and Sinclair 1988), whether wildebeest (*Conochaetes* spp.) in the Serengeti, caribou (*Rangifer tarandus*) in the Arctic, or pronghorn in Wyoming. Spring migrations represent a nutritional opportunity to "surf the green wave" of emerging vegetation along latitudinal or elevational gradients, allowing animals to recuperate from the nutritional deficits of winter (Merkle et al. 2016, Aikens et al. 2017), while autumn migrations reconnect animals with their winter ranges. Given that some migratory populations spend more than 100 days of the year migrating, managers are beginning to view migration corridors as critical seasonal range that complements winter and summer ranges (Sawyer et al. 2016).

Recognizing that migration promotes individual fitness and, in turn, population abundance, wildlife managers and stakeholders are motivated to ensure that migration routes remain connected and functional in the face of expanding wind energy development. At certain densities, wind energy infrastructure along ungulate migration routes could increase movement rates and reduce use of stopovers, where animals linger to gain energy reserves (Sawyer and Kauffman 2011, Lendrum et al. 2012, Sawyer et al. 2013, Skarin et al. 2015). We need a better understanding of how much energy development like wind energy migratory herds can withstand before routes are severed or movement rates increase such that nutritional benefits are lost and ungulate fitness is diminished (Sawyer et al. 2013).

Direct Mortality

Although direct and indirect habitat loss are primary considerations for wind energy impacts on ungulates, direct sources of mortality are worth noting. Energy infrastructure of all kinds creates road networks that can increase wildlife-vehicle collisions (Forman and Alexander 1998, Dzialak et al. 2011). Fences, including those bordering road rights-of-way or property lines, are well-documented sources of direct mortality for mule deer and, especially, pronghorn; pronghorn are unable to traverse fence lines unless they can crawl under them (Harrington and Conover 2016).

Cumulative Impacts

Given the rapidly growing human footprint, especially in landscapes of the western United States (Leu et al. 2008), any negative effects that ungulate populations experience from wind development are additive to the impacts they experience from other land-use changes and disturbances. As such, wind energy development should be considered in the context of cumulative impacts (Arnett and May 2016), requiring a perspective that goes beyond individual wind project boundaries. Such cumulative impact analyses are required by the National Environmental Policy Act of 1969 (42 US Code §4321 et seq.) in the form of environmental impact assessments for projects on federal lands or those that use federal funds (see Chapter 10, Arnett). More meaningful and substantive cumulative impact analyses will be increasingly important as residential development, multiple forms of energy development, climate change, and other major landscape drivers affect ungulates.

Several studies have demonstrated cumulative impacts of multiple anthropogenic landscape changes on ungulates. In Colorado, for instance, Johnson et al. (2016) showed that energy development's role in reduced recruitment of mule deer on their winter range was additive to the substantial impacts of rural residential development. In Norway, Nellemann et al. (2003) concluded that project-level impact analysis failed to address the cumulative impacts of progressive industrial development (e.g., roads, reservoirs, powerlines) that led to much-reduced reproductive success and permanently altered habitat use in reindeer (*Rangifer tarandus*). Such impacts can be easy to miss at the small spatial scales typical of site-level analyses (Vistnes and Nellemann 2008, Skarin et al. 2015). Added to multiple forms of industrial and residential development, cumulative impacts include those from habitat alterations caused by climate change, disease, and hunting pressure, among others (Dzialak et al. 2011, Christie et al. 2015).

Observed Impacts

The state of the science on wind energy impacts to ungulates is weak, though a few studies have examined wind energy impacts on North American ungulates (Arnett et al. 2007, Johnson and Stephens 2011, Northrup and Wittemyer 2013; Table 7.1). Notably absent from the literature are studies of wind energy impacts to mule deer, which have experienced population declines in many places (summarized by Johnson et al. 2016) and are known to avoid energy infrastructure (Northrup et al. 2015, Sawyer et al. 2017); mule deer have substantial range over-

Table 7.1. Peer-reviewed studies of wind energy effects on North American ungulate species

Species	Study	Processes Examined	Impacts	Study Design	Study Duration	Location	Number of Turbines	Sample Size
Reindeer	Colman et al. 2012	Abundance, movement across wind facility	No impact	Control-treatment, direct observations mapped to GIS	1–5 monthly observations/year for 5 years (1 year construction)	Northern Norway	17 (2.3 MW)	601–2,247 observations/year
	Colman et al. 2013	Abundance, avoidance of infrastructure	Possible road avoidance during construction, no avoidance of wind facility, higher density of reindeer on treatment site (related to habitat quality)	Control-treatment, direct observation, fecal pellet transects	6 years fecal pellet transects (pre-, during, and post-construction), 5 years of observations (during and post-construction)	Northern Norway	17 (2.3 MW)	16 transects wind, 4 transects control
	Flydal et al. 2009	Area use, vigilance of penned reindeer	No impact	Control-treatment, behavioral observations of semidomestic reindeer in enclosures adjacent to turbine or no turbine	Two falls: first season a pilot study with experimental manipulation of one turbine	Mid-Norway	5	3–5 female reindeer
	Skarin et al. 2015	Corridor and stopover use, step length	Reduced use of movement and stopover habitat and increased step length during construction	Before-after, GPS collar data	2 years pre-construction, 2 years during construction	Sweden	2 wind facilities: 8 and 10 turbines	80 females
	Tsegaye et al. 2017	Population density, habitat use	Avoidance during wind farm construction in the calving season, no effect during operation	Before-after, direct observation and GPS collar data	9 years observational data (1×/month), 3 years GPS data, pre-, during, and post-construction	Vannøy Island, Norway	18	14 GPS-collared females
Elk	Walter et al. 2006	Home range, forage quality from pellets	No impact	Before-after, VHF collar data	2 years: 2 months before, 10 months during, 12 months after construction	Southwestern Oklahoma	45 (1.65 MW)	10 elk
Pronghorn	Taylor et al. 2016	Adult female survival	No impact of turbines, decreased winter survival with proximity to major roads	Distance to wind infrastructure, GPS collar data	3 winters (1 during construction)	South-central Wyoming	74 (1.5 MW)	47 adult females

lap with abundant wind resources in the western United States. Also missing are wind energy impact studies for white-tailed deer (*Odocoileus virginianus*), moose (*Alces alces*), and bighorn sheep (*Ovis canadensis*). Where multiple studies have examined energy impacts to an ungulate species, reported impacts are often inconsistent (Hebblewhite 2011).

Below, we summarize the few wind-specific impact studies completed for North American ungulates (or their European counterpart, in the case of caribou). In some cases, we provide examples of research that assessed other energy infrastructure impacts to ungulates, recognizing that some aspects of wind development, such as spinning turbine blades and a larger spatial footprint, are unique (Jones and Pejchar 2013).

Reindeer

A push for expanded wind energy development in northern Europe, including in reindeer habitat in Sweden and Norway, led to some of the first studies of wind energy impacts on ungulates. Although North American populations of reindeer, known as caribou, currently are not exposed to wind development, studies of wind energy impacts to European reindeer are worth considering, given the nascency of research on wind energy impacts to North American ungulates.

Reindeer appear to shift habitat use during the construction phase of new wind facilities and associated infrastructure. A 9-year study of a resident reindeer population on the Fakken peninsula of northern Norway found reindeer reduced their use of areas close to (< 500 m) turbines by 50% and close to (< 250 m) major roads by 75% (Tsegaye et al. 2017). The construction phase also was associated with more direct reindeer mortalities from vehicle collisions (Tsegaye et al. 2017). Likewise, reindeer were sensitive to activity during construction of a high-voltage power line in Norway; they showed a 10% reduction in habitat use within 6 km of construction activity during the calving season (Eftestøl et al. 2016). After calving, during summer and au-

tumn, reindeer showed a 12% to 13% reduction in habitat use within 3.5 km of construction activities.

Wind facilities also can alter reindeer migration patterns during the construction phase. During construction of 2 small wind facilities in Sweden, migratory reindeer reduced their use of migration corridors across the site by 76%, spending more time in a "holding pattern" along the edges of the wind facilities (Skarin et al. 2015). Concurrently, reindeer increased step length within 5 km of the wind facility, indicating a faster rate of movement consistent with less time spent in stopover areas necessary to fulfill nutritional requirements during transitional seasons (Skarin et al. 2015).

Unlike displacement observed during construction of wind facilities, reindeer habitat use does not appear to be affected during the operational phase of wind facilities (Eftestøl et al. 2016, Tsegaye et al. 2017). Likewise, relative abundance did not decline for reindeer occupying habitat near operating wind facilities (Colman et al. 2013), and associated roads and power lines did not impede reindeer movements (Colman et al. 2012; but see Skarin et al. 2015). These results are consistent with observations of penned, semidomesticated reindeer that showed no significant changes in vigilance or restlessness when an adjacent 39 m-tall turbine was operational (Flydal et al. 2009).

Elk

Only 1 study has examined elk response to wind energy development. Walter et al. (2006) used movement data collected from VHF collars to examine home-range size, effects of distance from turbines, and diet quality of individuals in a nonmigratory and unhunted elk herd before, during, and after construction of a 45-turbine wind facility in Oklahoma. Home range size did not change, and nitrogen and carbon isotope analysis of fecal pellets showed no difference in dietary quality after construction of the wind facility. Two metrics of avoidance showed inconsistent results, although the authors concluded that elk had shifted the center of their home ranges away from the facility following construction. The study was complicated by small sample size, short time frame, small geographic scope, and, perhaps most importantly, a nearby crop of winter wheat that augmented native forage and likely drove elk habitat selection.

Pronghorn

Initial studies on wind energy impacts to pronghorn have reported that mortality and abundance were not affected by nearby wind facilities. One study of pronghorn and wind energy development tracked winter mortality factors for 47 female pronghorn in southeastern Wyoming (Taylor et al. 2016). The study spanned 3 winters, with the first winter including construction of infrastructure, and the second and third winters including the operation phase. The authors showed that pronghorn mortality risk did not vary with proximity to the 74-turbine wind facility. Also in Wyoming, Johnson and Stephens (2011) recorded similar numbers of pronghorn during surveys before and after construction of a wind facility, indicating no reduction in habitat use. Similarly, 2 studies have shown pronghorn do not avoid oil and gas well pads within otherwise high-value habitat (Beckmann et al. 2012, Christie et al. 2017).

Unlike their behavior with well pads and wind turbines, pronghorn avoid and are killed by roads and fences, which also are part of the infrastructure of renewable and conventional energy development. In a North Dakota oil field, pronghorn abundance was negatively correlated with road density (Christie et al. 2015), and in a Wyoming wind development study, pronghorn mortality risk decreased by 20% for every 1 km distance from major roads present before wind development (Taylor et al. 2016). These results are consistent with a study showing higher vigilance and lower foraging times for pronghorn near high-traffic roads, particularly in spring (Gavin and Komers 2006). Fences, including right-of-way fences alongside roads, cause direct mortality (e.g., Harrington and Conover 2016) and impede movement of pronghorn (e.g., Gates et al. 2012).

Prairie-Chickens and Sage-Grouse
Emerging Issues

Prairie grouse population declines and growing anthropogenic pressures have prompted multiple petitions and listings for Endangered Species Act protection of prairie grouse species. The US Fish and Wildlife Service (USFWS) is currently reviewing the lesser prairie-chicken (*Tympanuchus pallidicinctus*) for listing under the Endangered Species Act and, after determining in 2015 a listing for greater sage-grouse was not warranted, will review the species' status again in 2020. Prairie grouse species already listed include the endangered Attwater's prairie-chicken (*Tympanuchus cupido attwateri*) and the threatened Gunnison sage-grouse (*Centrocercus minimus*). Declining populations, shrinking habitat extent, sensitivity to anthropogenic disturbance, and current or potential federal protections for each of these species warrant a review of potential wind energy impacts.

Potential Impacts
Direct Mortality

Unlike many other bird species, collision with wind turbine blades is unlikely for low-flying prairie grouse (Kuvlesky et al. 2007). Although some individual prairie grouse may be killed by wind turbine towers and other infrastructure, collisions are unlikely to have population-level impacts on prairie grouse populations (Johnson and Holloran 2010).

Habitat Loss

Habitat loss—both direct (i.e., destruction) and indirect (e.g., behavioral avoidance of energy infrastructure)—is the primary concern for wind energy impacts on prairie grouse. Turbine pads, roads, and transmission lines cause direct habitat loss, fragment remaining habitat, and may promote the spread of non-native vegetation (Kuvlesky et al. 2007). Habitat loss and fragmentation from wind energy is additive to other sources of habitat loss that have resulted in widespread population declines for many prairie grouse species. For instance, sagebrush ecosystems that provide habitat for greater sage-grouse and Gunnison sage-grouse, and for some Columbian sharp-tailed grouse (*Tympanuchus phasianellus columbianus*) populations, have been reduced dramatically in the last century (Braun et al. 2002). Much of what remains has been degraded and fragmented through conventional energy and residential development, agricultural cultivation, changes in fire regime, cheatgrass (*Bromus tectorum*) or other weed invasions, and climate change (Braun et al. 2002, Connelly et al. 2004, Miller et al. 2011). Similar factors, plus changes in grazing regimes, have caused widespread loss and fragmentation of the short, mixed, and tallgrass prairies relied on by prairie-chicken species (Samson and Knopf 1994, Samson et al. 2004, Balch et al. 2013).

Direct loss of native prairie and sagebrush ecosystems may be exacerbated by indirect habitat loss represented by prairie grouse avoidance of disturbance and infrastructure associated with energy development (e.g., Robel et al. 2004, Pitman et al. 2005, Holloran et al. 2010), which has led some researchers to suspect that wind energy facilities could displace prairie grouse species (National Research Council 2007, Pruett et al. 2009a, 2009 b, Wyoming Game and Fish Department 2010). Tall structures associated with wind energy facilities may elicit avoidance by prairie grouse because such structures are novel to these species, which evolved in a treeless environment (Manville 2004) or because tall structures are associated with perching raptor species that hunt grouse (Pruett et al. 2009b). Regardless of the mechanism, greater and lesser prairie-chickens avoided powerlines in a study in Kansas, so they may avoid wind turbines as well (Pruett et al. 2009b). Concern for wind energy impacts on greater sage-grouse stems from studies showing the species' sensitivity to oil and gas infrastructure, including reductions in lek attendance by males (Doherty et al. 2010), brood success (Aldridge and Boyce 2007), nest initiation (Lyon and Anderson 2003), survival (Holloran et al. 2010), and habitat use near infrastructure (Doherty et al. 2008). Several studies have demonstrated a lag time between oil and gas development and decreased male lek attendance by greater sage-grouse, suggesting that short-term studies could miss some impacts of wind development on this and other prairie grouse species (Holloran 2005, Walker et al. 2007, Harju et al. 2010).

Disturbance

Most prairie grouse are lek-mating species; the males use booming vocalizations to communicate with females and other males during courtship displays (Sparling 1981, Blickley and Patricelli 2012). Noise associated with industrial development potentially could interfere with courtship communications of prairie grouse (Smith et al. 2016a). For instance, experimental evidence has shown decreased male lek attendance when greater sage-grouse were exposed to sounds from natural gas development (Blickley and Patricelli 2012). Lek attendance also decreased with exposure to intermittent sounds (e.g., traffic) more so than from continuous sounds (e.g., drilling noise), and males were more affected by noise than females (Blickley et al. 2012). Wind developments generate noise during the construction phase, from associated roads, and from the turbines themselves (Mockrin and Gravenmier 2012), which could interfere with intraspecific communications in lek-mating species that rely on booming vocalizations (Smith et al. 2016a).

Moving turbine blades may also represent a disturbance to prairie grouse. Both shadow flicker and reflection of sunlight off blades can create a visual disturbance on the ground (National Research Council 2007, Lovich and Ennen 2013). In turn, shadow flicker and reflected sunlight off blades might be visually perceived as a threat by ground-dwelling birds, including prairie grouse (Wyoming Game and Fish Department 2010).

Predation

Prairie grouse are prey for a range of mammalian and avian predators, and predation accounts for the majority of mortalities of individuals in most populations (Connelly et al. 2000a, Schroeder and Baydack 2001, Winder et al. 2014a). Given the importance of female survival in determining the population growth rate of long-lived prairie grouse species (e.g., McNew et al. 2012, Taylor et al. 2012, Dahlgren et al. 2016), changes to predation that may result from wind energy development are an important consideration for population performance. The indirect effect of wind development on prairie grouse through changes to predator communities depends on whether predators increase or decrease in response to new development (see Mammalian Carnivores section below for more detail).

Cumulative Impacts

As with ungulates, impacts of wind energy development on prairie grouse should be considered in a broader context with other conservation threats. Energy development impacts are additive to other stressors, including disease (e.g., West Nile virus; Walker and Naugle 2011), habitat loss, predation, and hunting (Johnson and Holloran 2010), which can cumulatively reduce survival or reproduction of species (Aldridge and Boyce 2007). Particularly because prairie grouse species require a complex of connected habitat types for lekking, nesting, brood-rearing, summer range, and wintering range, wind energy impacts should be considered across the entire life cycle and range of these species (Crawford et al. 2004).

Observed Impacts

Effects of wind energy development on North American prairie grouse were reviewed by Johnson and Stevens (2011). Here, we focus on more recent literature on greater prairie-chicken and greater sage-grouse (Table 7.2). The studies we review below are restricted to a few, relatively small wind facilities, making generalizations to other systems and wind facilities a challenge. Research on other imperiled grouse species is ongoing, but to our knowledge, no studies of wind energy impacts to sharp-tailed grouse or lesser prairie-chickens have been published.

Greater Prairie-Chicken

Female greater prairie-chickens have shown inconsistent responses to wind energy facilities. Wind facilities have changed habitat selection by female greater prairie-chickens, but without any apparent demographic consequences. At a 67-turbine wind development in northcentral Kansas, researchers collected 2 years of preconstruction data, followed by three years of post-construction data (McNew et al. 2014, Winder et al. 2014a, 2014b, 2015). During the breeding season, female greater prairie-chickens selected habitat farther from wind turbines and doubled their home range size in the post-construction period, suggesting indirect habitat loss from the wind facility for at least 3 years post-construction (Winder et al. 2014b). However, female annual survival was greater in the post-construction period (0.57) relative to the preconstruction period (0.32), leading the authors to speculate that the wind facility reduced avian and mammalian predator populations, which, in turn, reduced prairie-chicken mortality (Winder et al. 2014a). Thus, females tended to avoid habitat that actually conferred greater fitness—an example of a "perceptual trap" (Patten and Kelly 2010). In contrast, 2 studies have shown that nest-site selection and nest survival did not vary with distance to wind turbines (McNew et al. 2014) or along a disturbance gradient (Harrison et al. 2017) at wind facilities in Kansas and Nebraska, respectively.

Male greater prairie-chickens also appear to have mixed responses to wind energy development. Males in the Kansas study were twice as likely to abandon leks within 1 km of a wind turbine (probability of lek persistence = 0.5) relative to leks 3 to 8 km from a turbine (probability of lek persistence = 0.90 to 0.95), although overall lek persistence was not different between pre- and post-construction periods (Winder et al. 2015). Males also showed a decrease in body mass (2.4%) in the post-construction period, but the

Table 7.2. Peer-reviewed studies of wind energy effects on North American prairie grouse species

Species	Study	Processes Examined	Impacts	Study Design	Study Duration	Location	Number of Turbines	Sample Size
Greater sage-grouse	LeBeau et al. 2014	Nest and brood success, adult female survival	Decreased nest and brood survival near turbines, no impact on female survival (but see Le Beau et al. 2017b)	Distance to turbine, VHF radio transmitters, nest monitoring	2 years	Southeastern Wyoming	79 (1.5 MW)	116 females, 95 nests, 31 broods
	LeBeau et al. 2017a	Male lek counts	No impact	Before-after control-impact (BACI)	11 years (3 years pre- and 8 years post-construction)	Southeastern Wyoming	79 (1.5 MW)	5 treatment leks, 9 control leks
	LeBeau et al. 2017b	Habitat selection, nest, brood, and adult female survival	Females shifted brood rearing and summer habitat use away from surface disturbance, no impact on survival rates	Control-treatment, VHF radio transmitters	5 years	Southeastern Wyoming	79 (1.5 MW)	346 females
Greater prairie-chicken	Harrison et al. 2017	Nest site selection, daily nest survival	No impact	Disturbance gradient	2 years post-construction	North-central Nebraska	36 (1.65 MW)	64 females, 91 nests
	McNew et al. 2014	Nest site selection, daily nest survival	No impact	Modified BACI, distance to turbine	5 years (2 years pre- and 3 years post-construction)	North-central Kansas	67 (3.0 MW)	264 nests
	Smith et al. 2016	Female lek attendance, male lekking behavior	No effect on female lek attendance, males close to facility spent less time in nonbreeding behaviors	Distance to turbine	3 months	North-central Nebraska	36 (1.65 MW)	15 leks, 15,689 behavioral observations
	Smith et al. 2017	Daily female survival, predation risk	No impact on survival, lower mammalian predation risk near facility	Distance to facility, VHF radio transmitters	2 years, 5 months each	North-central Nebraska	36 (1.65 MW)	62 females

(continued)

Table 7.2. (*continued*)

Species	Study	Processes Examined	Impacts	Study Design	Study Duration	Location	Number of Turbines	Sample Size
Greater prairie-chicken (continued)	Winder et al. 2014a	Female habitat use	Female avoidance of turbines, breeding season home range size increased post-construction	Before and after construction, VHF radio transmitters	5 years (2 years pre- and 3 years post-construction)	North-central Kansas	67 (3.0 MW)	14–102 birds/group, 181-bird seasons
	Winder et al. 2014b	Adult female survival	Female survival increased post-construction	Before and after construction, VHF radio transmitters	5 years (2 years pre- and 3 years post-construction)	North-central Kansas	67 (3.0 MW)	220 females
	Winder et al. 2015	Lek persistence, male body mass	Decreased lek persistence with proximity to turbines, male body mass lower post-construction	Before and after construction, distance to turbine	5 years (2 years pre- and 3 years post-construction)	North-central Kansas	67 (3.0 MW)	23 leks, 408 males

decrease was unrelated to distance from wind turbines (Winder et al. 2015). In a behavioral study of greater prairie-chickens near a 36-turbine wind facility in the Nebraska Sandhills, greater prairie-chicken lekking behavior was not negatively affected by proximity to wind turbines (Smith et al. 2016a). Although the authors predicted changes to male lekking behavior due to acoustic interference from wind turbine noise, distance to wind turbine had no effect on time spent in courtship behaviors (i.e., booming, flutter-jumping) during 1 season of monitoring (Smith et al. 2016a). Instead, males farther from the wind facility spent more time exhibiting nonbreeding behaviors, possibly due to increased vigilance for predators.

Greater prairie-chicken predators may be influenced by wind energy development. In the Nebraska Sandhills, the avian predator community was similar near and far from (> 2 km) wind turbines (Smith et al. 2017); however, mammalian predators were less likely to encounter camera traps used in the study closer to the wind facility. Fewer mammalian predators may indicate reduced predation risk to greater prairie-chickens from predators nearer to wind energy facilities. However, greater prairie-chicken adult daily female survival did not vary with distance to wind turbines, suggesting that predator avoidance of the wind facility did not translate into improved demographic performance of prairie-chickens (Smith et al. 2017).

Greater Sage-Grouse

Similar to greater prairie-chickens, greater sage-grouse have shown some shifts in habitat selection without evidence of lasting demographic consequences after wind development. Following installation of a 79-turbine wind facility in Wyoming, radio-collared greater sage-grouse females shifted summer habitat use away from wind turbines (distance to wind turbines was 2.2 to 5.3 km following development), and distance from wind turbines increased annually throughout the study period, suggesting a possible lag time between construction and behavioral changes in this species (LeBeau et al. 2017b). A short-term study showed some evidence of greater nest and brood failure for greater sage-grouse near wind turbines (LeBeau et al. 2014), but effects on nest-site selection, nest survival, and brood survival were not apparent after additional years of study (LeBeau et al. 2017a). Similar to that of female greater prairie-chickens (Winder et al. 2014a), adult female greater sage-grouse survival increased with greater surface disturbance (e.g., turbine pads and access roads). Male lek attendance did not differ significantly between treatment leks near the wind facility and control leks (LeBeau et al. 2017a).

Deep Dive. Mitigation by Avoidance: Greater Sage-Grouse Core Area Policies

In the mitigation hierarchy, avoidance is the most effective means of mitigating wind energy development impacts to sensitive prairie-chicken and sage-grouse species. Avoidance refers to avoiding siting wind energy infrastructure where there are species of conservation concern. Although federal nondisturbance buffer zones for wind development are only recommendations (Manville 2004, USFWS 2012), state-level "core area" policies have been enacted to avoid disturbance from wind and other development in sage-grouse habitat.

With the largest greater sage-grouse population of any western state, Wyoming was one of the first states to offer a policy to limit new development and disturbance in areas with substantial sage-grouse populations. The Greater Sage-Grouse Core Area

Figure 7.1   High wind development potential areas (black shading) and greater sage-grouse (*Centrocercus urophasianus*) core-population areas (version 4 of the core-area map; gray polygons) map used to predict the effects of development on grouse populations in Wyoming. *Modified from USDA NRCS 2014. Courtesy of Holly Copeland.*

Protection Policy (Wyoming Executive Order 2015-4) effectively reduces new wind facilities inside designated core areas with high sage-grouse abundances by imposing caps on the number and extent of disturbances from wind development (Figure 7.1). An instructional memorandum from the Bureau of Land Management and revised resource management plans have extended the reach of core area policy to federal lands, and conservation easements have been used to protect some private lands in core areas (Copeland et al. 2013).

Wyoming's Core Area Policy Is Ambitious, But Is It Enough to Conserve Sage-Grouse?

Initial research suggests both promise and limits for effectiveness of current policy. By projecting future wind, oil and gas, and residential development, Copeland et al. (2013) estimated sage-grouse would decline 11% to 24% within core areas in the absence of the core area policy, but just 6% to 9% in core areas with the current core area policy. Sage-grouse declines could be further reduced through additional strategic targeting of conservation easements on private lands.

The core area policy represents an effort to balance greater sage-grouse conservation with economic development desires; some development is allowed inside core areas, and new energy developments are still allowed in non-core areas that support sage-grouse populations. Development still allowed under the state's core area policy could extirpate an estimated 6% of sage-grouse leks statewide (Copeland et al. 2013).

Reducing anthropogenic disturbance is intended to stabilize or improve sage-grouse population performance in core areas, but results have been mixed. An analysis of male lek attendance following implementation of Wyoming's core area policy showed no difference between core and non-core areas during a period of statewide sage-grouse population decline (Gamo and Beck 2017). During a period of population increase, lek attendance was greater in some portions of the state but not others. A legacy of extensive coalbed natural gas development prior to the core area policy could be cause for reduced effectiveness of the policy in some portions of Wyoming (Gamo and Beck 2017).

Wyoming's core area map was based on the distribution of nesting cover relative to mapped leks (Doherty et al. 2011), but lek-centered habitat prioritization did not adequately capture year-round habitat use for some sage-grouse populations (Smith et al. 2016b). Only one-third of medium- and high-use winter ranges for females was protected from development through inclusion in the core area policy, leaving some high-priority populations vulnerable to development for a significant portion of the species' annual life cycle. Version 4 of the core area map includes two sage-grouse winter concentration areas not included in previous versions, and new research likely will identify additional important winter ranges.

Wyoming's core area policy has been in place for only a short time, and it is too soon to know whether it will stabilize sage-grouse populations in the state. But evidence suggests it does reduce new energy development in sage-grouse areas of high conservation priority (Gamo and Beck 2017), and that is the crux of the mitigation hierarchy's avoidance goals. But early evidence also suggests that the core area policy is just the first step in effective sage-grouse conservation. Expansion of the core area map to protect additional sensitive winter ranges will further reduce development impacts on priority sage-grouse populations, including the impacts of wind energy (Smith et al. 2016b). And projected declines in sage-grouse populations, even with conservation on private lands and the core area policy (Copeland et al. 2013), mean that reducing wind energy impacts on sage-grouse outside of core areas could further reduce statewide population declines.

Insects

Wind facilities, by and large, have not been studied for their impacts on invertebrates. What little we do know is in relation to terrestrial insects, as opposed to aquatic invertebrates or non-insect invertebrates (e.g., spiders). However, the effects of wind development on insects—and vice versa—are worthy of attention. Indeed, this overlooked wildlife group can reduce wind turbine efficiency and may contribute to turbine-related bat fatalities.

Emerging Issues

An increasing number of petitions for federal protection of insects could change the regulatory

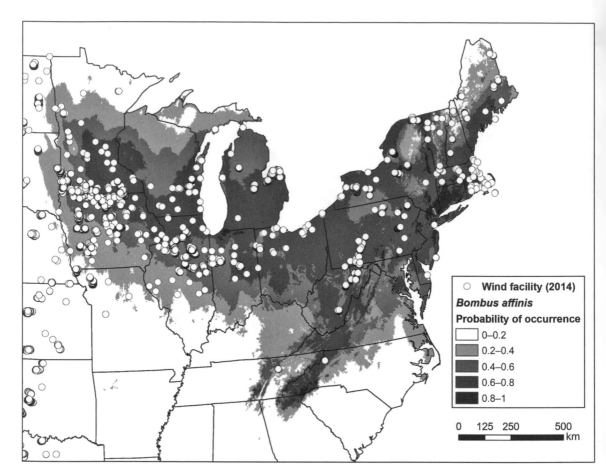

Figure 7.2   Predicted distribution of the rusty patched bumble bee (*Bombus affinis*) and the locations of existing wind facilities within the United States (as of March 2014). The rusty patched bumble bee was broadly distributed in eastern North America before population declines. *Source: Species distribution model, Williams et al. 2014; wind facility data, Diffendorfer et al. 2017.*

landscape for wind developers. For example, in 2017 the rusty patched bumble bee (*Bombus affinis*) became the first continental US hymenopteran to be listed as endangered. Although habitat loss and pesticides are the main causes of the bee's decline, the high degree of overlap between its range and locations of current and potential wind development elicits potential for conflict (Figure 7.2). Additionally, the migratory monarch butterfly (*Danaus plexippus*) is due for a listing decision under the Endangered Species Act in 2019, but impacts on this species from wind energy development are just now being explored. Because we know so little about wind energy impacts on bumble bees and

butterflies—or, for that matter, any other invertebrates—it remains to be seen whether developers will be required to conduct baseline surveys and monitoring for insect species or what additional requirements, if any, will be imposed for proposed and existing wind facilities.

Potential Impacts

Most potential wind energy–related impacts to insects are not unique to wind development, but a few are distinct. For example, direct insect mortality from collisions with wind turbine blades may be substantial, but have not been measured. While we do

Figure 7.3   Insect collisions with turbine blades can be substantial and reduce power generation. *Used with permission from BladeCleaning.com.*

not know about the quantity or kinds of insects that are directly affected by collisions with wind-turbine blades, we do know that collisions are greatest under low wind conditions, when insects most often fly. Insect carcasses can build up on blades enough to reduce power generation by 25% to 50% (Corten and Veldkamp 2001; Figure 7.3). These mortalities may reduce already-declining pollinator populations, for example. Wind turbines alter surface air currents, which, in turn, affects local climate characteristics like temperature and hydrology (Elzay et al. 2017). Changes to microclimate may affect insects, but this idea with respect to wind energy remains untested. Other potential effects may result from increased dust and chemical pollution, introduction of invasive species, and changes in trophic interactions stemming from wind energy development (Elzay et al. 2017).

Observed Impacts

Insects may not only inhabit areas coincident with turbines, but may actually be drawn to them. In a simple experiment measuring visitations to different colors of paper, insects were most attracted to the white or light gray hues typical of wind turbine structures (Long et al. 2011). Researchers suggested that painting structures a color less attractive to insects, such as purple, could reduce collisions. Insects also may seek out heat generated by warm rotor hubs of turbines (Elzay et al. 2017). Whatever may be attracting them, researchers hypothesize that insects congregate at wind facilities and, in turn, attract insectivorous bats (Rydell et al. 2010, Grodsky et al. 2012). The impacts of wind energy development on monarch butterflies are being studied at 1 wind facility in Mexico (Parque Eolico Coahuila). Although results of monitoring studies are not yet publicly available, a risk assessment concluded that butterflies were unlikely to collide with turbines (WEST, Inc. 2014). Most potential consequences of wind facilities for insects have been identified from other forms of disturbance or development. For example, evidence indicates that roads may cause habitat fragmentation for some ground beetles (Keller and Largiader 2003, Keller et al. 2004), although it is not clear what the causal mechanism for this change may be (e.g., collisions with vehicles, road avoidance, changes in host plants).

Other Taxonomic Groups

Literature on wind energy impacts on other terrestrial animals is scant but taxonomically diverse (Table 7.3). Some studies come from abroad or were conducted on common species, but findings may nevertheless be relevant to related taxa of conservation concern in North America. Studies on rodents, carnivores, and reptiles highlight some of the potential indirect impacts of wind development on lesser-studied taxonomic groups that may contain some species of conservation concern.

Rodents

Small mammal populations may incur fitness costs resulting from behavioral shifts in response to wind development. California ground squirrels (*Otospermophilus beecheyi*) at wind facilities spent more time engaged in antipredator behaviors than those at control sites (Rabin et al. 2006). Turbine noise may disrupt communication between individual ground squirrels, resulting in increased vigilance and other behavioral changes. Increased antipredator behaviors in the absence of an actual increase in predation risk could lead to a reduction in time spent for other important behaviors, such as foraging and mating. While these effects are likely of limited consequence for this common species, a similar response in a declining species like the Mohave ground squirrel (*Xerospermophilus mohavensis*) could have important fitness consequences (Inman et al. 2016).

A study on small mammals in Spain did not find discernible changes in wood mice (*Apodemus sylvaticus*) abundance at transects placed up to ~120 m from turbines (DeLucas et al. 2005). In a more detailed study of small mammals in Poland, community composition and population parameters of the most common rodent species did not differ between wind facilities and control sites (Łopucki and Mróz 2016). It is difficult to generalize these findings to cover implications for other small mammals, except to say that social rodents and those occurring in restricted ranges may be disproportionately affected by wind development.

Mammalian Carnivores

A small body of work on mammalian carnivores illustrates potentially important changes in space use and physiology in response to wind energy development. European badgers (*Meles meles*) close to wind facilities had chronically elevated levels of the stress hormone cortisol compared to control sites even 4 years post-construction (Agnew et al. 2016), perhaps due to turbine noise and vibrations. Prolonged, elevated cortisol levels are known to suppress immune function, growth, and reproduction in some mammals, although is believed to be adaptive in others (Boonstra 2013).

Several lines of evidence indicate that mammalian carnivores may avoid wind turbines. No studies have yet examined wind development effects on North American wolves, but a study of endangered Iberian wolves (*Canis lupus signatus*) in Portugal reported changes in space use from wind development, with possible consequences for reproduction (Torres and Fonseca 2016). Although wolves avoided wind facilities during construction, they used sites at pre-construction levels once wind turbines were operational (Álvares et al. 2011). However, wolves decreased their use of or abandoned high-quality breeding sites within wind facilities in exchange for low-quality sites away from wind facilities. The authors suggested that packs in these low-quality areas were less likely to reproduce, potentially exacerbating population declines (Helldin et al. 2017). Using visits to Agassiz's desert tortoise (*Gopherus agassizii*) burrows as a metric of habitat use, Agha et al. (2017) determined that mesopredators—bobcat (*Lynx rufus*), gray fox (*Urocyon cinereoargenteus*), coyote (*Canis latrans*), and western spotted skunk (*Spilogale gracilis*)—were more likely to use areas near roads at wind facilities, while avoiding wind turbines themselves. In a study of mammalian predators of prairie-chickens in Nebraska, researchers calculated an overall visitation rate for all species of potential predators (i.e., coyotes; badgers, *Taxidea taxus*; raccoons, *Procyon lotor*; and opossums, *Didelphis virginiana*) and reported that the capture index was positively related to distance from wind turbine, suggesting some wind turbine avoidance (Smith et al. 2017).

Desert Tortoise

Wind development appears compatible with the maintenance of local populations of the federally threatened Agassiz's desert tortoise. In southern California, individuals selected burrow locations close

Table 7.3. Peer-reviewed studies of wind energy effects on North American ground squirrels, carnivores, and reptiles

Species	Study	Processes Examined	Impacts	Study Design	Study Duration	Location	Number of Turbines	Sample Size
California ground squirrel	Rabin et al. 2006	Antipredator behavior	Increased vigilance behavior at turbine site	Control-treatment, conspecific call playback experiment	~1 summer month	Northern California	6-turbine string within 6,250-turbine project area (100 kW and other turbine types)	42 individuals at wind farm, 52 control
Carnivores (bobcat, coyote, gray fox, western spotted skunk)	Agha et al. 2017	Counts from camera traps	Counts decreased close to turbines, increased close to roads	Distance to roads and turbines	5.5 months (7,968 trap nights)	Southern California	460	Camera traps at 46 tortoise burrows
Carnivores (American badger, coyote, northern raccoon, striped skunk, Virginia opossum)	Smith et al. 2017	Capture index from camera traps, coyote occupancy	Capture index for all mammals lower near turbines, no effect on coyote occupancy	Distance to turbines	2 years, 5 months each (4,715 trap nights)	North-central Nebraska	36 (1.65 MW)	31 camera trap locations
Desert tortoise	Agha et al. 2015	Annual survival, size of activity center	Survival higher at wind farm, no effect on size of activity center	Control-treatment	18 years	Southern California	460	27 individuals at wind farm, 27 control
	Lovich and Daniels 2000	Burrow location	Burrows located closer to roads and turbines than random locations	Distance to roads and turbines of burrows and random points	2 years	Southern California	460	32 burrows
	Lovich et al. 2011a	Mortality due to culverts	One documented mortality, possibly linked to being entrapped in culvert	Anecdotal observation	N/A	Southern California	460	N/A
	Lovich et al. 2011b	Growth, sex ratio, mortality	No impact	Control-treatment (control was a closely related species in Arizona location without wind facility)	14 years	Southern California	460	69 individuals at wind farm

to wind energy infrastructure (Lovich and Daniels 2000). A study comparing Agassiz's desert tortoises at wind facilities to those of nearby intraspecific populations and those of a closely related species (*Gopherus morafkai*) showed no negative effects of wind development on tortoise population structure or growth (Lovich et al. 2011). A study that incorporated 18 years of data at the same site also reported that apparent survival estimates were slightly greater for desert tortoises occupying a wind facility (0.96) than at a nearby wilderness area (0.92; Agha et al. 2015). The authors hypothesized that greater survival could be attributable to fewer fatalities from vehicle collisions, enhanced resource availability because of the desirable conditions created by roads and turbine pads, and fewer predators compared to the wilderness area.

Mitigating Effects on Underrepresented Species

Even considering the limitations of existing science, a few themes emerge that can inform efforts to mitigate impacts of wind development on underrepresented species. First, building infrastructure causes habitat loss that can reduce carrying capacity, particularly for species such as ungulates or prairie grouse that show high site fidelity to specific ranges. Second, for species that tend to avoid infrastructure and human activity, indirect habitat loss further reduces

available habitat and the number of animals an area can support. And third, wind energy infrastructure, including roads, fences, and turbine pads, can fragment habitat and impede movement and migration for ungulates and other wide-ranging species. For some species such as mule deer and greater sage-grouse, migration connects populations to the seasonal resources needed to sustain robust populations (Fedy et al. 2015, Sawyer et al. 2017).

Broad-level impacts of wind development on wildlife can be matched with opportunities within the mitigation hierarchy—a nationally recognized approach to sequentially avoid, minimize, and compensate for impacts from disturbance such as wind energy development (Jakle 2012; Figure 7.4). The challenge is to define mitigation that is not excessively stringent but is appropriately precautionary, given emerging science on impacts to ungulates, prairie grouse, and insects, among other wildlife groups.

Avoidance

Avoidance is typically the most effective way to protect wildlife populations from known and unknown impacts of wind energy development. Avoidance typically involves *not* siting wind infrastructure where there are species of conservation concern. Even within a single project area, a developer may use micro-siting of infrastructure, including roads,

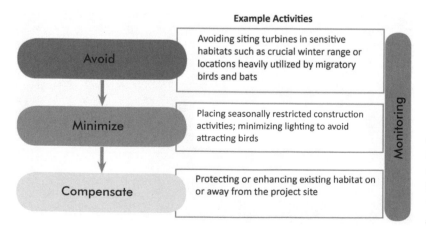

Figure 7.4 Steps of the mitigation hierarchy and example activities for mitigating potential impacts of wind energy production on wildlife. *Modified from Jakle 2012.*

buildings, and turbines, to avoid impacts on species of concern (Arnett and May 2016).

Another form of avoidance is siting wind facilities in marginalized landscapes (e.g., agricultural, urban) instead of in more undisturbed locations that are likely to provide wildlife habitat (Kiesecker et al. 2011, Fargione et al. 2012, Jones and Pejchar 2013). Such an approach could be implemented through county- or state-level energy planning that prioritizes wind energy in developed areas, while discouraging development in high-value wildlife habitat. For ungulates specifically, a precautionary approach would avoid siting intensive wind development in (1) state wildlife agency–designated crucial or vital habitat, such as geographically limited crucial winter range or parturition areas; and (2) migration corridors, which are critical to the year-round nutrition of migratory populations.

Studies showing shifts in habitat use by female greater sage-grouse and greater prairie-chicken and increased lek abandonment by male greater prairie-chickens close to wind turbines suggest that siting could help ameliorate wildlife impacts. At the project level, nondevelopment buffer zones can separate infrastructure from lekking, nesting, and brood-rearing areas, and from summer and winter resources required by prairie grouse. For greater sage-grouse, the most common approach is to use the lek as the center of a proposed nondevelopment buffer zone because leks are thought to be the approximate center of all annual resources. However, this may not be the case for migratory populations (Connelly et al. 2000b) and may underrepresent critical winter range for sage-grouse (Smith et al. 2016b).

For prairie grouse, appropriate size for buffer zones is species-specific and depends on both the intensity of development and the amount and quality of habitat around the lek (Walker et al. 2007, Winder et al. 2014a, LeBeau et al. 2017a, 2017b). In its most recent guidelines for wind development, the USFWS (2012) cited the need for more research to determine appropriate buffer zones, but earlier guidelines (USFWS 2003, Manville 2004) proposed an 8 km buffer

zone around prairie grouse leks. LeBeau et al. (2017a) recommended a 1.5 km nondevelopment buffer zone around active greater sage-grouse leks and a 1.2 km zone around nesting, brood-rearing, and summer range for wind energy development of similar intensity and in landscapes similar to those studied in southeastern Wyoming. Wind development effects on habitat use and lek attendance of greater prairie-chicken led researchers to support an 8 km nondisturbance buffer zone around active leks (Winder et al. 2015).

Siting wind facilities to avoid insect migration routes (e.g., those of monarch butterflies) may be one way to mitigate insect mortalities and other indirect impacts. Insect migration routes often follow wind currents and thus likely overlap with wind facilities and may further be coincident with bat migration paths (Rydell et al. 2010). Wind infrastructure siting to avoid negative consequences to bats would likely benefit migratory and nonmigratory flying insects as well.

Minimization

After siting wind facilities to avoid impacts to wildlife, minimizing remaining wildlife impacts is the next step in the mitigation hierarchy. For instance, for species that avoid human activity (Gavin and Komers 2006, Sawyer et al. 2009, Buchanan et al. 2014, Tsegaye et al. 2017), such as many ungulates, minimizing disturbances during the busy construction phase and reducing traffic and other disturbances during the operational phase are opportunities to reduce wildlife impacts. Wind developments can minimize impacts to wildlife by first timing construction and decommissioning activities to avoid wintering, parturition, and other sensitive life history stages for ungulates and breeding seasons of prairie grouse (Connelly et al. 2000b, Hagen et al. 2004, Johnson and Holloran 2010); developers also may spatially phase construction activities to avoid human disturbance throughout the entire project area at once. Second, to the extent possible, minimize the road network (USFWS 2012) and human activity within the wind facility and reduce vehicle traffic and speeds to lessen collisions with wildlife (LeBeau et al. 2017b). Restoring unneeded roads with native vegetation (USFWS 2012) and co-locating roads, transmission lines, and other linear features can further reduce wind facility footprints.

Impacts to ungulate migration routes may be reduced by minimizing fencing and using wildlife-friendly fencing, when necessary (Harrington and Conover 2016). For mule deer and elk, maintaining or enhancing refugia and structural elements, such as woody vegetation or rocky outcrops, can provide cover (Lendrum et al. 2012, Buchanan et al. 2014). Wind energy impacts to prairie grouse can be further minimized by reducing raptor perching opportunities by burying low- and medium-voltage power lines and using monopole wind turbine construction instead of lattice structures (Pruett et al. 2009b, Johnson and Holloran 2010, USFWS 2012).

For existing wind facilities, increasing the cut-in speed of turbines—the minimum wind speed at which blades begin to rotate—could reduce collisions with flying species. Bats, birds, and insects avoid flying at high wind speeds, and most wildlife-turbine collisions occur at lower wind speeds. Using a cut-in speed between 5.0 and 5.5 m/s substantially reduced the number of bat mortalities with minimal financial losses to the wind operator (Baerwald et al. 2009, Arnett et al. 2011). Lastly, altering the color of wind infrastructure, perhaps by reducing UV reflectance (Long et al. 2011), may be an inexpensive way to reduce insect attraction to wind turbines.

Compensation

Compensation is the last step of the mitigation hierarchy, and it is used to offset impacts, known as "debits," that occur after avoidance and minimization efforts are implemented. Compensatory mitigation offsets impacts by generating "credits" through protection, restoration, and enhancement of habitat away from the location of the wind facility. For instance, a developer may offset unavoidable impacts from a wind energy facility by purchasing credits from a USFWS-authorized conservation bank, managed for the protection of a threatened or endangered species (Hansen et al. 2013). Likewise, habitat exchanges offset impacts on species of concern that are not federally designated as threatened or endangered (Hansen et al. 2013).

Compensatory mitigation for wind development has potential to offset impacts on a range of species (Doherty et al. 2010, Cole 2011, Hansen et al. 2013) and could, for instance, become an important tool for landscape-scale mitigation, including reconnecting disjunct grouse populations (Pruett et al. 2009a) or protecting diverse seasonal ranges needed to sustain robust ungulate populations. Yet, the true value of conservation banks in offsetting impacts to prairie grouse and ungulates, for example, has not been validated (Arnett and May 2016). For some ungulates, such as mule deer that show high site fidelity to traditional winter ranges, parturition areas, and migration routes, off-site compensatory mitigation may not offset impacts to the affected population. Likewise, fidelity of prairie grouse species to lekking, wintering, and other seasonal ranges suggests that off-site mitigation may not compensate for impacts to the target population either. A final caveat is that credits (improvements) and debits (impacts) typically are calculated in terms of habitat units rather than wildlife population measures (Davis et al. 2009). Although preferable to the alternative of having no off-site mitigation, population-level benefits of habitat improvements to wildlife are not easily quantified or isolated from other drivers of wildlife population change (Doherty et al. 2010), making the true conservation value of off-site mitigation for affected wildlife populations uncertain.

Conservation banks have been used to conserve a range of endangered species but are just now being used to offset impacts from wind energy. For instance, the USFWS has authorized the Sweetwater River Conservancy, a 55,000-acre (22,000 ha) greater sage-grouse conservation bank in central Wyoming, to sell credits to offset industrial energy development (USFWS 2017), including a proposed wind development in southeastern Wyoming more than 150 km away (Nickerson 2011). Similarly, Common Ground Capital operates a conservation bank for mitigating wind energy and other impacts to lesser prairie-chicken (Common Ground Capital 2017a). Conservation banks have been used to conserve endangered insect species such as the American burying beetle (*Nicrophorus americanus*; Common Ground Capital 2017b) but, to our knowledge, have not been used to mitigate impacts to insects from wind energy development specifically.

Conclusions

The science of understanding wind energy impacts to ungulates, prairie grouse, insects, and other underrepresented taxa is limited. In ungulate research, 2 studies have shown minimal impacts of wind facilities on pronghorn (Johnson and Stephens 2011, Taylor et al. 2016), and too little is known about wind energy and elk to draw conclusions. Absent from the literature are studies of wind energy impacts on other ungulates, including mule deer, which are known to be sensitive to other forms of energy development. Given the economic and cultural importance of these ungulate species throughout the western United States, understanding wind energy impacts on a diversity of ungulate species should be a high priority.

Recent studies of greater sage-grouse and greater prairie-chicken have not borne out early predictions of dire consequences of wind energy development for prairie grouse. Well-designed studies have shown some shifts in habitat use and lek attendance, with no discernable consequences to population performance; however, studies are limited to a few wind facilities and do not include prairie grouse species of greatest conservation concern (e.g., Attwater's or lesser prairie-chicken). Well-designed before-after-control-impact (BACI) studies at wind facilities of different sizes and in a range of systems would help

managers predict population-level impacts of wind development, if any, on these avian species.

For insects, mammalian carnivores, and rodents, wind energy impacts are mostly unknown. For these and other underrepresented species, a lack of evidence of impacts is not evidence that there are no impacts. As the footprint of wind energy development grows in the United States, we have an opportunity to study effects on a variety of wildlife, especially for species of conservation concern. Greater transparency and sharing of industry-funded studies could fill the science gaps more quickly. In the meantime, a precautionary approach to mitigation can reduce potential impacts on underrepresented wildlife groups.

Even as more research is published to elucidate impacts of wind energy on wildlife, the very nature of wind development is changing. Many studies of effects of wind energy on wildlife have been conducted at facilities that are small relative to upcoming developments such as the Chokecherry-Sierra Madre wind facility in southcentral Wyoming, which is projected to site 1,000 turbines on a mix of federal and private land (US Department of the Interior 2017). The scale of this development is at least an order of magnitude greater than most of the studies reviewed in this chapter (Tables 7.1, 7.2, and 7.3). The unique impacts, if any, of such large wind facilities on ungulate, prairie grouse, and other wildlife populations are unknown.

As with all forms of development, potential impacts of wind facilities on wildlife should be evaluated in terms of trade-offs. Wind facilities add to an ever-growing human footprint from residential and other energy development, an expanding human population, and climate change, all of which affect wildlife. Relative to oil and natural gas development, wind energy has a larger physical footprint (Jones and Pejchar 2013), creating viewshed and other non-wildlife issues. But, as we individually and collectively decide whether biological and social costs of wind development are worth it, we must simultaneously consider the benefits of developing wind and other renewable energy resources relative to continued use of finite fossil fuels. Reducing greenhouse gases and mitigating climate change through renewable energy investments can benefit the very wildlife populations that are affected by wind facilities, as well as other species, including our own (Allison et al. 2014). Impacts of wind development should be minimized and weighed against consequences of climate change if the world does not transition to a new energy future that embraces energy conservation and efficiency while thoughtfully siting renewable energy sources.

ACKNOWLEDGMENTS

We thank Hall Sawyer and Chad LeBeau for reviewing parts of this chapter, and the editors for their helpful feedback. Thanks to Holly Copeland for contributing the Wyoming wind potential and greater sage-grouse core area map. Thanks to Leif Richardson for providing the species distribution model for the rusty patched bumble bee. Thank you to José Luis Ruiz of BladeCleaning for providing the image in Figure 7.3.

LITERATURE CITED

Adams, G. M. 2006. Bringing green power to the public lands: The Bureau of Land Management's authority and discretion to regulation wind-energy developments. Journal of Environmental Law and Litigation 21:445–508.

Agha, M., J. E. Lovich, J. R. Ennen, B. Augustine, T. R. Arundel, M. O. Murphy, K. Meyer-Wilkins, et al. 2015. Turbines and terrestrial vertebrates: Variation in tortoise survivorship between a wind energy facility and an adjacent undisturbed wildland area in the desert southwest (USA). Environmental Management 56: 332–341.

Agha, M., A. L. Smith, J. E. Lovich, D. Delaney, J. R. Ennen, J. Briggs, L. J. Fleckenstein, et al. 2017. Mammalian mesocarnivore visitation at tortoise burrows in a wind farm. Journal of Wildlife Management 81:1117–1124.

Agnew, R. C. N., V. J. Smith, and R. C. Fowkes. 2016. Wind turbines cause chronic stress in badgers (*Meles meles*) in Great Britain. Journal of Wildlife Diseases 52:459–467.

Aikens, E. O., M. J. Kauffman, J. A. Merkle, S. P. H. Dwinnell, G. L. Fralick, and K. L. Monteith. 2017. The greenscape shapes surfing of resource waves in a large migratory herbivore. Ecology Letters 20:741–750.

Aldridge, C. L., and M. S. Boyce. 2007. Linking occurrence and fitness to persistence: Habitat-based approach for endangered greater sage-grouse. Ecological Applications 17:508–526.

Allison, T. D., T. L. Root, and P. C. Frumhoff. 2014. Thinking globally and siting locally: Renewable energy and biodiversity in a rapidly warming world. Climatic Change 126:1. https://doi.org/10.1007/s10584-014-1127-y.

Álvares, F., H. Rio-Maior, S. Roque, M. Nakamura, D. Cadete, S. Pinto, and F. Petrucci-Fonseca. 2011. Assessing ecological responses of wolves to wind power plants in Portugal: Methodological constraints and conservation implications. Conference on Wind Energy and Wildlife Impacts. Norwegian Institute for Nature Research, May 2–5, Trondheim, Norway.

Arnett, E. B., D. B. Inkley, R. P. Larkin, S. Manes, A. M. Manville, J. R. Mason, M. L. Morrison, M. D. Strickland, and R. Thresher. 2007. Impacts of wind energy facilities on wildlife and wildlife habitat. Wildlife Society Technical Review 07-2. Bethesda, MD: Wildlife Society.

Arnett, E. B., M. M. P. Huso, M. R. Schirmacher, and J. P. Hayes. 2011. Altering turbine speed reduces bat mortality at wind-energy facilities. Frontiers in Ecology and the Environment 9:209–214.

AWEA (American Wind Energy Association). 2017. Public lands and wind energy. http://www.awea.org/public -lands. Accessed June 14, 2017.

Baerwald, E. F., J. Edworthy, M. Holder, and R. M. R. Barclay. 2009. A large-scale mitigation experiment to reduce bat fatalities at wind energy facilities. Journal of Wildlife Management 73:1077–1081.

Balch, J. K., B. A. Bradley, C. M. D'Antonio, and J. Gómez-Dans. 2013. Introduced annual grass increases regional fire activity across the arid western USA (1980–2009). Global Change Biology 19:173–183.

Beckmann, J. P., K. Murray, R. G. Seidler, and J. Berger. 2012. Human-mediated shifts in animal habitat use: Sequential changes in pronghorn use of a natural gas field in Greater Yellowstone. Biological Conservation 147:222–233.

Blickley, J. L., and G. L. Patricelli. 2012. Potential acoustic masking of greater sage-grouse (*Centrocercus urophasianus*) display components by chronic industrial noise. Ornithological Monographs 74:23–35.

Blickley, J. L., K. R. Word, A. H. Krakauer, J. L. Phillips, S. N. Sells, C. C. Taff, J. C. Wingfield, and G. L. Patricelli. 2012. Experimental chronic noise is related to elevated fecal corticosteroid metabolites in lekking male greater sage-grouse (*Centrocercus urophasianus*). PLOS One 11:e50462. doi:10.1371/journal.pone.0050462.

Boonstra, R. 2013. Reality as the leading cause of stress: Rethinking the impact of chronic stress in nature. Functional Ecology 27:11–23.

Braun, C. E., O. O. Oedekoven, and C. L. Aldridge. 2002. Oil and gas development in western North America: Effects on sagebrush steppe avifauna with particular emphasis on sage-grouse. Transactions of the North American Wildlife and Natural Resources Conference 67:337–349.

Buchanan, C. B., J. L. Beck, T. E. Bills, and S. N. Miller. 2014. Seasonal resource selection and distributional response by elk to development of a natural gas field. Rangeland Ecology and Management 67:369–379.

Christie, K. S., W. F. Jensen, J. H. Schmidt, and M. S. Boyce. 2015. Long-term changes in pronghorn abundance index linked to climate and oil development in North Dakota. Biological Conservation 192:445–453.

Christie, K. S., W. F. Jensen, and M. S. Boyce. 2017. Pronghorn resource selection and habitat fragmentation in North Dakota. Journal of Wildlife Management 81:154–162.

Cole, S. G. 2011. Wind power compensation is not for the birds: An opinion from an environmental economist. Restoration Ecology 19:147–153.

Colman, J. E., S. Eftestol, D. Tsegaye, K. Flydal, and A. Mysterud. 2012. Is a wind-power plant acting as a barrier for reindeer Rangifer tarandus tarandus movements? Wildlife Biology 18:439–445.

Colman, J. E., S. Eftestøl, D. Tsegaye, K. Flydal, and A. Mysterud. 2013. Summer distribution of semi-domesticated reindeer relative to a new wind-power plant. European Journal of Wildlife Research 59:359–370.

Common Ground Capital. 2017a. Lesser prairie chicken. http://commongroundcapital.com/lesser-praire-chicken/. Accessed July 13, 2017.

Common Ground Capital. 2017b. American burying beetle. http://commongroundcapital.com/american-burying-beetle/. Accessed July 13, 2017.

Connelly, J. W., A. D. Apa, R. B. Smith, and K. P. Reese. 2000a. Effects of predation and hunting on adult sage grouse Centrocercus urophasianus in Idaho. Wildlife Biology 6:227–232.

Connelly, J. W., M. A. Schroeder, A. R. Sands, and C. E. Braun. 2000b. Habitat and management guidelines to manage sage grouse populations and their habitats. Wildlife Society Bulletin 28:967–985.

Connelly, J. W., S. T. Knick, M. Schroeder, and S. J. Stiver. 2004. Conservation assessment of greater sage-grouse and sagebrush habitats. Unpublished Report. Cheyenne, WY: Western Association of Fish and Wildlife Agencies.

Cook, J. G., R. C. Cook, R. W. Davis, and L. L. Irwin. 2016. Nutritional ecology of elk during summer and autumn in the Pacific Northwest. Wildlife Monographs 195:1–81.

Copeland, H. E., A. Pocewicz, D. E. Naugle, T. Griffiths, D. Keinath, J. Evans, and J. Platt. 2013. Measuring the effectiveness of conservation: A novel framework to quantify the benefits of sage-grouse conservation policy and easements in Wyoming. PLOS One 8:e67261. https://doi.org/10.1371/journal.pone.0067261.

Corten, G. P., and H. F. Veldkamp. 2001. Insects can halve wind-turbine power. Nature 412:41–42.

Crawford, J. A., R. A. Olson, N. E. West, J. C. Mosley, M. A. Schroeder, T. D. Whitson, R. F. Miller, M. A. Gregg, and C. S. Boyd. 2004. Ecology and management of sage-grouse and sage-grouse habitat. Rangeland Ecology and Management 57:2–19.

Dahlgren, D. K., M. R. Guttery, T. A. Messmer, D. Caudill, R. D. Elmore, R. Chi, and D. N. Koons. 2016. Evaluating vital rate contributions to greater sage-grouse population dynamics to inform conservation. Ecosphere 7:e01249. doi:10.1002/ecs2.1249.

Davis, A., T. Weis, K. Halsey, and D. Patrick. 2009. Enabling progress: Compensatory mitigation scenarios for wind energy projects in the U.S. 2009. San Rafael, CA: Solano Partners.

DeLucas, M., G. F. E. Janss, and M. Ferrer. 2005. A bird and small mammal BACI and IG design studies in a wind farm in Malpica (Spain). Biodiversity and Conservation 14:3289–3303.

Doherty, K. E., D. E. Naugle, B. L. Walker, and J. M. Graham. 2008. Greater sage-grouse winter habitat selection and energy development. Journal of Wildlife Management 72:187–195.

Doherty, K. E., D. E. Naugle, and J. S. Evans. 2010. A currency for offsetting energy development impacts: Horse-trading sage-grouse on the open market. PLOS One 5:e10339. doi:10.1371/journal.pone.0010339.

Doherty, K. E., D. E. Naugle, H. E. Copeland, A. Pocewicz, and J. M. Kiesecker. 2011. Greater sage-grouse: Ecology and conservation of a landscape species and its habitats. In Studies in Avian Biology, edited by S. T. Knick and J. W. Connely, 505–516. 38th ed. Berkeley: University of California Press.

Dzialak, M. R., S. L. Webb, S. M. Harju, J. B. Winstead, J. P. Mudd, J. J. Wondzell, and L. D. Hayden-Wing. 2011. The spatial pattern of demographic performance as a component of sustainable landscape management and planning. Landscape Ecology 26:775–790.

Eftestøl, S., D. Tsegaye, K. Flydal, and J. E. Colman. 2016. From high voltage (300 kV) to higher voltage (420 kV) power lines: Reindeer avoid construction activities. Polar Biology 39:689–699.

Elzay, S., L. Tronstad, and M. E. Dillon. 2017. Terrestrial invertebrates. In Wildlife and Wind Farms: Conflicts and Solutions, Volume 1, Onshore: Potential Effects, edited by M. Perrow. Exeter, UK: Pelagic Publishing.

Fargione, J., J. Kiesecker, M. J. Slaats, and S. Olimb. 2012. Wind and wildlife in the Northern Great Plains: Identifying low-impact areas for wind development. PLOS One 7:e41468. doi:10.1371/journal.pone.0041468.

Fedy, B. C., C. P. Kirol, A. L. Sutphin, and T. L. Maechtle. 2015. The influence of mitigation on sage-grouse habitat selection within an energy development field. PLOS One 10:1–19.

Flydal, K., L. Korslund, E. Reimers, F. Johansen, and J. E. Colman. 2009. Effects of wind turbines on area use and behaviour of semi-domestic reindeer in enclosures. International Journal of Ecology 2009:1–14.

Forman, R. T. T., and L. E. Alexander. 1998. Roads and their major ecological effects. Annual Review of Ecology and Systematics 29:207–231.

Fryxell, J. M., and A. R. E. Sinclair. 1988. Causes and consequences of migration by large herbivores. Trends in Ecology and Evolution 3:237–241.

Gamo, R. S., and J. L. Beck. 2017. Effectiveness of Wyoming's sage-grouse core areas: Influences on energy development and male lek attendance. Environmental Management 59:189–203.

Gates, C. C., P. Jones, M. Suitor, A. Jakes, M. S. Boyce, K. Kunkel, and K. Wilson. 2012. The influence of land use and fences on habitat effectiveness, movements and distribution of pronghorn in the grasslands of North America. In Fencing for Conservation: Restriction of Evolutionary Potential or a Riposte to Threatening Processes?, edited by M. J. Somers and M. Hayward, 277–294. New York: Springer.

Gavin, S. D., and P. E. Komers. 2006. Do pronghorn (Antilocapra americana) perceive roads as a predation risk? Canadian Journal of Zoology 84:1775–1780.

Grodsky, S. M., C. S. Jennelle, D. Drake, and T. Virzi. 2012. Bat mortality at a wind-energy facility in southeastern Wisconsin. Wildlife Society Bulletin 36:773–783. http://doi.wiley.com/10.1002/wsb.191.

Grodsky, S. M., R. B. Iglay, C. E. Sorenson, and C. E. Moorman. 2015. Should invertebrates receive greater inclusion in wildlife research journals? Journal of Wildlife Management 79:529–536.

Hagen, C. A., B. E. Jamison, K. M. Giesen, and T. Z. Riley. 2004. Guidelines for managing lesser prairie-chicken populations and their habitats. Wildlife Society Bulletin 32:69–82.

Hansen, K., A. Jakle, and M. Hogarty. 2013. Market-based wildlife mitigation in Wyoming: A primer. Laramie, WY: Ruckelshaus Institute of Environment and Natural Resources. https://www.uwyo.edu/haub/_files/_docs / . . . /2013-market-based-mitigation.pdf.

Harju, S. M., M. R. Dzialak, R. C. Taylor, L. D. Hayden-Wing, and J. B. Winstead. 2010. Thresholds and time lags in effects of energy development on greater sage-grouse populations. Journal of Wildlife Management 74:437–448.

Harrington, J. L., and M. R. Conover. 2016. Characteristics of ungulate behavior and mortality associated with wire fences. Wildlife Society Bulletin 34:1295–1305.

Harrison, J. O., M. B. Brown, L. A. Powell, W. H. Schacht, and J. A. Smith. 2017. Nest site selection and nest survival of greater prairie-chickens near a wind energy facility. Condor 119:659–672.

Hebblewhite, M. 2011. Effects of energy development on ungulates. In Energy Development and Wildlife Conservation in Western North America, edited by D. E. Naugle, 71–94, Washington, DC: Island Press.

Helldin, J. O., A. Skarin, W. Neumann, M. Olsson, J. Jung, J. Kindberg, N. Lindberg, and F. Widemo. 2017. Terrestrial mammals. Wildlife and Wind Farms, Conflicts and Solutions. Volume 1, Onshore: Potential Effects, edited by M. Perrow, 222–240. Exeter, UK: Pelagic Publishing.

Holloran, M. 2005. Greater sage-grouse (Centrocercus urophasianus) population response to natural gas field development in western Wyoming. Dissertation, University of Wyoming, Laramie.

Holloran, M. J., R. C. Kaiser, and W. A. Hubert. 2010. Yearling greater sage-grouse response to energy development in Wyoming. Journal of Wildlife Management 74:65–72.

Inman, R. D., T. C. Esque, K. E. Nussear, P. Leitner, M. D. Matocq, P. J. Weisberg, and T. E. Dilts. 2016. Impacts of climate change and renewable energy development on habitat of an endemic squirrel, Xerospermophilus mohavensis, in the Mojave Desert, USA. Biological Conservation 200:112–121.

Jakle, A. 2012. Wind development and wildlife mitigation in Wyoming: A primer. Laramie, WY: Ruckelshaus Institute of Environment and Natural Resources.

Johnson, G., and M. Holloran. 2010. Greater sage-grouse and wind energy development: A review of the issues. Report commissioned by Renewable Northwest Project. Cheyenne, WY: Western EcoSystems Technology.

Johnson, G. D., and S. E. Stephens. 2011. Wind power and biofuels: A green dilemma for wildlife conservation. In Energy Development and Wildlife Conservation in Western North America, edited by D. E. Naugle, 131–155. Washington, DC: Island Press.

Johnson, H. E., J. R. Sushinsky, A. Holland, E. J. Bergman, T. Balzer, J. Garner, and S. E. Reed. 2016. Increases in residential and energy development are associated with reductions in recruitment for a large ungulate. Global Change Biology 23:578–591.

Jones, N. F., and L. Pejchar. 2013. Comparing the ecological impacts of wind and oil and gas development: A landscape scale assessment. PLOS One 11:e81391. doi:10.1371/journal.pone.0081391.

Keller, I., and C. R. Largiader. 2003. Recent habitat fragmentation caused by major roads leads to reduction of gene flow and loss of genetic variability in ground beetles. Proceedings of the Royal Society B: Biological Sciences 270:417–423.

Keller, I., W. Nentwig, and C. R. Largiadèr. 2004. Recent habitat fragmentation due to roads can lead to significant genetic differentiation in an abundant flightless ground beetle. Molecular Ecology 13:2983–2994.

Kiesecker, J. M., J. S. Evans, J. Fargione, K. Doherty, K. R. Foresman, T. H. Kunz, D. Naugle, N. P. Nibbelink, and N. D. Niemuth. 2011. Win-win for wind and wildlife: A vision to facilitate sustainable development. PLOS One 4:e17566. doi:10.1371/journal.pone.0017566.

Korfanta, N. M., M. L. Mobley, and I. C. Burke. 2015. Fertilizing western rangelands for ungulate conservation: An assessment of benefits and risks. Wildlife Society Bulletin 39:1–8.

Kuvlesky, W. P., L. A. Brennan, M. L. Morrison, K. K. Boydston, B. M. Ballard, and F. C. Bryant. 2007. Wind energy development and wildlife conservation: Challenges and opportunities. Journal of Wildlife Management 71:2487–2498.

LeBeau, C. W., J. L. Beck, G. D. Johnson, and M. J. Holloran. 2014. Short-term impacts of wind energy development on greater sage-grouse fitness. Journal of Wildlife Management 78:522–530.

LeBeau, C. W., J. L. Beck, G. D. Johnson, R. M. Nielson, M. J. Holloran, K. G. Gerow, and T. L. McDonald. 2017a. Greater sage-grouse male lek counts relative to a wind energy development. Wildlife Society Bulletin 41:17–26.

LeBeau, C. W., G. D. Johnson, M. J. Holloran, J. L. Beck, R. M. Nielson, M. E. Kauffman, E. J. Rodemaker, and T. L. Mcdonald. 2017b. Greater sage-grouse, habitat selection, survival, and wind energy infrastructure. Journal of Wildlife Management 81:690–711.

Lendrum, P. E., C. R. Anderson, R. A. Long, J. G. Kie, and R. T. Bowyer. 2012. Habitat selection by mule deer during migration: Effects of landscape structure and natural-gas development. Ecosphere 9:82. http://dx.doi.org/10.

Leu, M., S. E. Hanser, and S. T. Knick. 2008. The human footprint in the West: A large-scale analysis of anthropogenic impacts. Ecological Applications 18:1119–1139.

Long, C. V., J. A. Flint, and P. A. Lepper. 2011. Insect attraction to wind turbines: Does colour play a role? European Journal of Wildlife Research 57:323–331.

Łopucki, R., and I. Mróz. 2016. An assessment of non-volant terrestrial vertebrates response to wind farms: A study of small mammals. Environmental Monitoring and Assessment 188:1–9.

Lovich, J. E., and R. Daniels. 2000. Environmental characteristics of desert tortoise (Gopherus agassizii) burrow locations in an altered industrial landscape. Chelonian Conservation and Biology 3:714–721.

Lovich, J. E., and J. R. Ennen. 2013. Assessing the state of knowledge of utility-scale wind energy development and operation on non-volant terrestrial and marine wildlife. Applied Energy 103:52–60.

Lovich, J. E., J. R. Ennen, S. Madrak, K. Meyer, C. Loughran, C. Bjurlin, T. R. Arundel, W. Turner, C. Jones, and G. M. Groenendaal. 2011. Effects of wind energy production on growth, demography, and survivorship of a desert tortoise (Gopherus agassizii) population in southern California with comparisons to natural populations. Herpetological Conservation and Biology 6:161–174.

Lyon, A. G., and S. H. Anderson. 2003. Potential gas development impacts on sage grouse nest initiation and movement. Wildlife Society Bulletin 31:486–491.

Manville, A. M. 2004. Prairie grouse leks and wind turbines: U.S. Fish and Wildlife Service justification for a 5-mile buffer from leks; additional grassland songbird recommendations. Arlington, VA: USFWS, Division of Migratory Bird Management.

Mason, A., C. Driessen, J. Norton, and C. Strom. 2011. First year soil impacts of well-pad development and reclamation on Wyoming's sagebrush steppe. Natural Resources and Environmental Issues 17:41.

McNew, L. B., A. J. Gregory, S. M. Wisely, and B. K. Sandercock. 2012. Demography of greater prairie-chickens: Regional variation in vital rates, sensitivity values, and population dynamics. Journal of Wildlife Management 76:987–1000.

McNew, L. B., L. M. Hunt, A. J. Gregory, S. M. Wisely, and B. K. Sandercock. 2014. Effects of wind energy development on nesting ecology of greater prairie-chickens in fragmented grasslands. Conservation Biology 28:1089–1099.

Merkle, J. A., K. L. Monteith, E. O. Aikens, M. M. Hayes, K. R. Hersey, A. D. Middleton, B. A. Oates, H. Sawyer, B. M. Scurlock, and M. J. Kauffman. 2016. Large herbivores surf waves of green-up in spring. Proceedings of the Royal Society B 283:20160456.

Miller, R. F., S. T. Knick, D. A. Pyke, C. W. Meinke, S. E. Hanser, M. J. Wisdom, and A. L. Hild. 2011. Characteristics of sagebrush habitats and limitations to long-term conservation. In Studies in Avian Biology, edited by S. Knick and J. Connelly, 145–184. Berkeley: University of California Press.

Minnick, T. J., and R. D. Alward. 2015. Plant-soil feedbacks and the partial recovery of soil spatial patterns on abandoned well pads in a sagebrush shrubland. Ecological Applications 25:3–10.

Mockrin, M. H., and R. A. Gravenmier. 2012. Synthesis of wind energy development and potential impacts on wildlife in the Pacific Northwest, Oregon and Washington. USDA Forest Service General Technical Report PNW-GTR. https://www.fs.usda.gov/treesearch/pubs/40988.

Monteith, K. L., T. R. Stephenson, V. C. Bleich, M. M. Conner, B. M. Pierce, and R. T. Bowyer. 2013. Risk-sensitive allocation in seasonal dynamics of fat and protein reserves in a long-lived mammal. Journal of Animal Ecology 82:377–388.

National Research Council. 2007. Ecological effects of wind-energy development. In Environmental Impacts of Wind-Energy Projects, 67–139. Washington, DC: National Academies Press.

Nellemann, C., I. Vistnes, P. Jordhøy, O. Strand, and A. Newton. 2003. Progressive impact of piecemeal infrastructure development on wild reindeer. Biological Conservation 113:307–317.

Nickerson, G. 2011. Pathfinder developers hope to offset impacts of energy development. WyoFile. http://www.wyofile.com/banking-on-the-environment-pathfinder-developers-hope-to-offset-impacts-of-energy-development/. Accessed October 18, 2016.

Northrup, J. M., and G. Wittemyer. 2013. Characterising the impacts of emerging energy development on wildlife, with an eye towards mitigation. Ecology Letters 16:112–125.

Northrup, J. M., C. R. Anderson, and G. Wittemyer. 2015. Quantifying spatial habitat loss from hydrocarbon development through assessing habitat selection patterns of mule deer. Global Change Biology 21:3961–3970.

Patten, M. A., and J. F. Kelly. 2010. Habitat selection and the perceptual trap. Ecological Applications 20:2148–2156.

Pitman, J. C., C. Hagen, R. J. Robel, T. M. Loughin, and R. D. Applegate. 2005. Location and success of lesser prairie-chicken nests in relation to vegetation and human disturbance. Journal of Wildlife Management 69:1259–1269.

Pocewicz, A., H. Copeland, and J. Kiesecker. 2011. Potential impacts of energy development on shrublands in western North America. Natural Resources and Environmntal Issues 17:105–110.

Pruett, C. L., M. A. Patten, and D. H. Wolfe. 2009a. It's not easy being green: Wind energy and a declining grassland bird. BioScience 59:257–262.

Pruett, C. L., M. A. Patten, and D. H. Wolfe. 2009b. Avoidance behavior by prairie grouse: Implications for development of wind energy. Conservation Biology 23:1253–1259.

Rabin, L. A., R. G. Coss, and D. H. Owings. 2006. The effects of wind turbines on antipredator behavior in California ground squirrels (*Spermophilus beecheyi*). Biological Conservation 131:410–420.

Robel, R., J. Harrington Jr., C. Hagen, J. C. Pitman, and R. R. Reker. 2004. Effect of energy development and human activity on the use of sand sagebrush habitat by lesser prairie-chickens in southwestern Kansas. Transactions of the North American Natural Resources Conference 69:15.

Rydell, J., L. Bach, M. J. Dubourg-Savage, M. Green, L. Rodrigues, and A. Hedenström. 2010. Mortality of bats at wind turbines links to nocturnal insect migration? European Journal of Wildlife Research 56:823–827.

Samson, F., and F. Knopf. 1994. Prairie conservation in North America. BioScience 44:418–421.

Samson, F. B., F. L. Knopf, and W. R. Ostlie. 2004. Great Plains ecosystems: Past, present, and future. Wildlife Society Bulletin 32:6–15.

Sawyer, H., and M. J. Kauffman. 2011. Stopover ecology of a migratory ungulate. Journal of Animal Ecology 80:1078–1087.

Sawyer, H., R. M. Nielson, F. Lindzey, and L. L. McDonald. 2006. Winter habitat selection of mule deer before and during development of a natural gas field. Journal of Wildlife Management 70:396–403.

Sawyer, H., M. J. Kauffman, and R. M. Nielson. 2009. Influence of well pad activity on winter habitat selection patterns of mule deer. Journal of Wildlife Management 73:1052–1061.

Sawyer, H., M. J. Kauffman, A. D. Middleton, T. A. Morrison, R. M. Nielson, and T. B. Wyckoff. 2013. A framework for understanding semi-permeable barrier effects on migratory ungulates. Journal of Applied Ecology 50:68–78.

Sawyer, H., N. M. Korfanta, R. M. Nielson, K. L. Monteith, and D. Strickland. 2017. Mule deer and energy development: Long-term trends of habituation and abundance. Global Change Biology 23:4521–4529.

Schroeder, M. A., and R. K. Baydack. 2001. Predation and the management of prairie grouse. Wildlife Society Bulletin 29:24–32.

Schuster, E., L. Bulling, and J. Köppel. 2015. Consolidating the state of knowledge: A synoptical review of wind energy's wildlife effects. Environmental Management 56:300–331.

Skarin, A., C. Nellemann, L. Rönnegård, P. Sandström, and H. Lundqvist. 2015. Wind farm construction impacts reindeer migration and movement corridors. Landscape Ecology 30:1527–1540.

Smith, J. A., C. E. Whalen, M. Bomberger Brown, and L. A. Powell. 2016a. Indirect effects of an existing wind energy facility on lekking behavior of greater prairie-chickens. Ethology 122:419–429.

Smith, K. T., J. L. Beck, and A. C. Pratt. 2016b. Does Wyoming's core area policy protect winter habitats for greater sage-grouse? Environmental Management 58:585–596.

Smith, J. A., M. B. Brown, J. O. Harrison, and L. A. Powell. 2017. Predation risk: A potential mechanism for effects of a windenergy facility on greater prairie-chicken survival. Ecosphere 6:e01835. doi:10.1002/ecs2.1835.

Sparling, D. W. 1981. Communication in prairie grouse, I: Information content and intraspecific functions of principal vocalizations. Behavioral and Neural Biology 32:463–486.

Taylor, K. L., J. L. Beck, and S. V Huzurbazar. 2016. Factors influencing winter mortality risk for pronghorn exposed to wind energy development. Rangeland Ecology and Management 69:108–116.

Taylor, R. L., B. L. Walker, D. E. Naugle, and L. S. Mills. 2012. Managing multiple vital rates to maximize greater sage-grouse population growth. Journal of Wildlife Management 76:336–347.

Tollefson, T. N., L. A. Shipley, W. L. Myers, D. H. Keisler, and N. Dasgupta. 2010. Influence of summer and autumn nutrition on body condition and reproduction in lactating mule deer. Journal of Wildlife Management 74:974–986.

Torres, R. T., and C. Fonseca. 2016. Perspectives on the Iberian wolf in Portugal: Population trends and conservation threats. Biodiversity and Conservation 25:411–425.

Tsegaye, D., J. E. Colman, S. Eftestøl, K. Flydal, G. Røthe, and K. Rapp. 2017. Reindeer spatial use before, during and after construction of a wind farm. Applied Animal Behaviour Science 195:103–111.

US Department of the Interior. 2017. Decision record (DR) for phase 1 wind turbine development: Chokecherry and Sierra Madre (CCSM). Rawlins, WY.

USFWS (US Fish and Wildlife Service). 2003. Interim guidelines to avoid and minimize wildlife impacts from wind turbines. Washington, DC: Department of the Interior.

USFWS (US Fish and Wildlife Service). 2012. U.S. Fish and Wildlife Service land-based wind energy guidelines. Washington, DC: Department of the Interior.

USFWS (US Fish and Wildlife Service). 2017. Sweetwater River Conservancy Conservation Bank frequently asked questions. https://www.fws.gov/greatersagegrouse /QandAs/20150308Sweetwater River Conservancy Conservation Bank FAQs FINAL.pdf. Accessed June 30, 2017.

Vistnes, I., and C. Nellemann. 2008. The matter of spatial and temporal scales: A review of reindeer and caribou response to human activity. Polar Biology 31:399–407.

Walker, B. L., and D. E. Naugle. 2011. West Nile virus ecology in sagebrush habitat and impacts on greater sage-grouse populations. In Greater Sage-Grouse: Ecology and Conservation of a Landscape Species and Its Habitat, 127–142. Studies in Avian Biology 38. Berkeley: University of California Press.

Walker, B. L., D. E. Naugle, and K. E. Doherty. 2007. Greater sage-grouse population response to energy development and habitat loss. Journal of Wildlife Management 71:2644–2654.

Walter, W. D., D. M. Leslie, and J. A. Jenks. 2006. Response of Rocky Mountain elk (Cervus elaphus) to wind-power development. American Midland Naturalist 156:363–375.

WEST, Inc. (Western EcoSystems Technology, Inc.) 2014. Anexo 5.3: Evaluación de riesgo para aves, murciélagos y mariposas monarca del Parque de Energía Eólica de Coahuila Coahuila, México. Prepared for EDP Renewables North America, Houston, Texas. Prepared by C. Gordon and A. Poe. Translated by Sofia Agudelo, Austin, Texas. June 27.

Winder, V. L., L. B. McNew, A. J. Gregory, L. M. Hunt, S. M. Wisely, and B. K. Sandercock. 2014a. Effects of wind energy development on survival of female greater prairie-chickens. Journal of Applied Ecology 51:395–405.

Winder, V. L., L. B. McNew, A. J. Gregory, L. M. Hunt, S. M. Wisely, and B. K. Sandercock. 2014b. Space use by female greater prairie-chickens in response to wind energy development. Ecosphere 1:3. http://dx.doi.org /10.1890/.

Winder, V. L., A. J. Gregory, L. B. McNew, and B. K. Sandercock. 2015. Responses of male greater prairie-chickens to wind energy development. Condor 117:284–296.

Wyoming Game and Fish Department. 2010. Wildlife protection recommendations for wind energy develop-ment in Wyoming. Cheyenne: Wyoming Game and Fish Department.

PART III SOLAR ENERGY, WATERPOWER, AND WILDLIFE CONSERVATION

8 Solar Energy

A Technology with Multi-Scale Opportunities to Integrate Wildlife Conservation

Brian B. Boroski

Introduction

Solar energy technologies represent a rapidly developing renewable energy source worldwide (REN21 2016). Technological improvements in efficiency of solar energy are decreasing the time needed to recover initial investments, making solar power a financially competitive renewable energy source (Fthenakis et al. 2009, Dincer 2011). In 2016, for the first time in the United States, solar energy represented the largest new source of electricity-generating capacity, exceeding both natural gas and wind (Wood Mackenzie and SEIA 2016). While California, Georgia, and Utah collectively led solar energy installation within the United States in 2016, a record 22 additional states each added more than 100 megawatts$_{DC}$ (MW, direct current) of photovoltaic (PV) solar (Wood Mackenzie and SEIA 2016). China, Germany, and Japan documented increased growth in solar energy, with each installing greater solar photovoltaic capacity in 2015 than the United States (REN21 2016). Utility-scale (i.e., ≥ 1 MW) solar energy development is expected to grow worldwide with the strong international consensus to transition away from fossil fuels. Such growth will require large quantities of land for development, which could degrade or destroy habitat for some wildlife (REN21

2016). Nationally, total installed solar PV capacity is expected to nearly triple the current capacity between 2018 and 2022, and more than 18 gigawatts$_{DC}$ (GW) of solar PV capacity is predicted to be installed annually (Wood Mackenzie and SEIA 2016); given current technologies and practices, this growth is estimated to require an additional 36,422 to 72,844 ha per year (BLM/DOE 2012, Ong et al. 2013). The US Department of Energy (DOE) SunShot Initiative envisions that utility-scale solar electricity will affect approximately 849,775 to 2,509,181 ha of land by 2050 (US DOE 2012). Because of the broad geographical extent of utility-scale solar energy, development is occurring in many areas with high biodiversity and rare animal species. Consequently, understanding the interface between utility-scale solar energy development and wildlife conservation is increasingly important as utility-scale solar facilities continue to be constructed worldwide.

Construction and operation of solar energy facilities can have direct and indirect effects on wildlife, which may be negative or positive depending on existing landscape conditions at the time of construction, and may affect multiple trophic levels as well as trophic-level interactions (Hernandez et al. 2014b, Grodsky et al. 2017). Positive effects are most likely to occur on previously altered (e.g., marginalized) landscapes with limited wildlife habitat value. Negative effects are most likely to occur and be most severe in landscapes with previously intact biological communities, especially those characterized by relatively high plant and animal diversity. Examples of negative effects may include direct mortality, broad-scale habitat modification and fragmentation, changes to microclimate within solar arrays, increased levels of light and noise, invasion of non-native plant or animal species, and increased risk of wildfire (Grodsky et al. 2017, Lovich and Ennen 2011, Moore-O'Leary et al. 2017). Landscape connectivity may be reduced or enhanced as a result of siting and development of solar facilities, depending on species involved, existing conditions and location of the site selected for development, operation and maintenance activities during the life of the project, and policy decisions regarding the long-term use of the property. Few peer-reviewed studies on effects of utility-scale solar energy on wildlife exist, making assessments of effects of these systems on wildlife challenging (Lovich and Ennen 2011).

More information on potential effects of utility-scale solar development on wildlife is critically needed and may assist in proactive wildlife conservation. In this chapter, I describe the process for siting, constructing, and operating utility-scale solar energy facilities, summarize the literature on effects of solar energy development on wildlife, and suggest multi-scale measures to enhance potential positive effects and mitigate potential negative effects of solar energy development on wildlife. I provide a Deep Dive that illustrates how topics discussed in the chapter related to a project in California—a state with an ambitious renewable energy goal of 50% renewable electricity generation by 2030 and that is home to 154 animals listed under state and federal Endangered Species Acts (California Department of Fish and Wildlife 2017).

Solar Energy Technologies

Two basic types of solar energy technologies are deployed at the utility scale: PV and concentrating solar power (CSP). Photovoltaic systems use cells to convert sunlight to electric current, whereas CSP systems use reflective surfaces to concentrate sunlight to heat a receiver. Utility-scale CSP systems may comprise CSP trough, CSP tower, parabolic dish, or linear Fresnel reflector technologies (Mendelsohn et al. 2012). Photovoltaic systems usually include either crystalline silicon or thin-film technologies. In 2015, global operating capacity of CSP systems was approximately 4.8 GW compared with approximately 227 GW from PV systems (REN21 2016). Because utility-scale PV projects represent the majority of solar energy generation facilities (Mendelsohn et al 2012), they are the primary focus of this chapter. However, many of the topics discussed herein pertain to PV and CSP systems alike.

Photovoltaic technology comprises individual solar panels, referred to as PV modules, which are joined together in a series of increasing voltage to form an array (California Energy Commission 2001). Photovoltaic modules encompass the direct current power side of the overall solar power system, absorbing energy from the sun and converting it to electrical energy. PV modules may be fixed to a stationary rack or a moveable racking system, known as single-axis or dual-axis trackers, that enable panels to rotate to follow the sun. Because a variety of technologies are used (e.g., monocrystalline silicon, thin film), with different rated efficiencies and different variables for performance calculations, the size of arrays used to generate the same electrical output varies. Multiple arrays are connected together (i.e., in parallel) through a combiner box, and multiple combiner boxes feed into an inverter (California Energy Commission 2001). Inverters change the direct current unidirectional flow of electrons to alternating current (AC), and then a single transformer can convert the voltage to different levels. Multiple inverters can supply a transformer that is joined to a substation. From the substation, the electricity travels to the regional grid system.

Siting

Utility-scale solar energy development is projected to affect up to several million ha of land in the United States by 2050 (US DOE 2012). With PV systems, row spacing is important to maximize power density—kilowatt (KW)/ha—while minimizing panel shading, because power output can be greatly diminished by partial shading of a cell, panel, or array (Deline et al. 2011). Ong et al. (2013) reported that capacity-weighted averages for utility-scale solar energy ranged from a direct land-use requirement of approximately 3 ha/MW_{AC} (i.e., land directly occupied by solar arrays, access roads, substations, service buildings, and other infrastructure) to a total land-use area requirement of 3.6 ha/MW_{AC}, where total calculated area corresponds to all land enclosed by the solar site boundary. Moreover, suitability of a location to support a utility-scale solar energy facility depends on solar energy resources, topography, land cost, available transmission interconnections and capacity, and engagement and needs of utilities, among other factors (Hernandez et al. 2014a, Hartmann et al. 2016). Consequently, much effort has been allocated to studying siting of solar power facilities from an industry perspective, including development of software for siting tools (Macknick et al. 2014, Hernandez et al. 2015). A growing number of publications address solar facility siting from a broader land-use perspective, including identifying potential effects on wildlife habitat and areas of "least conflict" from a multidisciplinary perspective (Brown and Whitney 2011, BLM/DOE 2012, Hernandez et al. 2014a, 2015, BLM 2016, Pearce et al. 2016).

A fundamental principle for integration of wildlife conservation and utility-scale solar development is to avoid siting facilities in ecologically sensitive areas (Hernandez et al. 2015). Development of sites within or functionally linked to ecologically sensitive areas is likely to be less compatible with conservation objectives and requires more mitigation. Although the definition of what represents an ecologically sensitive area is broad, existing designations can aid in selection of sites that avoid conflicts with regional and national conservation goals, including designations of critical habitat (https://www.fws.gov/endangered/what-we-do/critical-habitats-faq.html), special protected areas (http://jncc.defra.gov.uk/page-162), and important bird and biodiversity areas (http://www.audubon.org/important-bird-areas, http://www.biodiversitya-z.org/content/important-bird-and-biodiversity-areas-iba), to name just a few. Compatibility of utility-scale solar development with conservation objectives should be assessed by ecologists familiar with local natural resource issues, core habitat areas, ecological interactions and functions (e.g., animal movement), and species of conservation concern (Guiller et al. 2017).

Co-location is an important concept pertaining to siting of utility-scale solar facilities that can reduce project costs, increase efficiency, and minimize negative effects on wildlife compared with development on more remote and undisturbed landscapes (Hernandez et al. 2014b). Identification of solar energy zones is a co-location strategy designed to encourage energy production and transmission in areas with reduced value for wildlife and natural resource conservation (BLM/DOE 2012). Siting decisions should give priority to previously degraded lands, including abandoned mines, landfills, and other sites where utility-scale solar installation provides an opportunity to increase vegetative cover and conserve soil and water resources (Hernandez et al. 2014b, Hoffacker et al. 2017). Two federal efforts to identify degraded lands appropriate for renewable energy are examples that encourage this type of development: (1) The Restoration Design Energy Project in Arizona uses American Recovery and Reinvestment Act (2009) funds to identify and assess degraded lands, including brownfields, abandoned mines, and landfills; and (2) the US Environmental Protection Agency's RE-Powering America's Land program evaluates the renewable energy potential of brownfields, Resource Conservation and Recovery Act sites, Superfund sites, and abandoned landfills and mines and describes considerations to address when integrating renewable energy development on such sites (US Environmental Protection Agency 2012). The solar development potential of these sites is estimated to be more than 920,000 MW. Hartmann et al. (2016) suggested that siting solar projects on certain types of previously used lands (e.g., formerly contaminated sites and other ecologically degraded lands) might provide the total land requirements to meet some national renewable energy program goals.

While avoiding impacts to rare species and their associated habitat will remain critical for wildlife conservation, avoidance alone is not likely to be enough to reach conservation goals. Siting criteria and compensatory mitigation offsets will continue to be important considerations (McDonald et al. 2009). A broader, multiple-use perspective of habitat conservation is likely to aid conservation efforts in the face of industrial expansion (Hernandez et al. 2014b, Francis 2015), including utility-scale solar development.

Utility-scale solar facilities possess unique features that other types of human structures lack. Whereas there is variability in the manner in which solar arrays are installed, the posts on which panels are mounted take up less than 1% of the land area of a facility; however, they are spaced apart to avoid shading of adjoining panels, which results in approximately 25% to 40% of the ground being shaded by raised panels (BRE 2014). BRE (2014) estimated that solar infrastructure, including inverters and other types of hardscape structures, typically overlay less than 5% of the ground. Because panels are raised (e.g., up to 2 m in height) and the distance between rows of panels can span from approximately 6 m when the trackers are horizontal to the ground to approximately 8 m when the trackers are tilted at 60 degrees, the area between and underneath arrays post-construction can sustain plant growth (Figure 8.1). With proper enhancement and management, including through complementary agricultural activities such as grazing, this area has the

Figure 8.1 A vegetated PV solar array in spring. *H. T. Harvey & Associates.*

potential to support habitat for a multitude of species (Hernandez et al. 2014b).

Maintaining electrical productivity and meeting wildlife conservation goals will be greatest when solar facilities are sited in areas with highly degraded habitat conditions. In some cases, even sites with moderate habitat value can be developed and contribute to broader conservation benefits through increasing landscape connectivity and habitat patch size (Guiller et al. 2017). In the San Joaquin Valley of California, where solar development potential is high, habitat for species of conservation concern has been extensively fragmented by agriculture and urban development (USFWS 1998, 2010, Cypher et al. 2013). In situations such as this, potential for improving conditions for species by providing habitat within solar arrays and, thereby, increasing habitat connectivity is high (Phillips and Cypher 2015). Cypher et al. (2007) recommended establishing corridors and improving connectivity for listed species in a portion of the San Joaquin Valley that included salt-impaired farmlands within the Westlands Water District. This area constitutes a large part of the 81,880 ha of "priority least-conflict" lands identified for utility-scale solar development in the San Joaquin Valley (Pearce et al. 2016), highlighting the great potential, in some circumstances, for solar energy development to contribute to regional conservation strategies and recovery efforts (USFWS 1998).

Wildlife Considerations during Construction and Operation

Wildlife professionals and the solar industry have taken actions to ensure that utility-scale solar projects are developed and operated in compliance with local, state, and national regulatory standards pertaining to wildlife. In 2011, Natural England (2011) published a technical information note with recommendations for siting solar projects, their potential impacts on wildlife, mitigation requirements for maintaining the natural environment, and opportunities for environmental stewardship. In 2012, the

BLM and DOE published a report, "Final Programmatic Environmental Impact Statement for Solar Energy Development in Six Southwestern States" (BLM/DOE 2012). The Renewable Energy Working Group of The Wildlife Society (www.wildlife.org /rewg) was established to facilitate the synthesis and dissemination of scientific information promoting responsible development of various renewable energy sectors that minimize impacts on wildlife through informed siting, project design, construction, operation, and decommissioning stages. More recently, Hartman et al. (2016) summarized emerging environmental impacts related to utility-scale solar development.

Solar projects should look beyond the traditional approach of avoidance, minimization, and mitigation during the preconstruction period to conserve wildlife (Council on Environmental Quality 1993). Through preconstruction surveys, wildlife professionals can help locate activities in less sensitive areas and minimize impacts to biodiversity, thereby integrating wildlife conservation into utility-scale solar development. Along these lines, the BRE National Solar Center in the United Kingdom developed a biodiversity guidance document for solar developers (BRE 2014); the document recommended a management plan be developed for solar facilities to describe site-wide objectives for biodiversity and how those objectives will be achieved throughout the life of the project. The plan should identify potential impacts arising from the site's development and outline mitigation, which often is a regulatory requirement. The management plan also should include land management activities (e.g., control of invasive weeds and vegetation management). A benefit of a management plan is the early consideration that it initiates for all phases of the project (e.g., preconstruction design and mitigation considerations, avoidance measures during construction and operational activities), which facilitates planning and budgeting efforts.

Examples are numerous of site-selection processes that rely on existing databases to identify elements such as prescreened renewable energy zones and areas that possess higher value for other purposes such as prime farmland or federally designated critical habitat (Macknick et al. 2014, Phillips and Cypher 2015, Pearce et al. 2016). Assorted databases provide different information; for instance, regional databases such as the California Gap Analysis Project and National Wetlands Inventory are often based on remotely derived data that are coarser than what may be obtained on a site-specific basis. Other databases, such as the California Natural Diversity Database, rely on volunteer data and rarely portray recent submittals because of the effort required to manage them. Data for public land might be fairly complete because of focused planning efforts (BLM 2015), whereas data for private lands might be absent because of a lack of prior access.

Conducting on-site baseline surveys in consultation with regulatory agencies is critical during the preconstruction phase of solar development to assess the presence or absence of natural resources, including those necessary for enhancement or restoration. The level of effort needed to establish a sufficient baseline for the purposes of designing the facility to achieve operational goals and avoid and minimize impacts on sensitive wildlife resources will vary. For species with high conservation value, multiyear surveys may be recommended by existing protocols (https://www.fws.gov/sacramento/es/Survey-Protocols -Guidelines/). In the western United States, however, many recommended protocols were established before utility-scale solar development was commonplace and were not designed with the thought of surveying hundreds to thousands of hectares, which may be necessary should extensive solar energy expansion arise.

During the preconstruction phase, survey data can be used to avoid siting solar in areas with high concentrations of focal wildlife. Multiyear surveys for species with population fluctuations, such as those exhibiting boom-and-bust cycles coincident with rainfall, may be important during the preconstruction phase of the project to capture year-to-year

variability. If wildlife populations are suppressed during baseline surveys, survey results can indicate potential areas where expansion could occur just prior to or during the construction phase and after favorable changes in environmental conditions, such as precipitation after a series of drought years (H. T. Harvey & Associates 2017b). Baseline documentation for resources that are relatively fixed, such as availability and extent of wetland resources, may take fewer years to establish.

Basic infrastructure associated with solar arrays and facility construction can have both negative and positive impacts on certain wildlife species. Unlike many forms of human development, PV solar generation facilities have relatively low levels of human activity after construction. During operation of the facility, inspections, testing, maintenance, and repairs are performed continually, with most routine activities occurring only once or twice per year for each tracker or major piece of equipment. Installed infrastructure requires minimal replacement of panels and equipment as a result of breakage or failure, and inverters currently require replacement approximately once every decade (H. T. Harvey & Associates 2015a). Most of this limited operational activity occurs during the day; however, limited nighttime work may occur while the system is de-energized for safety and to maintain energy production during the day. Most sites have a life span of 20 to 35 years or more. In some cases, commitments are made prior to construction to place the lands supporting a facility under a conservation easement, either when the facility becomes operational or after decommissioning, thereby providing the potential for wildlife conservation benefits in perpetuity (EMC Planning Group 2012, Quad Knopf 2014).

Areas underneath arrays are likely to be more heterogeneous than open fields owing to effects of shading and driplines (i.e., areas collecting rain runoff from panels). H. T. Harvey & Associates (2016a) reported establishment of 3 main vegetation zones at a solar facility in California: (1) the area directly beneath the panels; (2) the area along the dripline;

and (3) the area between the panels. In areas of the solar facility recovering from construction, H. T. Harvey & Associates (2016a) reported that, 3 years after construction, the vegetation community supported more native plant species than adjacent reference sites, but index values of diversity were similar.

Montag et al. (2016) compared wildlife diversity in solar facilities with nearby areas outside facilities under similar management regimes, such as livestock grazing. Overall, solar facilities had greater levels of plant diversity than other areas (Montag et al. 2016), possibly due to reseeding of species-rich wildflower mixes. Similarly, the study reported that abundance of butterflies and bumblebees was greater in solar facilities than outside solar facilities and that numbers of these pollinators were greatest where botanical diversity also was high. The researchers also documented a greater diversity of birds within the solar facilities and a greater abundance of birds within some of the solar facilities, when compared with paired adjacent areas.

Bird activity may change during and after solar facility construction. In California, development tended to displace most raptors, but establishment of new elevated perches from solar infrastructure in a grassland landscape benefited some raptor species, and use of solar facilities by some raptor species may increase as facilities transition to operational status (H. T. Harvey & Associates 2015b). Responses of other bird species to solar development are mixed. Potential reasons for positive responses of birds to project development may include (1) provision of structures that attract cavity or crevice nesters; (2) installation of solar panels that provide perch substrates and abundant shade in otherwise hot, dry landscapes; (3) provision of refuge for prey from predators (e.g., raptors) that tend to avoid construction activities and developed areas; (4) creation of new microhabitats that favor or concentrate some types of insect prey (e.g., in wind-sheltered areas behind solar panels); and (5) provision of opportunities for various human-adapted species to take advantage of ground-disturbing activities that routinely turn up insects, seeds, and other food resources (H. T. Harvey & Associates 2015b).

Other components of infrastructure can be problematic to wildlife, but solutions exist to minimize harm. Transmission lines often pose risks to wildlife species, a problem that similarly has been documented for other energy sectors. Site-specific assessments are recommended to design electrical infrastructure (Barrientos et al. 2011, APLIC 2012). Without modification, perimeter security fencing can represent a barrier that cause some species, such as the greater roadrunner (*Geococcyx californianus*), to be more prone to predation. Some fence designs are permeable to small and medium-sized wildlife (Cypher and Van Horn Job 2009) and others have been modified to facilitate passage of wildlife by raising the fence or modifying the standard weave to incorporate passageways spaced along the length of the fence (Cypher and Van Horn Job 2009, Cypher et al. 2012). Where solar arrays fragment landscapes used by migrating wildlife or those with large area requirements, arrays can be individually fenced and spaced apart to provide habitat linkages with the surrounding landscape.

Materials used in the construction process can also have a direct effect on wildlife but, similar to the situation with infrastructure described previously, solutions often exist to minimize harm. Hollow spherical piles used in the installation of solar arrays can easily number in the thousands or even tens of thousands, and each represents a potential trap for birds. In such cases, open ends can be capped or open-sided piles can be used to preclude avian mortality from entrapment. Some wildlife, including the San Joaquin kit fox (*Vulpes macrotis mutica*) in California, may use staged materials on the construction site for shelter. Depending on species-specific management goals, staging materials can be stored or stacked in a manner less attractive to wildlife or plugs and barriers can be installed to preclude wildlife use of the materials (Figure 8.2). Evaporation ponds, often used to store effluent from reverse osmosis

Figure 8.2   Example of piles plugged and barricaded to minimize use by wildlife. *H. T. Harvey & Associates.*

systems used to purify ground- or drain-water, may also present a risk to wildlife. Barriers can be designed to restrict access to the ponds by wildlife, and heavy-duty, textured pond liners may provide a means of escape for wildlife that become trapped. Basin configuration can be designed so as to deter use of the ponds by birds, bats, and invertebrate prey that may be susceptible to toxins such as selenium.

Protecting critical habitat elements, such as burrows, during the construction period may facilitate wildlife presence within arrays post-construction. Although barriers to and ramps inside open trenches are commonly employed to minimize impacts to wildlife at construction sites, the use of one-way door exclusion devices represents a specialized form of barrier more applicable to solar sites because of the habitat that can be maintained under the arrays for burrowing animals (Figure 8.3). These exclusion devices can be deployed within naturally occurring burrows or atypical dens, such as pipes or culverts, to preclude wildlife from entering the burrow while enabling them to escape. If the burrow occurs in the vicinity of a planned improvement but does not need to be filled to complete the construction process, the one-way door can be left in place during the construction period. When a significant percentage of burrows must be removed because of the installation of hardscape, installation of artificial burrows may facilitate the use of burrows in the solar facility without increased risk of predation while the natural occurrence of burrows is restored (Cypher et al. 2012).

Ongoing human activities during operation and maintenance of solar facilities can benefit or threaten wildlife resources. Solar facilities typically operate 7 days a week during daylight hours. Periodic monitoring and maintenance activities, both planned (e.g., routine inspection, repair, restoration, replacement, washing of solar panels, security surveys and actions, vegetation management, and modification of equipment and facilities) and unplanned (e.g., damage or failure of equipment or facilities), take place during daytime and nighttime hours. Management plans should address planned and unplanned maintenance activities throughout the life of the project. I summa-

Figure 8.3   Example of a one-way door exclusion device for San Joaquin kit fox. *H. T. Harvey & Associates.*

rize general measures that can be applied to minimize impacts on species and habitats from operational activities in Table 8.1.

The degree of habitat restoration or revegetation required following construction of a solar facility will depend on conditions prior to construction, site preparation, the practices used for construction, and desired long-term conservation goals for the site. A habitat restoration and revegetation plan that focuses on soil restoration and revegetation or restoration of areas disturbed during construction and operations is recommended. The plan should describe the quantitative monitoring of revegetation and restoration areas that will be conducted following planting. An annual status report summarizing site conditions and results of monitoring is recommended at the end of each year following implementation of the plan and should describe adaptive management measures to employ if vegetation cover success criteria were not met. The restoration goal should be clearly stated; it can be beneficial to relate the goal to conditions on adjacent undeveloped land that is concur-

rently sampled. In this way, variability in amount and timing of precipitation and its effect on vegetation growth can be used to inform progress toward meeting the success of revegetation criteria.

Appropriate vegetation management is vital to ensure habitat maintenance or enhancements achieve conservation goals. It is important to note that different areas within a utility-scale solar installation may be managed to meet different objectives. For instance, lower-growing native grasses and forbs may be managed within arrays whereas taller species, such as some milkweeds (*Asclepias* spp.) that support monarch butterflies (*Danaus plexippus*), are restored around the perimeter of arrays. In fact, establishment of native, pollinator-friendly plants between solar arrays is an increasingly popular approach for integrating conservation with renewable energy development. Wildfire risk is an important consideration within a vegetation management plan. Mowing to reduce or remove vegetative cover can be effective and is commonly used within smaller (less than 20 MW) solar arrays. Mowing can pose greater risk of injury to wildlife than the use of livestock grazing, so application of avoidance and minimization measures as described below are important.

Livestock grazing is a historical land use within many areas selected for solar development, and continued use of livestock grazing can serve as a primary vegetation management tool that enables co-location of solar energy and agricultural production. A grazing management plan should be developed to describe specific grazing approaches that will be used within arrays to manage vegetation, the rationale for these approaches, and how a monitoring program will be implemented within an adaptive management framework to assess the effectiveness of livestock grazing as a management tool.

The vegetation management component of the site management plan should address how to (1) minimize wildfire risk by managing vegetation height to reduce fuel loads while minimizing soil erosion to maintain basic resource values; and (2) maintain habitat

Table 8.1.   General measures that can be applied to minimize impacts of operational activities on species and their associated habitat

Normal operational activities

Provide an education program for site workers to cover the ecology of sensitive species that occur within the facility, how to identify and protect them, and the contact information for communicating when species are located.

Employ rodent control methods, including vegetation management, trapping, fumigation, or the careful use of toxic baits while observing label and other restrictions mandated by laws and regulations.

Restrict or prohibit use of erosion-control materials potentially harmful to wildlife, such as monofilament netting (erosion control matting) or similar material.

If design specifications allow, stow solar panels flat from sunset to sunrise to minimize the potential risk of bats striking the panels. Bats may sometimes perceive more vertically positioned smooth-surfaced panels as a void or open space, which could result in collisions.

Restrict access to the facility to emergency personnel and the operator's designees or employees, composed primarily of the operations and maintenance staff, panel-washing staff, environmental compliance specialists, researchers, and security personnel.

If nocturnal threatened or endangered species are present, operational activities should generally occur during daylight hours (sunrise to sunset); exceptions may involve, but are not limited to (1) capacitor bank wiring, connecting, and testing; (2) planned and unplanned maintenance activities that must occur after dark to ensure PV arrays are not energized; (3) interior use of the facility; (4) unanticipated emergencies (defined as an imminent threat to life or to a significant property interest), including forced outages and non-routine maintenance or repair requiring immediate attention; or (5) security patrols, which may occur 24 hours a day.

Transformers within arrays may contain food-grade oils that can be attractive to wildlife. If a transformer is suspected of leaking, the entire power conditioning station where the transformer is located should be enclosed with an exclusion barrier impermeable to wildlife until the leak can be fixed and cleaned up.

Vehicle/equipment operating and fueling procedures

Do not operate vehicles and equipment outside existing access roads within arrays except in compliance with the management plan or in the event of an emergency, and vehicles or equipment traveling between arrays should cross drainages only at improved crossings.

When threatened or endangered wildlife are present, restrict the speed limit on all roads within the facility to approximately 24 km per hour between sunrise and sunset and 16 km per hour between sunset and sunrise.

Stationary and mobile lighting, and fencing

Minimize light sources and kept to the lowest lumen/light intensity level possible, while still meeting minimum safety and security requirements.

Design solar array fencing to allow passage of wildlife compatible with operations. Fencing can be raised or have graduated openings from approximately 7 cm high to 15 or 18 cm high, and be installed with the larger openings at the bottom to allow wildlife passage.

Maintain fencing and gates around evaporation ponds to exclude terrestrial wildlife. The fence should be buried at least 0.6 m in the ground and extend at least 1.8 m above ground. Project personnel should photograph, document, and report to a qualified biologist the discovery of any dead wildlife at the evaporation pond to inform and improve adaptive management.

Weed control activities

Include weed control in the site management plan to prevent the spread of non-native, invasive species within the construction footprint and operational areas of the facility. The weed control plan should describe procedures to avoid the unintentional introduction of invasive species, required monitoring measures to detect invasive species, species-specific control measures that should be implemented if invasive species are detected, and the process by which the weed control plan should be implemented (e.g., the entity responsible for implementing it, funding mechanisms, and reporting procedures). The operator should monitor the facility each year for the presence of invasive weeds pursuant to the weed control plan.

Ground-disturbing or vegetation-disturbing activities

Limit disturbance to the smallest area practicable, schedule activities outside the breeding season, establish avoidance buffers, and thoroughly inspect excavated areas for trapped wildlife each morning before the onset of activities, before covering with plywood at the end of each workday, and before filling.

for wildlife species that may use arrays and respond to annual changes in climate and productivity of the vegetation community. Forb and grass biomass should be managed through grazing or mowing to encourage habitat that supports plant and animal diversity and manage residual dry matter (i.e., herbaceous plant material left standing or on the ground at the beginning of a new growing season). Intensity and duration of grazing, as well as the kinds and classes of livestock used for grazing, should be discussed, as well as when and how mowing may be used to meet vegetation goals, including mitigating wildfire risk. Numerous factors, including past vegetation management practices, climatic patterns, and wildfire affect the annual amount of grass and forb growth; therefore, monitoring often is required annually to adjust the intensity and duration of livestock grazing and identify areas that require mowing.

The Deep Dive discusses Phillips and Cypher's (2015) cited observations of 2 mammals listed under federal and state endangered species acts—the San Joaquin kit fox and the giant kangaroo rat (*Dipodomys ingens*)—using utility-scale solar facilities as well as 3 other species that may use solar facilities, given their similar habitat requirements.

Deep Dive: Meeting Wildlife Conservation Goals While Maintaining Electrical Productivity

The Renewable Energy Transmission Initiative is a California planning process to accommodate state renewable energy goals that has identified Competitive Renewable Energy Zones possessing the greatest potential for cost-effective and environmentally responsible development (http://www.energy.ca.gov/reti/). The California Valley Solar Ranch, located in the South Carrizo Competitive Renewable Energy Zone, is a 250 MW solar PV project on an 1,896 ha site in eastern San Luis Obispo County, California. When finalizing the initial site design, the developer had the option of determining where to locate the site within a 2,023 ha area composed of mostly grassland established following the cessation of dryland farming (Holland 1986). Within this area, the developer considered approximately 1,921 ha for siting of the solar facility. Ecological assessments and surveys within the area spanned 4 years and identified 7 plant species of conservation concern considered rare by the California Native Plant Society (http://www.cnps.org/cnps/rareplants/ranking.php). These plant species were broadly distributed across the project area and ranged from small, isolated populations of several plants to large, clearly defined polygons of land cover.

Surveys also identified 3 state and federally listed mammal species: (1) the federally and state-listed endangered giant kangaroo rat; (2) the federally and state-listed threatened San Joaquin kit fox; and (3) the state-listed threatened San Joaquin antelope squirrel (*Ammospermophilus nelsoni*; San Luis Obispo County 2010). Designated critical habitat for vernal pool species, the federally listed endangered longhorn fairy shrimp (*Branchinecta longiantenna*) and threatened vernal pool fairy shrimp (*Branchinecta lynchi*) also occurred within the project area (San Luis Obispo County 2010). Consequently, the initial design was revised through an iterative process to avoid construction near these resources, including the majority of giant kangaroo rat precincts (Figure 8.4).

The sections that follow present an overview of best management practices implemented during construction and operations at the facility from 2011 through 2017 for 2 focal species: San Joaquin kit fox and giant kangaroo rat. To mitigate habitat impacts on these species, permits required the permanent protection and management of approximately 3,706 ha, including a non-wasting endowment to cover the management, maintenance, and monitoring of these lands for the species in perpetuity.

Some giant kangaroo rats were translocated to surrounding on-site conservation lands using "soft release" methods to acclimate the

Figure 8.4 Distribution of giant kangaroo rat precincts relative to position of arrays of the California Valley Solar Ranch prior to construction. *H. T. Harvey & Associates.*

translocated animals to the release sites (H. T. Harvey & Associates 2011, 2013). A total of 427 precincts were mapped and trapped in an attempt to capture and relocated kangaroo rats before the sites were excavated (and, therefore, destroyed). More than 220 giant kangaroo rats were removed from construction areas, nearly all of which were successfully relocated to artificial burrow enclosures located within conservation lands (H. T. Harvey & Associates 2013).

In addition to the extensive redesign, avoidance of giant kangaroo rat precincts occurred during preconstruction and construction phases. Each individual or cluster of precincts identified as occurring within the solar facility was evaluated to identify which could be avoided. Consequently, 261 giant kangaroo rat precincts within planned construction impact areas were avoided through engineered redesign or by modifying construction methods (H. T. Harvey & Associates 2013).

San Joaquin kit fox use of the project site was carefully monitored. Measures for avoiding and protecting dens included restricted-entry buffers and use of one-way doors to protect unoccupied dens in place for future use. Excavation was a last resort for dens that would be directly affected by construction. Biologists closely monitored San Joaquin kit fox activity during construction. In 2013 alone, 416 sightings of San Joaquin kit foxes occurred in the construction area. By adhering to protective measures, no San Joaquin kit fox mortalities were documented during construction activities.

San Joaquin kit fox dens occurred broadly across the site (Figure 8.5). When a den required a one-way door, a minimum of 4 nights of camera data were collected before installation. All one-way doors were checked once per week in January and February and 5 times per week from March through October (i.e., pup rearing and dispersal period) for signs of digging around the boards, for general integrity and functionality, and to ensure that doors remained visible to construction crews. At the end of construction, one-way doors were removed and dens were available for reoccupation. Following one-way-door removal, reuse of dens was common and similar to levels recorded prior to construction.

During operations, sheep grazing was used in arrays to reduce vegetation height and biomass, with some supplemental mowing to reduce wildfire risk. Sheep grazing occurred in spring and usually included at least 1 rotation through each array. Bedding locations for sheep were kept away from sensitive biological resources, such as giant kangaroo rat precincts. Bedding and watering locations were identified by a qualified biologist, and maps of these approved locations were provided in advance to the sheep herder.

During the 2016–2017 rainy season, the project site received an above-average amount of precipitation, resulting in increased vegetation heights and biomass levels that required mowing in arrays, particularly along solar panel driplines. The design of the array fencing allowed passage of San Joaquin kit fox and giant kangaroo rats between solar array areas managed to perpetuate grassland conditions and adjacent conservation lands. Prior to mowing, biologists conducted surveys of kit fox and kangaroo rat activity in each array and within a buffer extending 15 m beyond the perimeter fence. The pre-mowing surveys documented active giant kangaroo rat precincts and San Joaquin kit fox dens.

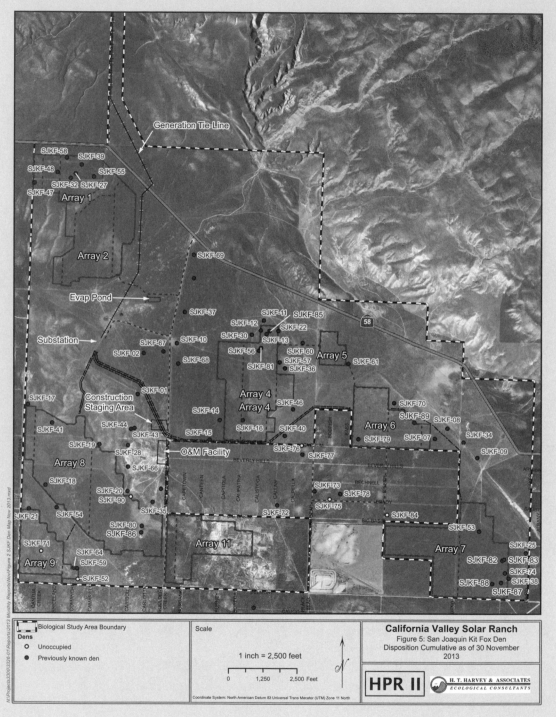

Figure 8.5 Cumulative distribution of San Joaquin kit fox dens between September 2012 and November 30, 2013, on the California Valley Solar Ranch. *H. T. Harvey & Associates.*

Table 8.2. Number of active giant kangaroo rat (*Dipodomys ingens*) precincts by array and in the buffer area of each array during 2017 surveys of California Valley Solar Ranch, San Luis Obispo County, California

Arrays	1	2	4	5	6	7	8	9	11	Total
Precincts	16	72	189	106	149	48	27	0	0	607
15 m buffer around arrays										
Precincts	5	122	106	132	62	26	25	26	47	551
Total	21	194	295	238	211	74	52	26	47	1,158

Table 8.3. Number of San Joaquin kit fox (*Vulpes macrotis mutica*) dens by array and type of den in 2017 at California Valley Solar Ranch, San Luis Obispo County, California

| Type of Den | Arrays | | | | | | | | | Total |
	1	2	4	5	6	7	8	9	11	
Historical	6	0	16	0	1	6	9	2	0	40
Natal	0	0	0	0	1	0	1	2	0	4
Escape	0	1	2	3	0	4	0	1	0	11
Total	6	1	18	3	2	10	10	5	0	55

The status of giant kangaroo rat precincts was determined using multiple indicators of aboveground activity, including suitably sized vertical shafts and horizontal openings, recently cropped vegetation, haystacks, pit-caching, tail drags, scat, and recent excavation activity. During surveys, biologists identified 1,158 active giant kangaroo rat precincts—607 within arrays and 551 precincts within 15 m buffers around the perimeter of array fencing. Precinct density in the buffer areas, which were not trapped and cleared during the construction period, was 11.54/ha compared with only 1.16/ha in the arrays, which were trapped and cleared of giant kangaroo rats via translocation between September 2009 and July 2013 (Table 8.2).

Fifty-five San Joaquin kit fox dens were located in arrays during surveys in 2017—40 inactive historical dens, 4 active natal dens, and 11 active escape dens (Table 8.3). Den use by San Joaquin kit foxes was identified through signs such as tracks, scat, and suitably sized dens, and by remote-sensing cameras. San Joaquin kit fox natal dens (Figure 8.6) were identified, monitored, and determined to be active within arrays at California Valley Solar Ranch in 5 out of 6 years. Prolonged drought and poor landscape condition appeared to cause low reproductive success for kit foxes in 2014.

The number of natal and escape dens within arrays indicated a stable use of arrays by San Joaquin kit fox, demonstrating the species' ability to use operating solar

Figure 8.6   San Joaquin kit fox adult with pups at a natal den within an array at the California Valley Solar Ranch.
H. T. Harvey & Associates.

facilities when properly designed, managed, and operated. Preliminary results from a San Joaquin kit fox telemetry study at California Valley Solar Ranch (H. T. Harvey & Associates in prep) support these inferences; radio-collared San Joaquin kit foxes used solar arrays and non-solar array areas.

The impact assessment used to permit the project (San Luis Obispo County 2011) assumed no future use of habitat by giant kangaroo rat within the arrays or within approximately 30.5 m beyond array perimeter fences. Results from the pre-mowing survey, however, demonstrated that giant kangaroo rats occupied some arrays in significant

numbers (Table 8.2) and did not avoid areas adjacent to arrays. A one-tailed t-test indicated that the estimated number of giant kangaroo rats occupying arrays in 2017 (based on the number of active precincts) was approximately 3 times greater (Table 8.4) than number of animals relocated from arrays during the construction period ($t = 2.28$, $p = 0.03$). Although data used for the t-test relied on a comparison of captured individuals versus active precincts, it is important to note that 27% of the animals relocated from the arrays were juveniles (H. T. Harvey & Associates 2013); the assumption is that an active precinct equates to one adult. If relocated juveniles we removed from the

Table 8.4. Comparison of giant kangaroo rat (*Dipodomys ingens*) numbers preconstruction (2011 through 2013) and post-construction (2017) at California Valley Solar Ranch, San Luis Obispo County, California

Array #	Preconstruction (2011 through 2013)	Post-construction (2017)
1	2	16
2	4	72
4	44	189
5	83	106
6	2	149
7	43	48
8	12	27
9	1	0
11	3	0
Total	194	607

analysis, the difference would be even greater.

The increase in number of giant kangaroo rat precincts within arrays did not mean that habitat within arrays was better than the pre-project conditions. Instead, increase in the number of giant kangaroo rats likely was in response to increased rainfall and vegetation productivity (Prugh et. al. 2015, H. T. Harvey & Associates 2017b). However, arrays did provide substantial habitat and added conservation benefits, supporting an estimated more than 600 individual giant kangaroo rats not anticipated to occupy the area post-construction.

Summary

Solar energy represents a rapidly developing and expanding renewable energy source (REN21 2016) that can influence wildlife habitat. Consequently, understanding the interface between utility-scale solar energy development and wildlife conservation is increasingly important. Construction and operation of solar energy facilities modifies wildlife habitat and can have direct and indirect effects on wildlife. Although siting of utility-scale solar facilities potentially affects landscape connectivity, the manner in which a solar facility is operated and maintained may also have positive or negative effects on landscape connectivity, because those activities can directly affect habitat values within the site during the life span of the project.

A fundamental principle for integrating wildlife conservation and utility-scale solar development is siting facilities to avoid ecologically sensitive areas. Facilities within or functionally linked to ecologically sensitive sites are likely to be less compatible with conservation objectives and require more mitigation. Compatibility of utility-scale solar development and conservation objectives should be assessed by ecologists familiar with key biodiversity issues in areas where construction may occur. Co-location of solar energy on marginalized lands is an important concept pertaining to the siting of utility-scale solar facilities and can reduce project costs, increase efficiency, and minimize wildlife conflicts compared with development on more remote and biologically diverse landscapes. Siting decisions should give priority to previously degraded lands, including abandoned mines, landfills, and other sites where utility-scale solar installation provides an opportunity to increase vegetative cover and conserve soil and water resources. However, avoidance is not likely to be enough to reach conservation goals, and a broader, multiple-use perspective of habitat conservation in and around solar facilities is likely to aid conservation efforts.

Utility-scale solar facilities possess unique features that other types of human development affecting wildlife lack; primarily, most of a solar facility site may remain in a condition that can sustain habitat capable of supporting a diverse array of plants and wildlife. Coordinated management of resources within a solar facility without impairment of electrical generation is a worthy goal. Benefits of approaching siting of utility-scale solar facilities with a goal of supporting multiple uses will be greatest when developments are sited in areas with highly degraded habitat conditions that can be enhanced post-construction and managed during operations.

ACKNOWLEDGMENTS

In alphabetical order, I thank those who contributed to this manuscript through thoughtful review and input: J. Babcock, W. Baker, M. Colon, R. Duke, S. Grodsky, R. Kelly, C. Moorman, H. Ogston, S. Rupp, and L. Tarasevic. I extend my sincere appreciation to all of the individuals who I have been blessed to work with on renewable energy projects. The number is too great to list here, and I fear offending someone through accidental omission. You will recognize your contribution to the manuscript, as I serve as the humble messenger of a small portion of our collective knowledge and understanding. Space is limited and there is much more to learn and share. Some aspects of this manuscript may be obsolete by the time it is published, as we learn more each day. Consequently, any errors or omissions that remain in the manuscript are my own, and I encourage others to join me in the discussion and dissemination of information on methods to integrate wildlife conservation and renewable energy development.

LITERATURE CITED

American Recovery and Reinvestment Act of 2009. Pub. L. No. 111-5 (2009). https://www.govinfo.gov/content/pkg/PLAW-111publ5/pdf/PLAW-111publ5.pdf.

APLIC (Avian Power Line Interaction Committee). 1994. Mitigating bird collisions with power lines: The state of the art in 1994. Washington, DC: Edison Electric Institute.

APLIC (Avian Power Line Interaction Committee). 2006. Suggested practices for avian protection on power lines: The state of the art in 2006. Washington, DC / Sacramento, CA: Edison Electric Institute/California Energy Commission.

APLIC (Avian Power Line Interaction Committee). 2012. Reducing avian collisions with power lines: The state of the art in 2012. Washington, DC: Edison Electric Institute and APLIC.

Barrientos, R., J. C. Alonso, C. Ponce, and C. Palacin. 2011. Meta-analysis of the effectiveness of marked wire in reducing avian collisions with power lines. Conservation Biology 25:893–903.

BLM (Bureau of Land Management). 2015. Desert renewable energy conservation plan: Proposed land use plan amendment and Final Environmental Impact Statement. Sacramento, CA: US Department of the Interior.

BLM (Bureau of Land Management). 2016. Regional mitigation strategy for the Dry Lake Valley North Solar Energy Zone. Environmental Science Division, Argonne National Laboratory. Washington, DC: US Department of the Interior. http://blmsolar.anl.gov/documents/docs/Final_DLVN_SRMS_2-25-16.pdf.

BLM/DOE (Bureau of Land Management/Department of Energy). 2012. Final programmatic environmental impact statement (PEIS) for solar energy development in six southwestern states. Report No. FES 12-14; DOE/EIS-0403. Washington, DC: US Department of the Interior.

BRE (Building Research Establishment). 2014. Biodiversity guidance for solar developments. Edited by G. E. Parker and L. Greene. https://app.croneri.co.uk/feature-articles/biodiversity-guidance-solar-developments.

Brown, P., and G. Whitney. 2011. U.S. renewable electricity generation: Resources and challenges. No. 7-5700 R41954. Washington, DC: Congressional Research Service.

California Department of Fish and Wildlife. 2017. California natural diversity database: State and federally listed endangered and threatened animals of California. Sacramento. https://www.wildlife.ca.gov/Data/CNDDB.

California Energy Commission. 2001. A guide to photovoltaic (PV) system design and installation. Consultant report prepared by Endecon Engineering with Regional Economic Research, Inc. Version 1.0. June 14. 500-01-020. https://www.energy.ca.gov/reports/2001-09-04_500-01-020.PDF.

Council on Environmental Quality. 1993. Incorporating biodiversity considerations into environmental impact analysis under the National Environmental Policy Act. Washington, DC: Executive Office of the President. https://ceq.doe.gov/docs/ceq-publications/Incorporating_Biodiversity_1993.pdf.

County of Los Angeles. 2016. Ordinance amending Title 22 of the Los Angeles County Code to Establish or Amend Requirements for Certain Renewable Energy Systems and Facilities, Wineries and Tasting Rooms, and Minor

Conditional Use Permits. December 13. Los Angeles: Office of the County Counsel.

Cypher, B. L., and C. Van Horn Job. 2009. Permeable fence and wall designs that facilitate passage by endangered San Joaquin kit foxes. Turlock: California State University, Stanislaus Endangered Species Recovery Program.

Cypher, B. L., C. Van Horn Job, and S. E. Phillips. 2012. Conservation strategies for San Joaquin kit foxes in urban environments. Prepared for the U. S. Bureau of Reclamation, Agreement No. R11AP20502. Turlock: California State University, Stanislaus Endangered Species Recovery Program.

Cypher, B. L., S. E. Phillips, and P. A. Kelly. 2013. Quantity and distribution of suitable habitat for endangered San Joaquin kit foxes: Conservation implications. Canid Biology and Conservation 16:25–31.

Deline, C., B. Marion, J. Granata, and S. Gonzalez. 2011. A performance and economic analysis of distributed power electronics in photovoltaic systems. Golden, CO: National Renewable Energy Laboratory. http://www.nrel.gov/docs/fy11osti/50003.pdf.

Devabhaktuni, V., M. Alam, S. S. S. R. Depuru, R. C. Green II, D. Nims, and C. Near. 2013. Solar energy: Trends and enabling technologies. Renewable and Sustainable Energy Reviews 19:555–564.

Dincer, F. 2011. The analysis on photovoltaic electricity generation status, potential and policies of the leading countries in solar energy. Renewable and Sustainable Energy Reviews 15:713–720.

EMC Planning Group. 2012. Final EIR Quinto Solar PV Project. State Clearinghouse #2010121039. Prepared for Merced County Planning and Community Development Department. September 13. Merced, CA: EMC.

Francis, C.D. 2015. Habitat loss and degradation: Understanding anthropogenic stressors and their impact on individuals, populations, and communities. Chapter 5 in Wildlife Habitat Conservation: Concepts, Challenges, and Solutions, edited by M. L. Morrison and Heather A. Mathewson. Baltimore, MD: John Hopkins University Press.

Fthenakis, V., J. E. Mason, and K. Zweibel. 2009. The technical, geographical, and economic feasibility for solar energy to supply the energy needs of the US. Energy Policy 37:387–399.

Grinnel, J. 1932. Habitat relations of the giant kangaroo rat. Journal of Mammalogy 13:305–320.

Grodsky, S. M., K. A. Morre-O'Leary, and R. R. Hernandez. 2017. From butterflies to bighorns: Multi-dimensional species-species and species-process interactions may inform sustainable solar energy development in desert ecosystems. Proceedings of the 31st Annual Desert Symposium, edited by R. L. Reynolds. April 14–15, California State University Desert Studies Center, Zzyzx, CA.

Guiller, C., L. Affre, M. Deschamps-Cottin, B. Geslin, N. Kaldonski, and T. Tatonia. 2017. Impacts of solar energy on butterfly communities in Mediterranean agro-ecosystems. Environmental Progress and Sustainable Energy. doi:10.1002/ep.12626.

H. T. Harvey & Associates. 2011. California Valley Solar Ranch Project: Plan for the relocation of giant kangaroo rats (Dipodomys ingens); Project 3103-01, San Luis Obispo County, California. https://www.energy.gov/sites/prod/files/2014/04/f14/CVSR_BA_11_08_10_Final.pdf.

H. T. Harvey & Associates. 2013. California Valley Solar Ranch Project giant kangaroo rat (Dipodomys ingens) relocation final report; Project 3103-01. Fresno, CA. Prepared for High Plains Ranch II, LLC, Carlsbad, CA. https://www.harveyecology.com/.

H. T. Harvey & Associates. 2015a. California Valley Solar Ranch site management plan: Project 3326-05. Prepared for High Plains Ranch II, LLC, Carlsbad, CA. https://www.harveyecology.com/.

H. T. Harvey & Associates. 2015b. California Valley Solar Ranch, San Luis Obispo County, California: Avian activity surveys final report, October 2011–October 2014: Project 3326-03. Prepared for High Plains Ranch II, LLC, Carlsbad, CA. https://www.harveyecology.com/.

H. T. Harvey & Associates. 2015c. Semiannual status report January–June 2015 for the California Valley Solar Ranch Project, San Luis Obispo County, California: Project 3326-05. Prepared for High Plains Ranch II, LLC, Carlsbad, CA. https://www.harveyecology.com/.

H. T. Harvey & Associates. 2016a. California Valley Solar Ranch, San Luis Obispo County, California: Year 3 (2016) onsite vegetation monitoring report: Project 3326-09. Prepared for High Plains Ranch II, LLC, Carlsbad, CA. https://www.harveyecology.com/.

H. T. Harvey & Associates. 2016b. Semiannual status report July–December 2015 for the California Valley Solar Ranch Project, San Luis Obispo County, California: Project 3326-10. Prepared for High Plains Ranch II, LLC, Carlsbad, CA. https://www.harveyecology.com/.

H. T. Harvey & Associates. 2016c. California Valley Solar Ranch conservation lands habitat mitigation and monitoring plan year 4 annual report. San Luis Obispo County, California: Project 3326-04. Prepared for High Plains Ranch II, LLC, Santa Margarita, CA. https://www.harveyecology.com/.

H. T. Harvey & Associates. 2017a. 2016 ITP Annual status report for the California Valley Solar Ranch Project, San Luis Obispo County, California: Project 3326-07.

Prepared for High Plains Ranch II, LLC, Santa Margarita, CA. https://www.harveyecology.com/.

H. T. Harvey & Associates. 2017b. Panoche Valley Solar Project: Revised technical assessment of additional effects on giant kangaroo rat, San Joaquin kit fox, blunt-nosed leopard lizard, and California tiger salamander. Fresno, California: Project 3128–03. Prepared for Panoche Valley Solar, LLC, Paicines, CA. https://www.harveyecology.com/.

H. T. Harvey & Associates. In preparation. California Valley Solar Ranch San Joaquin kit fox monitoring report: Project 3326–10. Fresno, California. Prepared for HPR II LLC, Santa Margarita, CA. https://www.harveyecology.com/.

Hartmann, H. M., M. A. Grippo, G. A. Heath, J. Macknick, K. P. Smith, R. G. Sullivan, L. J. Walston, and K. L. Wescott. 2016. Understanding emerging impacts and requirements related to utility-scale solar development. No. ANL/EVS-16/9. Washington, DC: US Department of the Interior, Argonne National Laboratory, Environmental Science Division.

Hernandez, R. R., M. K. Hoffacker, and C. B. Field. 2014a. Land-use efficiency of big solar. Environmental Science and Technology 48:1315–1323.

Hernandez, R. R., S. B. Easter, M. L. Murphy-Mariscal, F. T. Maestre, M. Tavassoli, E. B. Allen, C. W. Barrows, et al. 2014b. Environmental impacts of utility-scale solar energy. Renewable and Sustainable Energy Reviews 29:766–779.

Hernandez, R. R., M. K. Hoffacker, M. L. Murphy-Mariscal, G. C. Wu, and M. F. Allen. 2015. Solar energy development impacts on land cover change and protected areas. Proceedings of the National Academy of Sciences 112: 13579–13584.

Holland, R. F. 1986. Preliminary descriptions of the terrestrial natural communities of California. Sacramento: State of California, Resources Agency, Department of Fish and Game. https://nrm.dfg.ca.gov/FileHandler.ashx?DocumentID=75893.

Lovich, J. E., and J. R. Ennen. 2011. Wildlife conservation and solar energy development in the desert Southwest, United States. BioScience 61:982–992.

Macknick, J., T. Quinby, E. Caulfield, M. Gerritsen, J. Diffendorfer, and S. Haines. 2014. Geospatial optimization of siting large-scale solar projects. Joint Institute for Strategic Energy Analysis. Technical Report NREL/TP-6A50-61375. https://www.nrel.gov/docs/fy14osti/61375.pdf.

McDonald, R., J. Fargione, J. Kiesecker, W. Miller, and J. Powell. 2009. Energy sprawl or energy efficiency: Climate policy impacts on natural habitat for the United States of America. PLOS One 8:e6802. doi:10.1371/journal.pone.0006802.

Mendelsohn, M., T. Lowder, and B. Canavan. 2012. Utility-scale concentrating solar power and photovoltaics projects: A technology and market overview. Technical Report NREL/TP-6A20-51137. https://www.nrel.gov/docs/fy12osti/51137.pdf.

Montag, H., G. Parker, and T. Clarkson. 2016. The effects of solar farms on local biodiversity: A comparative study. Blackford, UK: Clarkson and Woods / Wychwood Biodiversity.

Moore-O'Leary, K. A., R. R. Hernandez, S. Abella, D. Johnston, J. Kreitler, A. Swanson, and J. Lovich. 2017. Sustainability of utility-scale solar energy: Critical ecological concepts. Frontiers in Ecology and the Environment 15:385–394.

Natural England. 2011. Solar parks: Maximising environmental benefits. Technical Information Note TIN101. 1st ed. https://webarchive.nationalarchives.gov.uk/20150902172007/http://publications.naturalengland.org.uk/publication/32027.

Ong, S., C. Campbell, P. Denholm, R. Margolis, and G. Heath. 2013. Land use requirements for solar power plants in the United States. NREL/TP-6A20-56290. Golden, CO: National Renewable Energy Laboratory.

Pearce, D., J. Strittholt, T. Watt, and E. N. Elkind. 2016. A path forward: Identifying least-conflict solar PV development in California's San Joaquin Valley. Prepared by Conservation Biology Institute and Center for Law, Energy and the Environment. Berkeley, CA: UC Berkeley School of Law.

Phillips, S. E., and B. L. Cypher. 2015. Solar energy development and endangered upland species of the San Joaquin Valley: Identification of conflict zones. Prepared for the California Department of Fish and Wildlife. Agreement P1440003 00. Turlock: California State University, Stanislaus Endangered Species Recovery Program.

Prugh, L., J. Brashares, and K. Suding. 2015. Carrizo Plain Ecosystem Project combined 2014 and 2015 report. December. http://ecnr.berkeley.edu/vfs/PPs/PrughLauR/web/.

Quad Knopf. 2014. Maricopa Sun Solar Complex habitat conservation plan final environmental impact statement. Report 090160.02 Prepared for US Fish and Wildlife Service, Sacramento, CA.

REN21. 2016. Renewables 2016 global status report. www.ren21.net/wp-content/uploads/2016/05/GSR_2016_Full_Report_lowres.pdf.

San Luis Obispo County. 2011. California Valley Solar Ranch conditional use permit (DRC2008-0097) and Twisselman conditional use permit/reclamation plan (DRC2009-00004) final environmental impact report.

State Clearing House #2009021009. http://www
.sloplanning.org/EIRs/CaliforniaValleySolarRanch/.

San Luis Obispo County Public Works. 2016. Volunteer
precipitation gauge station monthly precipitation
report: California Valley CDF #175. http://slocounty
water.org. Accessed November 2016.

USDOE (US Departure of Energy). 2012. SunShot vision
study. Report No.DOE/GO-102012-3037. Washington,
DC: Office of Energy Efficiency and Renewable Energy,
USDOE.

US EPA(US Environmental Protection Agency). 2012.
Handbook on siting renewable energy projects while
addressing environmental issues. Washington, DC:
Office of Solid Waste and Emergency Response's Center
for Program Analysis.

USFWS (US Fish and Wildlife Service). 1998. Recovery
plan for upland species of the San Joaquin Valley,
California. Portland, OR: US Fish and Wildlife Service.

USFWS (US Fish and Wildlife Service). 2010. Giant
kangaroo rat (*Dipodomys ingens*) 5-year review:
Summary and evaluation. Sacramento, CA: US Fish and
Wildlife Service, Sacramento Fish and Wildlife Office.

Wiser, R., T. Mai, D. Millstein, J. Macknick, A. Carpenter, S.
Cohen, W. Cole, B. Frew, and G. A. Heath. 2016. On the
path to SunShot: The environmental and public health
benefits of achieving high penetrations of solar energy
in the United States. NREL/TP-6A20-65628. Golden,
CO: National Renewable Energy Laboratory. http://www
/nrel.gov/docs/fy16osti/65628.pdf.

Wood Mackenzie and SEIA (Solar Energy Industries
Association). 2016. U.S. solar market insight, 2016 year
in review. NREL/TP-6A20-65628. https://www.seia.org
/research-resources/solar-market-insight-report-2016
-year-review.

Zar, J. H. 1984. Biostatistical analysis. 2nd ed. Englewood
Cliffs, NJ: Prentice-Hall.

9

Waterpower

Hydropower and Marine Hydrokinetic Energy

HENRIETTE I. JAGER AND
LINDSAY M. WICKMAN

Introduction

Moving water is a cheap source of power that can be harnessed in regulated rivers (i.e., rivers that are fragmented by dams and can produce hydropower) and in coastal and marine waters (hydrokinetic energy). Hydropower capitalizes on increased head (i.e., change in elevation from above to below dam) and storage provided by dams. Reservoirs created by damming store water for multiple purposes (e.g., irrigation, municipal water supply, recreation, flood control). Marine hydrokinetic energy captures energy of tidal rivers, waves, and ocean currents; the technologies used include tidal barrages, wave power, and tidal-stream turbines.

Waterpower development can have direct consequences (e.g., entrainment in turbines) and indirect consequences (e.g., habitat fragmentation or conversion) for many wildlife taxa. Research on direct and indirect ecological effects of conventional hydropower has focused primarily on fish and aquatic insects. In this chapter, we review research on the potential impacts of hydropower to these taxa. In addition, we review interactions between hydropower and other wildlife, such as mammals, birds, and mussels. Next, we review the risks to wildlife associated with marine hydrokinetic energy technologies. We also highlight research directions and new technologies that may provide society with water-based renewable energy while simultaneously conserving habitat for wildlife.

Hydropower
Generating Electricity from River Flow

The United States is among the countries with the most regulated rivers, including regulation of all major rivers in the country (Figure 9.1). Systems of dams and reservoirs along rivers currently supply 15% of US energy, 6.5% of US electricity, and one-third to over half of renewable energy (US Energy Information Agency 2016).

Hydropower generation is not always the primary reason for damming rivers. In most cases, rivers are dammed to form reservoirs and increase predictability of water supplies. Stored water can be used later for irrigation, hydropower generation, and recreation. In temperate climates, flow is released from storage projects during winter to prepare for spring floods (i.e., draw down), thereby serving the function of flood control. An important exception in the United States is the Columbia River basin, where hydropower provides more than one-third of electricity from a series of run-of-river (non-storage) projects.

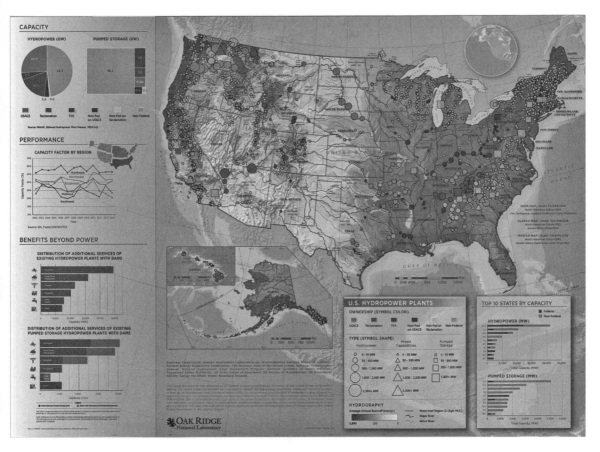

Figure 9.1 Statistical summaries of US hydropower produced for the Department of Energy's Water Power Technologies Office. *Oak Ridge National Laboratory 2016.*

From an energy perspective, hydropower has the advantage that a single drop of water can be used to generate energy multiple times as it passes downstream, but hydropower cannot be stored. Hydropower also cannot reliably meet baseload electricity demand in the way that fossil fuel sources do. To provide electricity when it is most needed, water must be stored and generation must be delayed. In some cases (e.g., pumped storage projects), this is accomplished by pumping water up to a reservoir using wind or solar power and then generating power by releasing reservoir water at times when electricity demand is high, allowing hydropower (and similar less-predictable renewables such as wind and solar power) to supplement baseloads to meet peak energy demand.

Most hydropower is generated by turbines installed in integral powerhouses, which are built into the structure of dams and raise the head of water that they impound. Integral powerhouses are typical of larger storage projects (i.e., one or more dams and associated infrastructure operated together). Smaller hydropower projects tend to have nonintegral powerhouses and generate electricity from water diverted downstream through penstocks (i.e., pipes) along the side of the river. These configurations have different kinds of ecological impacts, with larger dams and reservoirs often having more adverse effects on aquatic biota than smaller dams (Paish 2002). However, small water diversion projects can adversely affect downstream fish communities if too much water is diverted (Kibler and Tullos 2013,

Jager and McManamay 2014, Benejam et al. 2016, Lange et al. 2018).

River basins of the southeastern United States are notable for their biodiversity and support many aquatic species (e.g., amphibians, crayfishes, and fishes). The scientific literature related to hydropower effects on biota rarely extends beyond aquatic species. Yet, rivers support terrestrial communities by providing wildlife with a source of water and by supporting production of aquatic insects that contribute to terrestrial food chains. Many wildlife species are central-place foragers (or drinkers) that must consider access to water when setting boundaries for their home ranges. For example, beavers (*Castor canadensis*), waterbirds (e.g., herons, ducks), other birds (e.g., eagles), and some mustelids (e.g., river otter, *Lontra canadensis*; mink, *Neovison vison*) are closely associated with rivers and floodplain wetlands. In arid climates, rivers, springs, and arroyos (i.e., a small steep-sided watercourse or gulch—often dry—typically found in arid and semiarid areas of the US Southwest) are oases that provide water needed to support life. Aerial insectivores, including bats (Nummi et al. 2011) and birds, drink water and feed on insects while flying over rivers, and navigation along river corridors is common among many terrestrial species. Because access to freshwater is such a fundamental biological requirement, studies examining both where and how river regulation might influence diverse fish and wildlife communities are important, especially in biodiversity hotspots (Kano et al. 2016, Winemiller et al. 2016).

The initial inundation of reservoirs behind large dams has significant, adverse effects on wildlife living in riparian zones (Benchimol and Peres 2015b). Recent studies have sought to develop relationships between threshold distances up and downstream (e.g., typically 1,000 to 5,000 m) influenced by dam construction as a function of reservoir volume and land use (Zhao et al. 2013). Impoundment can create an archipelago of small islands resulting in habitat loss and fragmentation (Benchimol and Peres

2015b). As one might expect from principles of biogeography (MacArthur and Wilson 1967), the remaining small islands support fewer species than the larger contiguous terrestrial area did prior to flooding. One study estimated that, 26 years after flooding, 99% of the islands would harbor vertebrate communities with less than 80% of the original species diversity (Jones et al. 2016). In one example, the creation of the Balbino Hydroelectric Reservoir in Brazil inundated 312,900 ha of contiguous lowland tropical forest, creating an archipelago of 3,546 islands (Benchimol and Peres 2015b). Large vertebrates and species able to travel through water were found on more islands than less motile species (Benchimol and Peres 2015a). Similarly, inundation following reservoir filling along southern rivers in the United States may have reduced the availability of breeding habitat for the endangered gray bat (*Myotis grisescens*), which is selective in its use of limestone caves scattered along rivers (USFWS 1997).

Conversion from Lotic to Lentic Habitat

The effects of converting free-flowing, aquatic habitat (i.e., lotic) to reservoirs (i.e., lentic) does not result in a consistent increase or decrease in fish species richness (Bilotta et al. 2016). However, only a subset of lotic fish species (fish found in flowing waters) can reproduce and survive in reservoirs. Lotic fish specialists often are less prevalent in fragmented reaches (Guenther and Spacie 2006, Freedman et al. 2014). For example, in the United States, lower lotic fish diversity was reported in reservoirs of the Monongahela River (Pennsylvania) compared with free-flowing sections of the river, with greater-than-expected fish richness found in free-flowing rivers (Freedman et al. 2014). However, beyond these general results, there is a lack of consensus among studies, which stems, in part, from the methods used, with different results emerging from studies that compared reaches above and below dams than from studies that compared dammed and undammed reaches (Fencl et al. 2017). Sites above dams typically supported less biodiversity than those below dams (i.e., negative dam effect), whereas studies comparing similar reaches, one below a dam and the other not, report a negligible dam effect (Fencl et al. 2017). Differences between lotic and lentic species are also reflected in their abundances. For example, in one river system, lentic centrarchid species dominated reaches above and below impoundments, whereas obligate-stream (lotic) cyprinids were more abundant in nearby rivers without impoundments (Kashiwagi and Miranda 2009).

Storage reservoirs tend to have degraded water quality compared with run-of-river reservoirs (i.e., those with outflows equal to inflows) and free-flowing rivers, which can have a negative effect on fish diversity and abundance. Because stored water often supports irrigation, agricultural nutrients draining into storage reservoirs can stimulate the growth of algal blooms when temperatures are warm, causing eutrophication. When blooms sink to the bottom, they decay and contribute to periods of low dissolved oxygen, which can result in reduced fish species richness. In some cases, dam removal has resolved eutrophication concerns (Tuckerman and Zawiski 2007, Poff and Zimmerman 2010).

Shifts in flow regimes below dams alter downstream ecosystems both longitudinally (i.e., up and downstream) and laterally. The serial discontinuity concept (SDC; Ward and Stanford 1983) suggests that longitudinal habitat characteristics of tailwaters recover downstream, along with the aquatic communities they support. However, the SDC, in some ways, downplays the ecosystem effects of regulation by implying that the primary effects diminish within a short distance downstream of a dam. The SDC hypothesis has recently been evaluated and new results have emerged that paint a more complex picture of longitudinal recovery. For example, benthic invertebrate diversity can decrease below dams and may not recover quickly downstream (Ellis and Jones 2013). The ecosystem-transformative processes that depend on lateral connections between main channel and lateral slow-shallow floodplains are critically important to address in the context of habitat alteration, but they are far more difficult to study (Naiman et al. 2010).

Longitudinal Alterations of River Flow Regimes

Although many hydropower dams operate in run-of-river mode, in which inflows match outflows, larger storage reservoirs are designed to provide flexibility to modify and reshape downstream flow releases. When operating specifically for electricity generation, hydropower facilities release flows at times when electricity demand is high. Facilities that operate in peaking or load-following mode fluctuate flow releases within a day to provide electricity when demand is highest, which tends to be when air temperatures are extreme (winter or summer) and during business hours (Ruibal and Mazumdar 2008, Jager and Uria-Martinez 2012). Peaking at an upstream reservoir can propagate through downstream reaches and dams (Jager and Bevelhimer 2007).

Fluctuating Reservoir Elevation

Rapid changes in flow release rates (hydropeaking) can cause problems for wildlife both above and below dams. Above and below reservoirs, fluctuating water levels can dewater fish nests and strand fish (Nagrodoski et al. 2012). Reservoir fluctuations can influence non-fish species as well. For example, the western painted turtle (*Chrysemys picta*), a species of concern in Canada (Committee on the Status of Endangered Wildlife in Canada 2006), overwinters (hibernates) in muddy substrates along Arrow Lakes Reservoir, where fluctuating water levels during reservoir operations result in reduced wetland inundation, increased predation on turtle nests, and lower body weights for turtles in the drawdown zone (Committee on the Status of Endangered Wildlife in Canada 2006).

Downstream Hydropeaking

The effects of flow alterations on aquatic biota are often mediated by water temperature (Poff et al. 2010, Jager 2014, McManamay et al. 2015). Water temperature fluctuations associated with sudden changes in flow releases can shock aquatic biota below dams. Seasonal shifts in water temperature caused by upstream reservoirs include a general lag in seasonal patterns because of the large thermal mass of reservoirs and moderation of temperature extremes with a cold-bias created by hypolimnetic (i.e., lake-bottom) withdrawals. Thermal gradients recover much farther downstream (e.g., hundreds of kilometers) than other gradients (Ellis and Jones 2013). Nevertheless, the temporal and spatial extent of impacts can be extensive. In simulations of dam removal on the Klamath River, California, downstream changes in water temperature extended up to 200 km in the main stem (Bartholow et al. 2004). Although effects on biota tend to be site-specific, predictable shifts in thermal regimes have been documented across multiple hydropower projects in other rivers. A study in Canada described general changes in tailwater temperatures as a function of storage capacity (Maheu et al. 2016). Storage and peaking dams that impounded at least 10% of annual runoff reduced thermal variation and increased fall water temperatures (Maheu et al. 2016). For tailwater biota, diurnal shifts in water temperature in response to load-following or peaking operations (i.e., thermopeaking) are even more harmful than seasonal shifts because species are unable to acclimate gradually to thermal shocks. As with other effects of flow, hydropeaking is best studied in the context of other factors, such as thermopeaking and channel simplification (Metcher et al. 2017).

Another indirect effect of regulation is change in sediment dynamics. Changes in benthic invertebrate communities caused by river regulation tend to reflect indirect effects of flow on sediment and organic matter (Tupinambás et al. 2015). For example, an Index of Biological Integrity for macroinvertebrates showed differences between reference sites and the first site downstream of each of 9 dams in California; the differences were caused by lower variability in flows and coarsening of substrates below dams, with partial recovery in index scores farther downstream (Rehn 2009). Invertebrate composition, but not species richness, responded to flow regime in western US streams, especially base (i.e., minimum) flow and temperature (Chinnayakanahalli et al. 2011).

Invertebrate drift rates are affected by hydro- and thermopeaking operations of reservoirs. Macroinvertebrate guilds respond differently to peaking of upstream reservoirs; drift rates of rheophilic taxa (i.e., species that prefer fast-flowing water, like drift feeders) and interstitial taxa tend to decrease, whereas limnophilic taxa (i.e., those that prefer standing water) drift at increased rates (Schulting et al. 2016). In Great Britain, species evenness showed a significant change before and after construction of 22 run-of-river hydropower projects measured within 1 km of a dam, but other metrics—for example, family richness, an index that reflects flow (lentic to lotic gradient)—the proportion of sediment-sensitive invertebrates, and tolerance to organic pollution, did not (Bilotta et al. 2017).

Effects of flow fluctuations can be moderated by setting ramping-rate restrictions (i.e., limits on the rate of change in flow releases from dams) and by avoiding large-magnitude flow changes at night (Metcher et al. 2017). For example, effects on tailwater fish communities were modest below Austrian hydropower projects that adhered to limits on the rate of change in flow releases and avoided nighttime peaking, although benthic invertebrate drift rates decreased (Schmutz et al. 2015). Most studies of ecological effects of hydropower on wildlife have relied on hydrodynamic models to quantify the amount of habitat lost during periods of lowering (Boavida et al. 2013) or have used telemetry to measure changes in swimming effort of species of concern to avoid dewatering (Ruetz and Jennings 2000).

The signal of hydropower effects on biota can be difficult to detect against a backdrop of other watershed influences. At the regional scale, a study of

1,883 streams representing 9 regions across the conterminous United States determined that dam density and impoundment were less important in determining fish composition than was the percentage of urban and agriculture lands and natural attributes of watersheds (Hill et al. 2017). Dam density and impoundment influenced fish assemblages in 4 of 9 regions, with positive effects in the Northern Plains, but negative effects in the Northern Appalachian and xeric ecoregions of the United States (Hill et al. 2017). Farther north, cold-water, riffle-dwelling fishes and fish that are intolerant of environmental perturbations tended to be absent where annual flow variability was high and summer temperatures were low (Macnaughton et al. 2016). Consequently, the largest observed changes were increases in biomass of the remaining tolerant species in response to stabilizing flow and increasing summer temperatures. Although the effects of flow and temperature regimes on community composition were clear, effects of regulation of flow were not statistically significant (Macnaughton et al. 2016). Fragmentation by dams and flow alterations have had the largest anthropogenic impacts on riverine fauna across the conterminous United States (Cooper et al. 2017). Fish species adapted to lentic habitat increased with dams, whereas species that prefer fast-moving water (rheophilic), benthic species (lithophilic), and fishes intolerant to disturbance decreased (Cooper et al. 2017).

Many species of freshwater mussels depend on host fishes for reproduction and show poor tolerance to siltation and cold temperatures below dams. Mussels have declined in tailwaters below large, but not small, reservoirs (Hornbach et al. 2014). Mussel growth is faster below small dams than behind larger dams, possibly because productivity of small upstream reservoirs provides more food for filter-feeding mussel beds in the tailwaters (Hornbach et al. 2014). Below large dams, hypolimnetic releases of cold water likely make mussel development and reproduction difficult in tailwaters (Hornbach et al. 2014). Thermodynamic effects diminish below the dam as long as there is enough space between dams (Galbraith and Vaughn 2011). Dam construction can indirectly reduce mussel populations by harming the host fishes that disperse mussel larvae. Many mussels are long-lived, and some problems can be avoided through translocation above the dam or to a location with a more favorable thermal regime (Lessard and Hayes 2003). Mussel communities seem to recover with greater distance downstream of a dam. Immediately below 2 tailwaters in Texas (Lake Tawakoni and Toledo Bend Reservoir), opportunistic strategist mussel species (i.e., small, highly fecund species) were more abundant immediately downstream of dams, whereas periodic and equilibrium strategist mussel species (i.e., large, long-lived species) were more abundant farther downstream (Randklev et al. 2016).

Dam removals provide opportunities to study before-and-after effects of flow regulation. Recovery of some components of aquatic communities has been observed following dam removal (Burroughs et al. 2010). Following removal of one small dam, macroinvertebrates recovered quickly, but macrophytes recolonized slowly and mussels did not recover during the period of the study (Doyle et al. 2005). The migratory community of fishes below dams on the Penobscot River in Maine recovered quickly once lower dams were removed (Hogg et al. 2013). After removal of low-head dams in Wisconsin, species tolerant of degraded water quality decreased whereas intolerant species increased, and recolonizing species included large-river fishes known to migrate between spawning and overwintering habitat (Catalano et al. 2007).

Lateral Alterations of River Flow Regimes

River floodplains are among the world's most biologically diverse ecosystems. In a recent study of Missouri-Mississippi-Ohio River drainages, channel complexity provided the best surrogate measure for fish community health (Taylor et al. 2013). The wide variety of habitat types provided by complex channels

Figure 9.2   Wildlife associated with regulated rivers. (*A–C*) Three kingfishers; (*D*) imagery showing patterns in a meander-ing river; (*E*) cottonwood stakes planted at McKay Creek, BC, to support migratory songbirds at Revelstoke Reach, one year after planting; (*F*) western painted turtle; (*G*) yellow-rumped warbler; (*H*) river otters; (*I*) beaver dam. *Photo credits: A–C, Roger-Ashley-Photography-Belted Kingfisher, taken by Roger Ashley Photography at Chicamauga Lake, Tennessee; E, Cooper Beauchenesne & Associates Ltd. 2011; G, photo by Brian Sullivan, in Hunt and Flaspohler 1998; F–I, Creative Commons.*

and their floodplains (Figure 9.2D), in turn, sup-port a high diversity of species. Slow, shallow floodplains are highly productive ecosystems that support macrophytes and aquatic insects. They also serve as refuge from predators for juvenile and small-bodied species of fishes. Stanford et al. (1996) sug-gested that alluvial reaches support core populations of species needed to recolonize other satellite reaches and should therefore be considered first when river restoration is attempted.

Altered Sediment Regimes and River Channel Evolution

Dams and reservoirs simplify channels through a va-riety of mechanisms, often to the detriment of aquatic and riparian communities. Large dams pre-vent channel meandering and the formation of mul-tiple channels by reducing spring peak flows and the transport of coarse sediment and large woody debris that benefit aquatic communities (National Research Council 2002, Feld et al. 2011). Although fine sedi-

ments tend to reduce spawning survival of fish and mussels, coarse sediments benefit spawning survival of these wildlife by providing connections with the hyporheic zone—a zone of interaction between stream water and groundwater. Larger debris flows contribute to formation of channels into pool-riffle sequences that provide a matrix of heterogeneous conditions for aquatic biota (Roni et al. 2015). Flow diversions tend to move fine sediments even during base flows, but only a small fraction of coarse sedi-ments are transported during peak flows (Angelaki and Harbor 1995).

Sediment-depleted waters released from reser-voirs alter generation of sandbars and islands, but the magnitude of this effect and associated responses among wildlife species varies (Phillips 2001). For ex-ample, in the United States, the endangered least tern (*Sternula antillarum*) and piping plover (*Char-adrius melodus circumcinctus*) have declined in the Missouri River, in part because sandbars on which they nest are no longer generated. Flow regimes designed to recover these 2 birds and the endangered pallid sturgeon (*Scaphirhynchus albus*) require operating dams to mimic historical high spring flows and low summer flows.

Riparian Vegetation

One reason for restoring natural channel evolution is that it maintains a range of successional stages in downstream riparian vegetation (Ward and Stanford 1995, Kondolf 2011). Where flow regulation slows, the rate of creation of disturbed patches and recruitment of native floodplain species is reduced (Katz and Shafroth 2003). Typically, riparian woody species that colonize along rivers are pioneer species that are fast growing and are adapted to historical and seasonal patterns in water flow. Along rivers of the interior western United States, native riparian forests are dominated by cottonwood (*Populus* spp.) and willow (*Salix* spp.), pioneer species that rely on physical disturbance (e.g., seasonal flooding) for seed establishment. However, below dams, invasive riparian shrubs such as Russian olive (*Elaeagnus angustifolia*) and saltcedar (*Tamarix* spp.) have become established in the understory (Mortenson and Weisberg 2010). Though non-native and invasive, Russian olive provides habitat for insects and supports a diversity of bird and mammal species that may be comparable to that supported by native vegetation (Katz and Shafroth 2003).

Wildlife responses to flow regulation may be mediated by effects on riparian vegetation. Regulation has contributed to the estimated loss of 87% of riparian vegetation in the Columbia River Basin of the United States (Moody et al. 2007). In British Columbia, one-half of forest-dwelling vertebrate species use riparian vegetation during at least 1 life-history stage (Bunnell et al. 1999), and migratory songbirds follow valleys during annual migrations along riparian corridors. To protect riparian vegetation as habitat, British Columbia Hydropower planted cottonwood stakes to supplement perching and foraging sites for migrating songbirds (Figure 9.2E). For example, in the vicinity of Revelstoke Dam, British Co-

lumbia, impact of reservoir operations could be reduced for migrating American pipit (*Anthus rubescens*), yellow-rumped warbler (*Setophaga coronata*) (Figure 9.2G), and other songbirds by avoiding refill until after the spring migration season (van Oort et al. 2011).

In Colorado, regulated rivers supported greater densities of beavers than an unregulated one (Figure 9.2H) because, in part, regulation prevented the river from freezing over and retained more streamside willow (*Salix* spp.) patches used by beavers for dam construction and food (Breck et al. 2003b). However, flow regulation by dams prevented recruitment of riparian trees in other cases, and thereby reduced beaver forage (Breck et al. 2003a). In arid climates with more saline soils, non-native saltcedar has become more common, particularly in regulated rivers (Merritt and Poff 2010). Although saltcedar supports some wildlife, including yellow-billed cuckoo (*Coccyzus americanus*) and southwestern willow flycatcher (*Empidonax traillii extimus*) (Shafroth et al. 2005, van Riper et al. 2008), other avian species at risk, including cavity nesters, frugivores, and granivores, do not use saltcedar (Shafroth et al. 2005).

Flood-Pulse Dynamics

Large upstream dams can eliminate flood-pulse dynamics that inundate floodplain wetlands and other slow, shallow riparian systems historically high in aquatic biodiversity. Lateral flooding promotes abundance of aquatic insects (Benke et al. 2000) and provides rearing and escape cover for juvenile fish (Rood et al. 2005, Jager 2014). For example, when Yolo Bypass at the mouth of the Sacramento and Feather Rivers of California flooded, juvenile salmon rearing in the floodplain grew faster than those in the main channel (Sommer et al. 2001b). Managing flows such that over-bank pulse flows occur during seasons when wildlife require access to slow, shallow water may benefit juvenile fishes.

Flood-pulse dynamics also are important to taxa other than fish. For example, migrating shorebirds stopover in inundated floodplains (Sommer et al. 2001a, Feyrer et al. 2006). In the Amazon, changes in flood-pulse dynamics are a concern for the Amazonian manatee (*Trichechus inunguis*), which migrates just prior to the annual low-flow season (Arraut et al. 2017). Proposed dams will divide the current population of these manatees and could increase the unpredictability of when low-flow bottlenecks will occur.

River systems can be regulated in ways that maintain historical timing of pulse flows, sediment dynamics, channel evolution, and vegetative successional-stage diversity for wildlife. In general, the practice of letting rivers do the work of restoring downstream function seems to be best. A 3-pronged strategy is needed to restore natural river function. First, if a new dam is sited upstream of tributaries that provide an influx of sediment- and debris-laden waters, disruption to sediment dynamics will be lower. Second, protecting multistage floodplains at different heights and distances from the river ensures that development will not encroach, thus allowing the river to heal itself over time (Kondolf 2011). Finally, research is needed to overcome technological and financial obstacles associated with sediment bypasses (Kondolf et al. 2014), such as tunnels and vented dams (Shetkar and Mahesha 2011). Challenges to maintaining a sediment bypass include scouring damage to the structure and high maintenance cost. Dechannelizing rivers using a combination of the strategies above can have significant ecological benefits (Franklin et al. 2001, Hohensinner et al. 2004).

Fragmentation of Rivers by Hydropower Dams

Public concerns about reconnecting river reaches fragmented by dams are high because dams prevent movements by fishes and other aquatic biota. Dam removal has become a full-fledged environmental movement, and over 1,400 dams have been removed in the United States, although most of these were small, non-power dams. Longer distances between dams and the presence of tributaries between series of dams can help to protect fish biodiversity (da Silva et al. 2015, de Oliveira et al. 2016). Upstream of dams, tributaries provide free-flowing refuge from adverse conditions in reservoirs. Tributaries are used for spawning by many fishes, including flannelmouth (*Catostomus latipinnis*), blue suckers (*Catostomus discobolus*), and paddlefish (family Polyodontidae) (Pracheil et al. 2009, Fraser et al. 2017). Downstream of dams, tributaries moderate thermal effects of flow regulation. Furthermore, confluences are depositional zones with unique substrate and hyporheic flows and are used for spawning by sturgeon (family Acipenseridae) and possibly other fishes (Perrin et al. 2003).

Effects of Asymmetric Migration

An important consequence of dams is that smaller-bodied species and younger life stages of fish move easily downstream from natal areas, whereas older individuals or larger-bodied species are unable to move upstream to spawn or to return downstream safely. Thus, larvae produced upstream can be swept downstream, leaving no recruits in the upstream reach. As these juveniles grow in downstream reaches, they are unable to recolonize upstream reaches. This asymmetric migration leads to a higher risk of extirpation in upstream reaches (Jager et al. 2001).

Risk of Entrainment and Turbine Strike

Even when upstream passage is provided, this may add to the risk to wildlife species during downstream movement. Entrainment into turbine intakes, followed by turbine strike, is a primary risk caused by hydropower facing fishes and other aquatic biota. The probability of entrainment differs among species because of variation in their avoidance behaviors (Amaral et al. 2015). Although research has been funded to design so-called fish-friendly turbines, few are truly safe for larger fishes. Strike risk is somewhat lower for larger-diameter turbines and especially enclosed bulb turbines (Waters and Aggidis 2016). Unfortunately, bulb turbines are less efficient in generating electricity.

Effects of Hydropower on Fish Migration

Among the better-known effects of dams is the challenge to fishes with life histories that require long-distance migrations between different areas. Below, we describe the effects of hydropower and potential mitigation strategies for 3 groups of migratory fishes, some of which travel between marine and freshwaters: (1) anadromous (i.e., adults return to freshwater to breed and spawn and their offspring migrate to estuaries or ocean) salmon; (2) sturgeons, which show nearly every possible migratory strategy except catadromy; and (3) catadromous (i.e., adults breed in the ocean and their offspring follow ocean currents to estuaries or freshwater systems) eels.

ANADROMOUS SALMON

In the United States, the costs and engineering challenges involved in moving anadromous salmon and steelhead upstream to spawn in rivers are substantial. Fish ladders enable adults to move upstream past numerous dams. Hatcheries are operated to enhance early survival of salmon (Naish et al. 2008). After their parents breed, some juvenile offspring are transported downstream by trucks; others navigate through reservoirs aided by supplemental flows, thus completing the complex and highly managed round-trip from the ocean and back. For some species of conservation concern, these measures seem to be resulting in recovery. For example, the Snake River fall Chinook salmon (*Oncorhynchus tshawytscha*) was recently petitioned for delisting because the number of returning adults has remained high for longer than a generation. Others, such as sockeye salmon (*Oncorhynchus nerka*) are continuing to decline.

STURGEONS

Sturgeons are imperiled globally, in part because they are unable to thrive in regulated river systems. As members of a capital-breeding species that build energy reserves over multiple years before migrating to spawn, female sturgeons have high fecundity, and populations do best with access to different habitat types (e.g., deep pools, lakes, or estuaries for feeding, confluences or other specialized habitats for spawning). Some sturgeon are potamodromous (i.e., species that migrate within river), especially in the Mississippi River basin, whereas others are anadromous (e.g., Atlantic sturgeon, *Acipenser oxyrinchus oxyrinchus*), semi-anadromous (facultatively anadromous; e.g., white sturgeon, *Acipenser transmontanus*), or amphidromous (migrate from estuary to freshwater; e.g., shortnose sturgeon, *Acipenser brevirostrum*). For potamodromous species, dams should be spaced far enough apart to provide a variety of microhabitats, including free-flowing rivers, confluences, areas with substrate and cover for incubation and larval rearing, and deep pools or reservoirs. For anadromous and amphidromous species, it may be necessary to remove dams that no longer serve a purpose or where hydropower facilities operate at a loss. Where dams still serve a valuable purpose, it may be worthwhile to implement translocation or hatchery programs or to design passage facilities and bypass structures that allow passage of these large-bodied, bottom-oriented fishes (Jager et al. 2016).

CATADROMOUS EELS

Eels are present in coastal rivers around the world and are especially challenged during downstream migration through hydropower projects. The American eel (*Anguilla rostrata*) once represented over 40% of fish biomass in US rivers of the Atlantic Coast, but the biomass of eels has now declined dramatically (Machut et al. 2007). The eel is a catadromous fish (i.e., migrates to the ocean to breed) and an apex predator. Adults migrate to the Sargasso Sea to breed as one large, panmictic population, after which juveniles return via ocean currents to rivers of the Atlantic Coast. Although some individual eels grow from juveniles to adults in the estuary, a fraction swim upstream to grow and develop in rivers. Eels are adept at moving upstream over barriers and can pass above dams by climbing relatively inexpensive eel ladders (Jager et al. 2013). However, unlike salmon, eels are large adults when they migrate downstream to sea. Some individuals find a bypass or spill over the dam, but they experience a high rate of mortality when entrained into turbines. Researchers are exploring whether turbines can be shut down temporarily during ocean migration of eels and whether individual eels can locate exit routes through deep spillways (Eyler et al. 2016, Haro et al. 2016).

Deep Dive: Can Passage for Eels Help Restore Mussels?

Freshwater unionid mussels have suffered the most significant declines among freshwater taxa in the United States, partly because of damming. Lydeard et al. (2004) reported that 202 of 300 species of mussels in North America were presumed to be on the spectrum from vulnerable to extinct, with more than 37 species presumed extinct. In the Guntersville Lake section of the Tennessee River, 14 species of mussel were considered to have some level of risk, and 6 of these species are federally listed. The most diverse assemblage of mussel fauna was in an area named "Muscle" Shoals at a confluence of the Tennessee River in Alabama. Of 69 original species in the watershed, 32 have not been recorded since the construction of dams in the early 1900s; ten species that have since colonized the area are adapted to lentic conditions (Lydeard et al. 2004). Similar changes were recorded in the Caney Fork River, Kentucky, following the construction of Center Hill Dam (Layzer et al. 1993).

Dams can restrict the upstream movement of host fishes of mussels, and the host fishes of most mussel species remain unknown. To protect declining mussel species, studies are being developed to identify host fishes for species at risk. In addition, passage is being provided for migratory fish species that host mussel larvae, including the American eel. For example, eel ladders are being added to low-head dams in the Shenandoah and Potomac Rivers with the hope that American eels will recolonize upstream reaches (Eno et al. 2013). The goal is to facilitate the eel population and to bolster mussel species (e.g., eastern elliptio (*Elliptio coplanata*); (Lellis et al. 2013) (Figure 9.3).

Options for Restoring Connectivity in Regulated Rivers

Management practices at the frontier of wildlife-friendly hydropower all lead to restored functional connectivity of rivers. Dam removal is an option, but this management action often results in a complete loss of hydropower. Fish passage technologies may alleviate the deleterious effects of dams; a wealth of research on fish passage technology was not reviewed here because of space limitations (Brown et al. 2013, Silva et al. 2018, Wilkes et al. 2018). Water-vortex turbines are one type of system that may help minimize fish trauma during passage in the future (Timilsina et al. 2018). In general, we characterize these actions as incremental advances that are unlikely to revolutionize the ability of wildlife to coexist with hydropower. Below, we focus on what we consider to be more visionary options that are currently being studied, including dam-free hydropower.

Several options exist for generating electricity from water without a complete dam or impoundment. Partial dams can be

Figure 9.3   (*Top left*) eastern elliptio (*Elliptio coplanata*); (*Top right*) eel ladder at Millville Dam on the Shenandoah River; (*bottom*) American eel (*Anguilla rostrata*). *Photo credits (clockwise): Phillip Westcott; Karl Blankenship; US Great Lakes Environmental Research Laboratory.*

used to divert water to a downstream power-house through penstocks without fully blocking the river channel. In addition, river kinetic energy can be generated from free turbines installed in rivers, and there are several options for power generation with water by using turbines installed in pipes either to convey water for other purposes or as part of a conjunctive use of groundwater (i.e., aquifer-pumped storage).

Free Turbines

Free turbines can be installed in rivers, water-falls, irrigation canals, and natural flumes (Lahimer et al. 2012). Kinetic energy may be generated in larger flowing rivers using dam-free turbines once technical barriers are surmounted (Struthers et al. 2017). A study of fish responses to operation of the Canadian Hydrokinetic Turbine Testing Center on the Winnipeg River, Manitoba, determined that risk of blade strike would be greatest for lake sturgeon moving along the river bottom during spring (Struthers et al. 2017). We note that the energy produced from dam-free hydropower is considerably lower than that from conventional hydropower because power generation is proportional to the head, which is the difference in water elevation above and below the dam.

TURBINES IN PIPES

Turbines installed in irrigation or wastewater pipes can produce hydropower that is environmentally benign. Typically, the power is used locally (e.g., to pump water through a central irrigation pivot). Aquifer-pumped storage is a promising dam-free option for hydropower, particularly in arid climates. Because other renewables, such as wind and solar, are being integrated into electricity systems, there is an opportunity to pump water from aquifers during times when there is wind or solar radiation and then generate on that water when these sources are unavailable (Hernandez et al. 2014). Integrated with aquifer storage and recovery systems, the use of underground aquifers to store water in combination with a small aboveground reservoir does not require dams and hence avoids the negative ecological impacts of dams and reservoirs. Thus, it appears that adverse effects on wildlife could be avoided entirely.

Re-visioning Design of Regulated River Basins

Interest is growing in how regulated river basins can be better designed to accommodate wildlife. We envision these design elements operating at 2 temporal scales. At a slow scale, natural geomorphological processes through which river channels evolve produce the diversity of habitat conditions that aquatic and terrestrial (i.e., riparian) biota require. For example, disturbance regimes associated with wildfire, landslides, debris flows, sediment dynamics, and river flows produce riffle-pool sequences in upper reaches and meanders and oxbows in lower reaches (Hessburg and Agee 2003, Holden et al. 2012, Flitcroft et al. 2016). These disturbance regimes contribute to maintaining shallow water with slow flows that support high wildlife diversity, including aquatic invertebrates, juvenile fishes, wading birds, beavers, and other water-associated mammals. Hydropower developers provide slow moving, shallow water through consideration of the following: (1) release of floodplain inundation flows when juvenile fish are rearing; (2) sediment bypasses, such as vented dams, that pass sediment-laden water during event-flows (Shetkar and Mahesha 2011, Kondolf et al. 2014); and (3) purchase of wide riparian conservation easements to ensure that large woody debris is supplied and that 2-stage floodplains can be formed (National Research Council 2002). Together, these 3 management strategies maintain natural processes that control channel evolution downstream (Kondolf 2011).

At a faster temporal scale, fishes and other aquatic biota require access to dispersed food and cover resources at different times and for different life stages. Research on sustainable river basin design has produced a number of recommendations for dam siting that help ensure that migratory species can persist after the dam is built (Jager et al. 2015). Some design concepts include (1) concentration of dams within a subset of tributary watersheds and avoidance of downstream main stems of rivers; (2) dispersal of freshwater reserves among the remaining tributary catchments; (3) assurance that habitat between dams will support reproduction of aquatic biota and retain their offspring; and (4) recognition of spatial issues that affect wildlife at the scale of large river basins (Jager et al. 2015). Benefits to wildlife of implementing these strategies at short and long timescales should be quantified in the future. They can provide a starting place for regional-scale research on how best to accommodate the need for water storage and hydropower without adverse effects on wildlife.

Marine Hydrokinetic Energy

Marine hydrokinetic energy (MHK) technologies are in the early stages of development (Jimenez et al. 2015) and promise to eventually supply one-third of total US energy demand (Albernaz et al. 2012). Estimates indicate that areas along the Pacific Coast of the United States are especially rich in wave-energy potential (Figure 9.4). At this writing, however, there are no operating MHK facilities in the United States (US EIA 2016).

Generating Electricity from Marine Currents

Two categories of MHK are current-energy converters and wave-energy converters (Jimenez et al. 2015). Wave-energy converters take a multitude of forms, from floating structures (e.g., buoys, overtopping devices) to submerged pressure-differential devices and oscillating pendulums. Current-energy converters can be divided into 2 basic types: (1) tidal barrages or lagoons, which capture water in a holding area; and (2) tidal-stream turbines or "fences" that capture energy as water moves past the turbine, without the use of a holding area. The wide variety of MHK turbine designs used by current-energy converters in both of these types (see Figure 9.5) can be classified into 3 groups: (1) axial-flow turbines; (2) cross-flow turbines; and (3) oscillating hydrofoils (Belloni 2013).

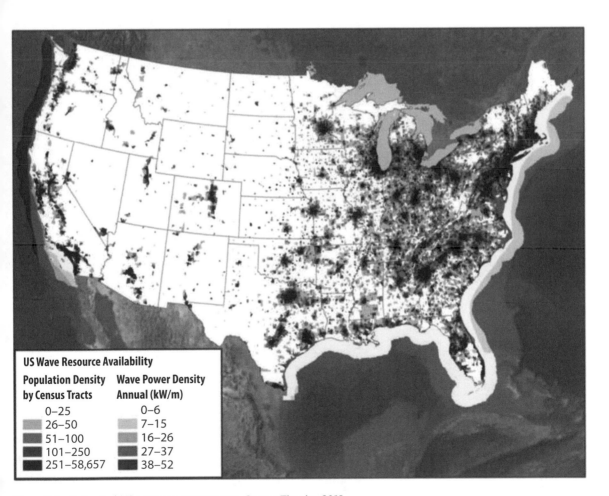

Figure 9.4   Estimated US wave energy resource. *Source: Thresher 2013.*

Figure 9.5   Examples of tidal and ocean current devices. (*A*) Atlantis AK1000 axial-flow turbine (1 MW); (*B*) Atlantis Solon-K ducted turbine (1 MW); (*C*) OpenHydro open-center turbine (250 kW); (*D*) Kepler transverse horizontal axis water turbine (THAWT), a kind of cross-flow turbine (4 MW); (*E*) Edinburgh vertical axis cross-flow turbine concept (100 MW); (*F*) Gorlov helical turbines before deployment at Cobscook Bay, Maine, US; (*G*) bioSTREAM (150 kW) oscillatory-hydrofoil turbine; (*H*) Minesto deep sea tidal kite; (*I*) Flumill Archimedes screw turbine. *Source: A, Borthwick 2016; F, image courtesy of Ocean Renewable Power Company.*

Tidal Barrages

Tidal barrages are the most established technology among marine hydrokinetic facilities worldwide (e.g., La Rance, France; Sihwa Lake, South Korea; and Annapolis, Canada). Axial-flow turbines in tidal barrages are like those used in hydroelectric dams, except that water flows in both directions. Turbines generate power during both parts of the tidal cycle, first as the barrage allows water to flow into a bay or estuary during high tide, and then again as the barrage releases the water during low tide. A new technology—the tidal lagoon—works similarly to the barrage. The Swansea Bay Tidal Lagoon in Wales is scheduled to be the first structure of this type. Unlike barrages, this tidal lagoon will comprise U-shaped breakwaters (i.e., jetty or sea wall) built out from the coast with a bank of turbines within. The man-made lagoon will fill up and empty as tides rise and fall, permitting generation of electricity 4 times per day. However, approval for Swansea was delayed when Natural Resources Wales claimed that 20%–25% of salmon and sea trout in the area could be killed as they migrate to and from local rivers (Youle 2016).

Tidal-Stream Turbines

Tidal-stream turbines are located where currents are strong (e.g., tidal channels with channel constrictions). Turbines vary in structure and design, but they are similar to those used to generate wind power (e.g., axial turbines; Figure 9.5A). Tidal-stream turbines are currently operating in Canada, Ireland, Scotland, Norway, the Netherlands, and possibly offshore areas of other countries. Several test installations have been deployed in the United States along the coast of Pacific Ocean, Alaska, and in the East River near New York City. A significant challenge to implementation of tidal-stream turbines has been building devices capable of withstanding strong currents.

Wave Power

The global resource of wave power is estimated to exceed 2 terawatts (Gunn and Stock-Williams 2012). Wave farms use wave energy converters (WECs) to produce electricity that is transmitted to the grid through submarine cables along the ocean floor (Khan et al. 2017). Importantly, when co-located with offshore wind facilities, wave farms can offset the intermittent nature of wind resources and provide a more consistent energy resource (Onea et al. 2017).

Wave energy converters have different designs depending on where they are deployed (e.g., nearshore versus offshore) and their position in the water column. Designs for wave energy differ in how they affect wildlife. Electromagnetic cables for WECs run along the ocean floor, which may disrupt navigation of some sensitive marine species. Floating structures (e.g., buoys, surface attenuators) and overtopping devices generate energy at the water surface. Oscillating water column devices use wave action to alternately compress and decompress trapped air to create electricity (Falcao and Henriques 2016). Designs that use flexible membranes to extract energy from submerged pressure differentials can be situated in the mid-water column, where interference with ship navigation and other surface interactions are avoided. Because WECs are secured to the seafloor by cables, entanglement in mooring lines is a concern for fish, marine mammals, and birds.

Changes in Marine Habitat

Marine hydrokinetic energy influences wildlife both during initial construction and during operation (Table 9.1). Habitat loss and change is likely greatest with the construction of tidal barrages. In the case of ebb-only generation, permanent inundation of areas upstream of the barrage threatens species that specialize on feeding in the upper-shore area by destroying nearshore environments, such as salt marshes, and it reduces the amount of time that tidal flats are exposed for use by foraging birds (Goss-Custard et al. 1991, Parsons Brinckerhoff Ltd 2008, Kidd et al. 2015).

Change in water currents can be significant near MHK installations, especially for tidal barrages. Algal blooms may appear upstream of barrages. Further, reduced current speed upstream of MHK facilities may increase water clarity (Clare et al. 2017). Clearer water increases light penetration and phytoplankton biomass, which, in turn, increases habitat for filter feeders and pelagic fish at higher trophic levels (Underwood 2010, Warwick and Somerfield 2010). Changes in the transport of nutrients accompany shifts in currents and mixing in the water column. Increased stratification from MHK, for example, may lead to toxic algal blooms (Roche et al. 2016), as observed at a UK barrage in Northumberland (Worrall and McIntyre 1998). Turbulent flows have been shown to emerge downstream of barrages and immediately upstream of the inflows (Clare et al. 2017).

Alteration of flow patterns near MHK installations may affect dispersal and colonization-extinction dynamics of marine larvae (Bulleri and Chapman 2010), subsequently affecting recolonization by fish and other animal communities postconstruction. In addition, altered flow fields can prevent animal movements between the areas occupied by energy installations and neighboring environments (Hooper and Austen 2013). These effects are thought to be greatest for tidal barrages and smallest and most temporary for tidal-stream turbines (Langhamer et al. 2010).

Table 9.1.   Potential effects of marine hydrokinetic energy (MHK) barrages and wave arrays on wildlife and suggestions for mitigation

Type of MHK Technology	Risk to Wildlife	Species / Functional Groups Affected	Mitigation
Tidal barrage, ebb-only generation	Loss of mud-flat feeding areas	Wading birds	Create offsite intertidal habitat Dual-cycle generation
Tidal barrage	Algal blooms	Filter feeders, pelagic fish	Design currents to mix waters and avoid stratification Nutrient reduction (offsite)
Tidal barrage, tidal-stream turbines	Advective dispersal / entrainment	Marine larvae	More research needed
Tidal barrage, tidal-stream turbines, wave arrays	Sedimentation and dredging	Benthic communities	Locate arrays far from shore (distance determined based on numeric modeling of local conditions)
Tidal barrage, tidal-stream turbines	Turbine mortality	Fish, marine mammals, turtles	Provide safe fish passage through turbine arrays; use slow rotation speeds; use wave energy devices and tidal devices that employ oscillating beams, bulb or other enclosed turbines
Wave array	Entanglement in cables	Sea birds, marine mammals, fish	More research needed
Tidal barrage, tidal-stream turbines, wave array	Noise and habitat displacement	Marine mammals, fish, benthic communities	More research needed

Benthic habitat can be influenced by MHK turbine operation (Sheehan et al. 2010). Seabed erosion and scour pits form in areas of high turbulence alongside turbines, causing increased sediment deposition elsewhere (Shields et al. 2011, Ahmadian et al. 2012). In the case of tidal barrages, changes in sediment transport can be extreme. For example, sedimentation caused by a barrage at Goolwa, near the mouth of the Murray River, South Australia, has resulted in shallow conditions at the inlet every other year (Walker 2003). Consequently, periodic dredging was required to keep the mouth of the river open. Operation of tidal-stream turbines also causes localized seabed erosion and scouring within 100 to 200 m of the turbine (Gasparatos et al. 2017).

Large quantities of suspended sediment potentially resulting from MHK operations may interfere with feeding and digestion of benthic marine life (e.g., crustaceans, oligochaetes, polychaetes, benthic fish, insects). Resuspended sediments resulting from MHK may remobilize sediment-bound contaminants

(e.g., Bock and Miller 1994, Gill 2005, Hooper and Austen 2013). Even if sedimentation does not cause smothering of biota directly, dredging to remove sediment deposited from MHK disturbs the community in the same way as does erosion and scouring.

Wave-energy generation in offshore arrays likely has imperceptible effects on wave structure because waves are naturally variable in the open ocean (e.g., Millar et al. 2007, Palha et al. 2010). Theoretically, effects could include reduced mixing in the upper water column and changes to coastal environments (Boehlert and Gill 2010, Frid et al. 2012). For example, Abanades et al. (2015) predicted that beach erosion could decrease by 20% when offshore MHK arrays were within 2 to 3 km of shore. Such effects could be avoided by locating wave arrays farther from the coast; however, they may be desired to protect coastal habitat from storm-induced erosion and sea-level encroachment.

Depending on where they are located, MHK installations can have variable effects on biodiversity.

Coastal areas are hot-spots for biodiversity and, therefore, nearshore coastal MHK installations are more likely to have negative impacts than offshore MHK installations. In some cases, changes in the environment caused by tidal barrages may alter fish assemblages and lead to reduced biodiversity (e.g., Shields et al. 2009, J. D. Yoon et al. 2016, J. Yoon et al. 2017). However, the area surrounding MHK devices will likely be closed off to bottom-trawl fishing, which may benefit benthic communities, fish, and other marine wildlife (Inger et al. 2009, Boehlert and Gill 2010, Langhamer 2010, 2012). In other words, MHK arrays may inadvertently become marine protected areas. In fact, it has been suggested that ocean zoning should seek opportunities for co-location for conservation, fisheries maintenance, and marine renewable energy (Yates et al. 2015).

Research is needed to understand how artificial structures that support MHK devices might be designed to enhance habitat for marine animals (Inger et al. 2009, Langhamer 2012). The structure and scour protection provided by MHK devices may create artificial reefs that attract benthic species and fish (Inger et al. 2009, Langhamer 2010, Langhamer 2012). Broadhurst and Orme (2014) determined that an artificial reef created by a tidal turbine supported higher overall biodiversity, including a greater proportion of predators, than a control site. Often, fish and epibenthic assemblages that colonize artificial reef materials (e.g., concrete, plastic, or metal) are substantially different from those that colonize nearby natural substrates (Bulleri and Chapman 2010). Incorporating artificial cover that mimics the surrounding areas and creating microhabitats that function as refuges against predators, wave action, desiccation, and scouring can increase the ability of larvae to settle and survive on MHK structures (Bulleri and Chapman 2010).

Risk of Wildlife-Turbine Collision

Blade-strike risk experienced by fish and other mobile taxa should be lower for MHK than for hydro-power because the water passing through turbines is not entrained. Hydrokinetic devices that capture energy without turbines (e.g., wave energy devices and tidal stream devices that employ oscillating beams) eliminate the risk of blade strike. Tidal barrages that use traditional turbines are much more likely to cause fish injury or death than tidal-stream turbines (Jacobson et al. 2012) or facilities that use enclosed (e.g., bulb) turbines. Unlike traditional turbines in barrages, hydrokinetic turbines usually produce only minor changes in shear, turbulence, and pressure (Jacobson et al. 2012). The greatest risks of collision with turbines likely occur in partially enclosed water bodies such as estuaries or bays (Dadswell and Rulifson 1994) or where turbines form an extensive barrier to movement (Gill 2005). Blade-strike mortality is less likely at slower rotational speeds (i.e., those of larger-diameter turbines), in clear water, and from turbines with fewer blades; it also varies depending on placement in the water column (Schweizer et al. 2011, Hammar et al. 2015).

The risk of wildlife-turbine collision depends on the natural history of the focal wildlife taxa. Species thought to be most at risk of collision with hydrokinetic turbines include larger species and apex predators, including marine mammals, sharks, and other large fishes (e.g., Gill 2005, Hammar et al. 2015). In addition, diving seabirds—for example, guillemots (family Alcidae) and cormorants (family Phalacrocoracidae)—may be susceptible to blade strike (Langton et al. 2011). Turbine collisions involving marine mammals are of conservation concern because their large size elevates risk of blade strike, and many populations already are threatened by other human impacts. Interestingly, however, no collisions of marine life with MHK devices have been recently reported (Baring-Gould et al. 2016).

Video cameras and acoustic devices have been used to observe fish behavior in the vicinity of tidal turbines (Hammar et al. 2013, Viehman and Zydlewski 2015, Bevelhimer et al. 2016). Fish avoidance of turbines and evasion of blades when passing through turbines has been documented (e.g., Hammar et al. 2013, Viehman and Zydlewski 2015, Bevelhimer et al. 2016). However, detection equipment cannot always determine whether blade strike occurred or the fate of a fish after it entered the turbine.

Because field monitoring can be challenging, physical or computer simulations often are substituted (Hammar et al. 2015). For example, Carlson et al. (2014) used a physical model to evaluate effects of blade strike on material similar to killer whale (*Orcinus orca*) tissue. Under a worst-case scenario (i.e., head being struck by a turbine spinning at 5 m/s), bruising and mild lacerations would likely have resulted. Using a different approach, Thompson et al. (2015b) simulated blade strike by performing collisions between a boat equipped with a horizontal bar (meant to act as a turbine blade) and previously frozen seal carcasses. No obvious injuries were observed, but damage to soft tissue could not be assessed.

Also, potential strike impacts of hydrokinetic turbines have been explored by simulation models of collision risk, and these have estimated a negligible number of strikes. For example, Romero-Gomez and Richmond (2014) predicted a survival rate of 87% to 99% for simulated fish that approached the turbine. However, more realistic simulation is possible by estimating the ability of individual fishes or marine mammals to avoid turbines. For fish, this may depend on how their swimming patterns are affected by increased current speed and whether they can avoid entrainment and collision with turbine blades (Hammar et al. 2015). Experimental evidence from flume studies of small fishes indicated that survival rates can exceed 95% (Normandeau Associates Inc. 2009, Jacobson et al. 2012, Amaral et al. 2015), because most fish avoided turbines by swimming upstream (in a flume) or being swept around the turbine.

Species with different traits vary in their susceptibility to turbine mortality. Fish with less maneuverable body shapes (e.g., compressiform) or of a large size have been observed avoiding a turbine at greater distances than smaller fish or those that have a more maneuverable, fusiform shape (Hammar et al. 2013). Hammar et al. (2013) also observed that "bolder" fish

species, like wrasses (family Labridae) and stumpnoses (family Sparidae), more closely approached turbines. In general, probability of strike following entrainment increases with animal size. Fish are better able to avoid turbines during the day (i.e., when visual cues are available) and when swimming in schools (Viehman and Zydlewski 2015). More research on blade-strike risk is needed to quantify the cumulative effects of turbine arrays, rather than a single turbine, and how species with different traits differ in their ability to avoid collisions (Hammar et al. 2015, Baring-Gould et al. 2016).

Mooring lines and cables associated with MHK structures could cause entanglement of turtles, large fish, seabirds, and marine mammals, especially baleen whales (e.g., Cada et al. 2007, Frid et al. 2012). This threat might be reduced by use of large, heavy cables that are less likely to cause entanglement (Baring-Gould et al. 2016). Low-flying birds may be at risk of colliding with floating structures that are sometimes associated with wave-energy devices (Hammar et al. 2017). This risk may be amplified if birds are attracted to prey that may aggregate around the artificial reefs created by such wave-energy devices (Langton et al. 2011).

Barriers to Wildlife Movement or Habitat Use

Migratory species, including sharks, seabirds, marine mammals, and eels, are among those at greatest risk of extinction (Jager et al. 2008). MHK structures that block or alter species migration routes or alter environments required to complete their life cycles are therefore a conservation concern. Species that migrate between marine and freshwater environments may be especially vulnerable to MHK projects located in coastal areas (e.g., Lucas et al. 2009, Piper et al. 2013, Silva et al. 2017, Yoon et al. 2017).

Tidal barrages can physically prevent fish and marine mammals from reaching feeding and breeding areas (Frid et al. 2012). In one case, a whale was entrapped in a MHK plant in Canada (Gasparatos et al.

Figure 9.6   Harbor seal outfitted to study how the tidal turbine at Strangford Lough might affect movement of the species. *Source: http://www.smruconsulting.com/tidal-turbine-affect-seal-movements/.*

2017). Harbor porpoises (*Phocoena phocoena*), harbor seals (*Phoca vitulina*), and gray seals (*Halichoerus grypus*) regularly were observed transiting past the world's first commercial-scale tidal turbine in Northern Ireland's Strangford and Lough (Savidge et al. 2014). There was no evidence for a barrier effect from this MHK installation, although the number of transits by marine mammals decreased by 20% when the turbine was operating (Sparling et al. 2017). Field observations also showed local wildlife avoidance of operating turbines (Keenan et al. 2011, Band et al. 2016). At Strangford Lough, a passive acoustic monitoring device was used to detect porpoises and dolphins within 50 m of turbines based on echolocation clicks (Figure 9.6). When protected species are detected approaching the array, turbine operation can be interrupted (Savidge et al. 2014).

Tidal-stream turbines have not yet been reported to change the distribution or numbers of bird and benthic species (Gasparatos et al. 2017). However, concerns remain for species that are transported by ocean currents between areas required for different parts of their life histories. For example, adults of American eel migrate from coastal rivers of North America to the Sargasso Sea and larval elvers (young eels) return by riding the Gulf Stream (Diaz et al. 2015). Placement of tidal stream MHK structures along the Gulf Stream, for example, might adversely affect this species. Such risks may be partially mitigated by engineering spaces for fish to pass through (i.e., fishways) or using navigation locks to allow fish to pass (Frid et al. 2012, Silva et al. 2017).

Effects of Noise on Wildlife

The loudest noises associated with MHK likely will occur during construction of turbine arrays.

Construction of tidal barrages is more extensive and will likely lead to more chronic exposure of animals to high noise levels (Roche et al. 2016); however, construction for MHK is probably quieter than that of offshore wind farms (Matuschek and Betke 2009, Copping et al. 2014). Loud impulse sounds from operations like those created by pile drivers are thought to have the most damaging effects on wildlife (Frid et al. 2012, Hammar et al. 2017). Behavioral observations during construction of tidal turbines indicated that harbor porpoises left the area during construction, but returned to near-baseline numbers once construction had ceased (Keenan et al. 2011). The construction phase for all MHK devices also includes sounds associated with increased vessel traffic (e.g., Thompson et al. 2015a, Gasparatos et al. 2017).

The effects of noise from MHK devices on wildlife is difficult to assess, as noise levels are influenced by a variety of factors, including design of the device, layout of the arrays, oceanographic conditions and background noise, as well as location (e.g., depth and topography; Patrício et al. 2009, Popper and Hastings 2009). Although short-term exposure to noise during normal MHK operation is not likely to cause physiological damage to most animal species, it could have behavioral effects on marine mammals and fish or disrupt underwater communication among marine animals (Keenan et al. 2011, Frid et al. 2012, Copping et al. 2014, Baring-Gould et al. 2016). Schramm et al. (2017) exposed 4 species of freshwater fish to broadcasted sounds from tidal turbines, but they were unable to provide strong evidence of significant behavioral changes caused by the noise. More research is needed to characterize the sound levels of MHK devices during operation and to evaluate the effects of noise produced by large, multi-turbine arrays on different species of marine animals (e.g., Patrício et al. 2009, Baring-Gould et al. 2016, Bevelhimer et al. 2016, Roche et al. 2016). In addition to the effects of sound, more information is needed to understand how MHK devices influence substrate vibration and particle motion, which, in turn, may affect some benthic species (e.g., crustaceans; Thomsen 2015, Baring-Gould et al. 2016).

Mitigating noise associated with MHK, although possible, may not always be advisable. Mitigation strategies include acoustic shielding or damping on devices, tuning devices to operate at different frequencies, and operating at different rotational speeds (Cada et al. 2007). However, if sound levels become low, mobile species may not detect MHK devices to avoid collision (Roche et al. 2016). Acoustic alarms to deter marine mammals have been proposed, but these may not be effective, and introducing another noise source into the environment may have negative consequences for other wildlife (Wilson and Carter 2013).

Exposure of Wildlife to Chemical Pollution and Electromagnetic Fields

Marine hydrokinetic operations may pollute the environment through chemical spills (e.g., hydraulic fluid) or through leaching of anti-fouling paints that are used to deter organisms (e.g., barnacles) from attaching to mooring lines (Dolman and Simmonds 2010, Henkel et al. 2014). Tidal barrages can also change the amount of suspended sediment and salinity and influence flushing of deoxygenated water, all of which may lead to mass mortality of fish species (Copping et al. 2013, Broadhurst and Orme 2014).

Electromagnetic fields produced by cables associated with marine hydrokinetic devices have the potential to disrupt species that use electromagnetic fields for navigation or finding prey. Fish that are most sensitive to electromagnetic fields include sharks, rays, and agnathans (Gill et al. 2014). Behavioral responses by fishes to electromagnetic fields generated by cables have been shown in a laboratory, but inferring effects of these cables in the field based on laboratory studies alone is still speculative (Öhman et al. 2007, Bevelhimer et al. 2013, Woodruff et al. 2013, Gill et al. 2014). Benthic species are likely to experience greater exposure to electromagnetic fields and thus experience more risk than pelagic species, as they are more likely to be near subsea cables (Woodruff et al. 2013, Gill et al. 2014). Effects of MHK are most likely in the immediate vicinity of grid-connection cables and attenuate within a few meters when MHK devices are on a small scale (i.e., fewer ~10 devices; Bevelhimer et al. 2013, Baring-Gould et al. 2016).

ACKNOWLEDGMENTS
We appreciate an excellent review of the conventional hydropower section by Rebecca Novello (Oak Ridge Institute for Science Education) and a thorough review of the marine hydrokinetic section provided by Dr. Mark Bevelhimer (Oak Ridge National Laboratory). This chapter was authored by UT-Battelle, LLC, under Contract No. DE-AC05-00OR22725 with the US Department of Energy. By accepting the article for publication, the publisher acknowledges that the United States government retains a nonexclusive, paid-up, irrevocable, worldwide license to publish or reproduce the published form of the manuscript, or allow others to do so, for United States government purposes. The Department of Energy will provide public access to these results of federally sponsored research in accordance with the DOE Public Access Plan (http://energy.gov/downloads/doe-public-access-plan).

LITERATURE CITED
Albernaz, A. L., R. L. Pressey, L. R. F. Costa, M. P. Moreira, J. F. Ramos, P. A. Assuncao, and C. H. Franciscon. 2012. Tree species compositional change and conservation implications in the white-water flooded forests of the Brazilian Amazon. Journal of Biogeography 39:869–883.

Amaral, S. V., M. S. Bevelhimer, G. F. Čada, D. J. Giza, P. T. Jacobson, B. J. McMahon, and B. M. Pracheil. 2015. Evaluation of behavior and survival of fish exposed to an axial-flow hydrokinetic turbine. North American Journal of Fisheries Management 35:97–113.

Angelaki, V., and J. M. Harbor. 1995. Impacts of flow diversion for small hydroelectric power plants on sediment transport, northwest Washington. Physical Geography 16:432–443.

Arraut, E. M., J. L. Arraut, M. Marmontel, J. E. Mantovani, and E. Novo. 2017. Bottlenecks in the migration routes of Amazonian manatees and the threat of hydroelectric dams. Acta Amazonica 47:7–17.

Bartholow, J. M., S. G. Campbell, and M. Flug. 2004. Predicting the thermal effects of dam removal on the Klamath River. Environmental Management 34:856–874.

Belloni, C. 2013. Hydrodynamics of ducted and open-centre tidal turbines. Trinity, UK: Oxford University.

Benchimol, M., and C. A. Peres. 2015a. Predicting local extinctions of Amazonian vertebrates in forest islands created by a mega dam. Biological Conservation 187: 61–72.

Benchimol, M., and C. A. Peres. 2015b. Widespread forest vertebrate extinctions induced by a mega hydroelectric dam in lowland Amazonia. PLOS One 10:e0129818. https://doi.org/10.1371/journal.pone.0129818.

Benejam, L., S. Saura-Mas, M. Bardina, C. Sola, A. Munne, and E. Garcia-Berthou. 2016. Ecological impacts of small hydropower plants on headwater stream fish: From individual to community effects. Ecology of Freshwater Fish 25:295–306.

Benke, A. C., I. Chaubey, G. M. Ward, and E. L. Dunn. 2000. Flood pulse dynamics of an unregulated river floodplain in the southeastern US coastal plain. Ecology 81:2730–2741.

Bilotta, G. S., N. G. Burnside, J. C. Gray, and H. G. Orr. 2016. The effects of run-of-river hydroelectric power schemes on fish community composition in temperate streams and rivers. PLOS One 11: e0154271. https://doi.org/10.1371/journal.pone.0154271.

Bilotta, G. S., N. G. Burnside, M. D. Turley, J. C. Gray, and H. G. Orr. 2017. The effects of run-of-river hydroelectric power schemes on invertebrate community composition in temperate streams and rivers. PLOS One 12:e0171634. https://doi.org/10.1371/journal.pone.0171634.

Boavida, I., J. M. Santos, M. T. Ferreira, and A. Pinheiro. 2013. Fish habitat-response to hydropeaking. Proceedings of the 35th Iahr World Congress, Vols. I and II:2512–2519.

Boehlert, G. W., and A. B. Gill. 2010. Environmental and ecological effects of ocean renewable energy development: A current synthesis. Oceanography 23:68–81.

Borthwick, A. G. L. 2016. Marine renewable energy seascape. Engineering 2:69–78.

Breck, S. W., K. R. Wilson, and D. C. Andersen. 2003a. Beaver herbivory and its effect on cottonwood trees: Influence of flooding along matched regulated and unregulated rivers. River Research and Applications 19:43–58.

Breck, S. W., K. R. Wilson, and D. C. Andersen. 2003b. Beaver herbivory of willow under two flow regimes: A comparative study on the Green and Yampa Rivers. Western North American Naturalist 63:463–471.

Brown, J. J., K. E. Limburg, J. R. Waldman, K. Stephenson, E. P. Glenn, F. Juanes, and A. Jordaan. 2013. Fish and hydropower on the US Atlantic coast: Failed fisheries policies from half-way technologies. Conservation Letters 6:280–286.

Bulleri, F., and M. G. Chapman. 2010. The introduction of coastal infrastructure as a driver of change in marine environments. Journal of Applied Ecology 47:26–35.

Bunnell, F. L., L. L. Kremsater, and E. Wind. 1999. Managing to sustain vertebrate richness in forests of the Pacific Northwest: Relationships within stands. Environmental Reviews 7:97–146.

Burroughs, B. A., D. B. Hayes, K. D. Klomp, J. F. Hansen, and J. Mistak. 2010. The effects of the Stronach Dam removal on fish in the pine river, Manistee County, Michigan. Transactions of the American Fisheries Society 139:1595–1613.

Catalano, M. J., M. A. Bozek, and T. D. Pellett. 2007. Effects of dam removal on fish assemblage structure and spatial distributions in the Baraboo River, Wisconsin. North American Journal of Fisheries Management 27:519–530.

Chinnayakanahalli, K. J., C. P. Hawkins, D. G. Tarboton, and R. A. Hill. 2011. Natural flow regime, temperature and the composition and richness of invertebrate assemblages in streams of the western United States. Freshwater Biology 56:1248–1265.

Clare, D. S., M. Spencer, L. A. Robinson, and C. L. J. Frid. 2017. Explaining ecological shifts: The roles of temperature and primary production in the long-term dynamics of benthic faunal composition. Oikos 126:1123–1133.

Committee on the Status of Endangered Wildlife in Canada. 2006. COSEWIC assessment and status report on the western painted turtle, Chrysemys picta bellii (Pacific Coast population, Intermountain-Rocky Mountain population and Prairie/Western Boreal–Canadian Shield population), in Canada. https://www.registrelep-sara registry.gc.ca/virtual_sara/files/cosewic/sr_Western% 20Painted%20Turtle_2016_e.pdf.

Cooper, A. R., D. M. Infante, W. M. Daniel, K. E. Wehrly, L. Wang, and T. O. Brenden. 2017. Assessment of dam effects on streams and fish assemblages of the conterminous USA. Science of the Total Environment 586:879–889.

Copping, A., H. Battey, J. Brown-Saracino, M. Massaua, and C. Smith. 2014. An international assessment of the environmental effects of marine energy development. Ocean and Coastal Management 99:3–13.

da Silva, P. S., M. C. Makrakis, L. E. Miranda, S. Makrakis, L. Assumpcao, S. Paula, J. H. P. Dias, and H. Marques. 2015. Importance of reservoir tributaries to spawning of migratory fish in the Upper Parana River. River Research and Applications 31:313–322.

de Oliveira, A. K., J. C. Garavello, V. V. Cesario, and R. T. Cardoso. 2016. Fish fauna from Sapucai-Mirim River, tributary of Grande River, upper Parana River basin, Southeastern Brazil. Biota Neotropica 16.

Diaz, S., S. Demissew, J. Carabias, C. Joly, M. Lonsdale, N. Ash, A. Larigauderie, et al. 2015. The IPBES Conceptual Framework: Connecting nature and people. Current Opinion in Environmental Sustainability 14:1–16.

Doyle, M. W., E. H. Stanley, C. H. Orr, A. R. Selle, S. A. Sethi, and J. M. Harbor. 2005. Stream ecosystem response to small dam removal: Lessons from the heartland. Geomorphology 71:227–244.

Ellis, L. E., and N. E. Jones. 2013. Longitudinal trends in regulated rivers: A review and synthesis within the context of the serial discontinuity concept. Environmental Reviews 21:136–148.

Eno, N. C., C. L. J. Frid, K. Hall, K. Ramsay, R. A. M. Sharp, D. P. Brazier, S. Hearn, et al. 2013. Assessing the sensitivity of habitats to fishing: From seabed maps to sensitivity maps. Journal of Fish Biology 83:826–846.

Eyler, S. M., S. A. Welsh, D. R. Smith, and M. M. Rockey. 2016. Downstream passage and impact of turbine shutdowns on survival of silver American eels at five hydroelectric dams on the Shenandoah River. Transactions of the American Fisheries Society 145:964–976.

Falcao, A. F. O., and J. C. C. Henriques. 2016. Oscillating-water-column wave energy converters and air turbines: A review. Renewable Energy 85:1391–1424.

Feld, C. K., S. Birk, D. C. Bradley, D. Hering, J. Kail, A. Marzin, A. Melcher, et al. 2011. From natural to degraded rivers and back again: A test of restoration ecology theory and practice. In Advances in Ecological Research, edited by G. Woodward, 119–209. Cambridge, MA: Academic Press.

Fencl, J. S., M. E. Mather, J. M. Smith, and S. M. Hitchman. 2017. The blind men and the elephant examine biodiversity at low-head dams: Are we all dealing with the same dam reality? Ecosphere 8:e01973.https://doi.org/10.1002 /ecs2.1973.

Feyrer, F., T. Sommer, and W. Harrell. 2006. Managing floodplain inundation for native fish: Production dynamics of age-0 splittail (Pogonichthys macrolepidotus) in California's Yolo Bypass. Hydrobiologia 573:213–226.

Flitcroft, R. L., S. L. Lewis, I. Arismendi, R. LovellFord, M. V. Santelmann, M. Safeeq, and G. Grant. 2016. Linking hydroclimate to fish phenology and habitat use with ichthyographs. PLOS One 11:e0168831. https://doi .org/10.1371/journal.pone.0168831.

Franklin, S. B., J. A. Kupfer, S. R. Pezeshki, R. A. Hanson, T. L. Scheff, and R. W. Gentry. 2001. A comparison of hydrology and vegetation between a channelized stream and a nonchannelized stream in western Tennessee. Physical Geography 22:254–274.

Fraser, G. S., D. L. Winkelman, K. R. Bestgen, and K. G. Thompson. 2017. Tributary use by imperiled flannel-mouth and bluehead suckers in the Upper Colorado River Basin. Transactions of the American Fisheries Society 146:858–870.

Freedman, J. A., B. D. Lorson, R. B. Taylor, R. F. Carline, and J. R. Stauffer. 2014. River of the dammed: Longitu-dinal changes in fish assemblages in response to dams. Hydrobiologia 727:19–33.

Galbraith, H. S., and C. C. Vaughn. 2011. Effects of reservoir management on abundance, condition, parasitism and reproductive traits of downstream mussels. River Research and Applications 27.

Gill, A. B., I. Gloyne-Phillips, J. A. Kimber, and P. Sigrary. 2014. Marine renewable energy, electromagnetic fields and EM-sensitive animals. In Humanity and the Sea: Marine Renewable Energy Technology and Environmen-tal Interactions, edited by M. Shields and A. Payne. Suffolk, UK: Springer, Halesworth.

Guenther, C. B., and A. Spacie. 2006. Changes in fish assemblage structure upstream of impoundments within the Upper Wabash River Basin, Indiana. Transactions of the American Fisheries Society 135:570–583.

Gunn, K., and C. Stock-Williams. 2012. Quantifying the global wave power resource. Renewable Energy 44:296–304.

Hammar, L., S. Andersson, L. Eggertsen, J. Haglund, M. Gullström, J. Ehnberg, and S. Molander. 2013. Hydroki-netic turbine effects on fish swimming behaviour. PLOS One 8:e84141. https://doi.org/10.1371/journal.pone .0084141.

Hammar, L., L. Eggertsen, S. Andersson, J. Ehnberg, R. Arvidsson, M. Gullstrom, and S. Molander. 2015. A probabilistic model for hydrokinetic turbine collision risks: Exploring impacts on fish. PLOS One 10: e0117756.

Haro, A., B. Watten, and J. Noreika. 2016. Passage of downstream migrant American eels through an airlift-assisted deep bypass. Ecological Engineering 91:545–552.

Hernandez, E. A., V. Uddameri, and S. Singaraju. 2014. Combined optimization of a wind farm and a well field for wind-enabled groundwater production. Environmen-tal Earth Sciences 71:2687–2699.

Hessburg, P. F., and J. K. Agee. 2003. An environmental narrative of inland northwest United States forests, 1800–2000. Forest Ecology and Management 178:23–59.

Hill, R., E. Fox, S. Leibowitz, A. Olsen, D. Thornbrugh, and M. Weber. 2017. Predictive mapping of the biotic condition of conterminous-USA rivers and streams. Ecological Applications 27:2397–2415.

Hogg, R., S. M. Coghlan, and J. Zydlewski. 2013. Anadro-mous sea lampreys recolonize a Maine coastal river tributary after dam removal. Transactions of the American Fisheries Society 142:1381–1394.

Hohensinner, S., H. Habersack, M. Jungwirth, and G. Zauner. 2004. Reconstruction of the characteristics of a natural alluvial river-floodplain system and hydromor-phological changes following human modifications: The Danube River (1812–1991). River Research and Applications 20:25–41.

Holden, Z. A., C. H. Luce, M. A. Crimmins, and P. Morgan. 2012. Wildfire extent and severity correlated with annual streamflow distribution and timing in the Pacific Northwest, USA (1984–2005). Ecohydrology 5:677–684.

Hornbach, D. J., M. C. Hove, H. T. Liu, F. R. Schenck, D. Rubin, and B. J. Sansom. 2014. The influence of two differently sized dams on mussel assemblages and growth. Hydrobiologia 724:279–291.

Hunt, P. D., and D. J. Flaspohler. 1998. Yellow-rumped Warbler (Setophaga coronata). Version 2.0 in The Birds of North America, edited by P. G. Rodewald. Ithaca, NY: Cornell Lab of Ornithology.

Inger, R., M. J. Attrill, S. Bearhop, A. C. Broderick, W. J. Grecian, D. J. Hodgson, C. Mills, et al. 2009. Marine renewable energy: Potential benefits to biodiversity? An urgent call for research. Journal of Applied Ecology 46:1145–1153.

Jacobson, P. T., S. Amaral, T. Castro-Santos, D. Giza, A. Haro, G. Hecker, B. McMahon, N. Perkins, and N. Pioppi. 2012. Environmental effects of hydrokinetic turbines on fish: Desktop and laboratory flume studies. https://www.osti.gov/servlets/purl/1084623.

Jager, H. 2014. Thinking outside the channel: Timing pulse flows to benefit salmon via indirect pathways. Ecological Modelling 273:117–127.

Jager, H. I., and M. S. Bevelhimer. 2007. How run-of-river operation affects hydropower generation and value. Environmental Management 40:1004–1015.

Jager, H. I., and R. A. McManamay. 2014. Comment on "Cumulative biophysical impact of small and large hydropower development in Nu River, China" by Kelly M. Kibler and Desiree D. Tullos. Water Resources Research 50:758–759.

Jager, H. I., and R. Uria-Martinez. 2012. Optimizing river flows for salmon and energy. Oak Ridge, TN: Oak Ridge National Laboratory.

Jager, H. I., J. A. Chandler, K. B. Lepla, and W. Van Winkle. 2001. A theoretical study of river fragmentation by dams and its effects on white sturgeon populations. Environmental Biology of Fishes 60:347–361.

Jager, H., K. A. Rose, and A. Vila-Gispert. 2008. Life history correlates and extinction risk of capital-breeding fishes. Hydrobiologia 602:15–25.

Jager, H. I., B. Elrod, N. Samu, R. A. McManamay, and B. T. Smith. 2013. ESA protection for the American eel: Implications for US hydropower. Oak Ridge, TN: Oak Ridge National Laboratory.

Jager, H., R. Efroymson, J. Opperman, and M. Kelly. 2015. Spatial design principles for sustainable hydropower development in river basins. Renewable and Sustainable Energy Reviews 45:808–816.

Jager, H. I., M. J. Parsley, J. J. Cech Jr., R. L. McLaughlin, P. S. Forsythe, R. F. Elliott, and B. Pracheill. 2016. Reconnecting fragmented sturgeon populations in North American rivers. Fisheries 41:140–148.

Jimenez, T., S. Tegen, and P. Beiter. 2015. Economic impact of large-scale deployment of offshore marine and hydrokinetic technology in Oregon coastal counties. National Renewable Energy Laboratory. Technical Report NREL/TP-5000-63506. March. https://www.boem.gov/BOEM-2015-018/.

Jones, I. L., N. Bunnefeld, A. S. Jump, C. A. Peres, and D. H. Dent. 2016. Extinction debt on reservoir land-bridge islands. Biological Conservation 199:75–83.

Kano, Y., D. Dudgeon, S. Nam, H. Samejima, K. Watanabe, C. Grudpan, J. Grudpan, et al. 2016. Impacts of dams and global warming on fish biodiversity in the Indo-Burma hotspot. PLOS One 11:e0160151. https://doi.org/10.1371/journal.pone.0160151.

Kashiwagi, M. T., and L. E. Miranda. 2009. Influence of small impoundments on habitat and fish communities in headwater streams. Southeastern Naturalist 8:23–36.

Katz, G. L., and P. B. Shafroth. 2003. Biology, ecology and management of Elaeagnus angustifolia L. (Russian olive) in western North America. Wetlands 23:763–777.

Khan, N., A. Kalair, N. Abas, and A. Haider. 2017. Review of ocean tidal, wave and thermal energy technologies. Renewable and Sustainable Energy Reviews 72:590–604.

Kibler, K. M., and D. D. Tullos. 2013. Cumulative biophysical impact of small and large hydropower development in Nu River, China. Water Resources Research 49:3104–3118.

Kondolf, G. M. 2011. Setting goals in river restoration: When and where can the river "heal itself"? In Stream Restoration in Dynamic Fluvial Systems: Scientific Approaches, Analyses, and Tools, edited by A. Simon, S. J. Bennett, and J. M. Castro, 29–43. Geophysical Monograph Series. Washington, DC: American Geophysical Union.

Kondolf, G. M., Y. X. Gao, G. W. Annandale, G. L. Morris, E. H. Jiang, J. H. Zhang, Y. T. Cao, et al. 2014. Sustainable sediment management in reservoirs and regulated rivers: Experiences from five continents. Earths Future 2:256–280.

Lahimer, A., M. Alghoul, K. Sopian, N. Amin, N. Asim, and M. Fadhel. 2012. Research and development aspects of pico-hydro power. Renewable and Sustainable Energy Reviews 16:5861–5878.

Lange, K., P. Meier, C. Trautwein, M. Schmid, C. T. Robinson, C. Weber, and J. Brodersen. 2018. Basin-scale effects of small hydropower on biodiversity dynamics. Frontiers in Ecology and the Environment 16:397–404.

Langhamer, O. 2010. Effects of wave energy converters on the surrounding soft-bottom macrofauna (west coast of Sweden). Marine Environmental Research 69:374–381.

Langhamer, O. 2012. Artificial reef effect in relation to offshore renewable energy conversion: State of the art. Scientific World Journal 2012:e386713.

Layzer, J. B., M. E. Gordon, and R. M. Anderson. 1993. Mussels, the forgotten fauna of regulated rivers: A case study of the Caney Fork River. Regulated Rivers-Research and Management 8:63–71.

Lellis, W. A., B. S. White, J. C. Cole, C. S. Johnson, J. L. Devers, E. V. Gray, and H. S. Galbraith. 2013. Newly documented host fishes for the eastern Elliptio mussel, Elliptio complanata. Journal of Fish and Wildlife Management 4:75–85.

Lessard, J. L., and D. B. Hayes. 2003. Effects of elevated water temperature on fish and macroinvertebrate communities below small dams. River Research and Applications 19:721–732.

Lydeard, C., R. H. Cowie, W. F. Ponder, A. E. Bogan, P. Bouchet, S. A. Clark, K. S. Cummings, et al. 2004. The global decline of nonmarine mollusks. Bioscience 54:321–330.

MacArthur, R. H., and E. O. Wilson. 1967. The theory of island biogeography. Princeton, NJ: Princeton University Press.

Machut, L. S., K. E. Limburg, R. E. Schmidt, and D. Dit-Rman. 2007. Anthropogenic impacts on American eel demographics in Hudson River tributaries, New York. Transactions of the American Fisheries Society 136:1699–1713.

Macnaughton, C. J., C. Senay, I. Dolinsek, G. Bourque, A. Maheu, G. Lanthier, S. Harvey-Lavoie, J. Asselin, P.

Legendre, and D. Boisclair. 2016. Using fish guilds to assess community responses to temperature and flow regimes in unregulated and regulated Canadian rivers. Freshwater Biology 61:1759–1772.

Maheu, A., A. St-Hilaire, D. Caissie, N. El-Jabi, G. Bourque, and D. Boisclair. 2016. A regional analysis of the impact of dams on water temperature in medium-size rivers in eastern Canada. Canadian Journal of Fisheries and Aquatic Sciences 73:1885–1897.

Matuschek, R., and K. Betke. 2009. Measurements of construction noise during pile driving of offshore research platforms and wind farms. Proceedings of the NAG/DAGA International Conference on Acoustics, 262–265. Rotterdam: NAG/DAGA.

McManamay, R. A., B. K. Peoples, D. J. Orth, C. A. Dolloff, and D. C. Matthews. 2015. Isolating causal pathways between flow and fish in the regulated river hierarchy. Canadian Journal of Fisheries and Aquatic Sciences 72:1731–1748.

Merritt, D. M., and N. L. Poff. 2010. Shifting dominance of riparian *Populus* and *Tamarix* along gradients of flow alteration in western North American rivers. Ecological Applications 20:135–152.

Metcher, A. H., T. H. Bakken, T. Friedrich, F. Greimel, N. Humer, S. Schmutz, B. Zeiringer, and J. A. Webb. 2017. Drawing together multiple lines of evidence from assessment studies of hydropeaking pressures in impacted rivers. Freshwater Science 36:220–230.

Moody, A., P. Slaney, and J. Stockner. 2007. Footprint impact of BC hydro dams on aquatic and wetland primary productivity in the Columbia Basin. AIM Ecological Consultants Ltd. in association with Eco-Logic Ltd. and PSlaney Aquatic Science Ltd. Unpublished report prepared for Columbia Basin Fish & Wildlife Compensation Program, Nelson, British Columbia. http://www.sgrc.selkirk.ca/bioatlas/pdf/FWCP-CB_Impacts_Summary.pdf.

Mortenson, S. G., and P. J. Weisberg. 2010. Does river regulation increase the dominance of invasive woody species in riparian landscapes? Global Ecology and Biogeography 19:562–574.

Nagrodoski, A., G. Raby, C. Hasler, M. Taylor, and S. Cooke. 2012. Fish stranding in freshwater systems: Sources, consequences, and mitigation. Journal of Environmental Management 103:133–141.

Naiman, R. J., J. S. Bechtold, T. J. Beechie, J. J. Latterell, and R. Van Pelt. 2010. A process-based view of floodplain forest patterns in coastal river valleys of the Pacific Northwest. Ecosystems 13:1–31.

Naish, K. A., J. E. Taylor, P. S. Levin, T. P. Quinn, J. R. Winton, D. Huppert, and R. Hilborn. 2008. An evaluation of the effects of conservation and fishery enhancement hatcheries on wild populations of salmon. Advances in Marine Biology 53:61–194.

National Research Council. 2002. Riparian areas: Functions and strategies for management. Washington, DC: The National Academies Press. https://doi.org/10.17226/10327.

Normandeau Associates Inc. 2009. An estimation of survival and injury of fish passed through the Hydro Green Energy Hydrokinetic System, and a characterization of fish entrainment potential at the Mississippi Lock and Dam No. 2 Hydroelectric Project (P-4306). https://tethys.pnnl.gov/sites/default/files/publications/Hastings_Fish_Passage_Report.pdf.

Nummi, P., S. Kattainen, P. Ulander, and A. Hahtola. 2011. Bats benefit from beavers: A facilitative link between aquatic and terrestrial food webs. Biodiversity and Conservation 20:851–859.

Oak Ridge National Laboratory. 2016. The 2016 National Hyropower Map: Statistics on U.S. existing hydropower assets. Oak Ridge, TN: Oak Ridge National Laboratory.

Onea, F., S. Ciortan, and E. Rusu. 2017. Assessment of the potential for developing combined wind-wave projects in the European nearshore. Energy and Environment 28:580–597.

Paish, O. 2002. Small hydro power: Technology and current status. Renewable and Sustainable Energy Reviews 6:537–556.

Perrin, C. J., L. L. Rempel, and M. L. Rosenau. 2003. White sturgeon spawning habitat in an unregulated river: Fraser River, Canada. Transactions of the American Fisheries Society 132:154–165.

Phillips, J. D. 2001. Sedimentation in bottomland hardwoods downstream of an east Texas dam. Environmental Geology 40:860–868.

Poff, N. L., and J. K. Zimmerman. 2010. Ecological responses to altered flow regimes: A literature review to inform the science and management of environmental flows. Freshwater Biology 55:194–205.

Poff, N. L., B. D. Richter, A. H. Arthington, S. E. Bunn, R. J. Naiman, E. Kendy, M. Acreman, C. Apse, B. P. Bledsoe, and M. C. Freeman. 2010. The ecological limits of hydrologic alteration (ELOHA): A new framework for developing regional environmental flow standards. Freshwater Biology 55:147–170.

Pracheil, B. M., M. A. Pegg, and G. E. Mestl. 2009. Tributaries influence recruitment of fish in large rivers. Ecology of Freshwater Fish 18:603–609.

Randklev, C. R., N. Ford, S. Wolverton, J. H. Kennedy, C. Robertson, K. Mayes, and D. Ford. 2016. The influence of stream discontinuity and life history strategy on mussel community structure: A case study from the Sabine River, Texas. Hydrobiologia 770:173–191.

Rehn, A. C. 2009. Benthic macroinvertebrates as indicators of biological condition below hydropower dams on west-slope Sierra Nevada streams, California, USA. River Research and Applications 25:208–228.

Roni, P., T. Beechie, G. Pess, and K. Hanson. 2015. Wood placement in river restoration: Fact, fiction, and future direction. Canadian Journal of Fisheries and Aquatic Sciences 72:466–478.

Rood, S. B., G. M. Samuelson, J. H. Braatne, C. R. Gourley, F. M. R. Hughes, and J. M. Mahoney. 2005. Managing river flows to restore floodplain forests. Frontiers in Ecology and the Environment 3:193–201.

Ruetz, C. R., and C. A. Jennings. 2000. Swimming performance of larval robust redhorse, *Moxostoma robustum*, and low-velocity habitat modeling in the Oconee River, Georgia. Transactions of the American Fisheries Society 129:398–407.

Ruibal, C. M., and M. Mazumdar. 2008. Forecasting the mean and the variance of electricity prices in deregulated markets. Transactions on Power Systems 23:25–32.

Savidge, G., D. Ainsworth, S. Bearhop, N. Christen, B. Elsaesser, F. Fortune, R. Inger, et al. 2014. Strangford Lough and the SeaGen Tidal Turbine. In Marine Renewable Energy Technology and Environmental Interactions, edited by M. A. Shields and A. I. L. Payne, 153–172. Dordrecht: Springer.

Schmutz, S., T. H. Bakken, T. Friedrich, F. Greimel, A. Harby, M. Jungwirth, A. Melcher, G. Unfer, and B. Zeiringer. 2015. Response of fish communities to hydrological and morphological alterations in hydropeaking rivers of Austria. River Research and Applications 31:919–930.

Schulting, L., C. K. Feld, and W. Graf. 2016. Effects of hydro- and thermopeaking on benthic macroinvertebrate drift. Science of the Total Environment 573:1472–1480.

Shafroth, P. B., J. R. Cleverly, T. L. Dudley, J. P. Taylor, C. Van Riper, E. P. Weeks, and J. N. Stuart. 2005. Control of *Tamarix* in the western United States: Implications for water salvage, wildlife use, and riparian restoration. Environmental Management 35:231–246.

Sheehan, E. V., T. F. Stevens, and M. J. Attrill. 2010. A quantitative, non-destructive methodology for habitat characterisation and benthic monitoring at offshore renewable energy developments. PLOS One 5:e14461. https://doi.org/10.1371/journal.pone.0014461.

Shetkar, R. V., and A. Mahesha. 2011. Tropical, seasonal river basin development through a series of vented dams. Journal of Hydrologic Engineering 16:292–302.

Silva, A. T., M. C. Lucas, T. Castro-Santos, C. Katopodis, L. J. Baumgartner, J. D. Thiem, K. Aarestrup, et al. 2018. The future of fish passage science, engineering, and practice. Fish and Fisheries 19:340–362.

Sommer, T. R., B. Harrell, M. Nobriga, R. Brown, P. Moyle, W. Kimmerer, and L. Schemel. 2001a. California's Yollo Bypass: Evidence that flood control can be compatible with fisheries, wetlands, wildlife, and agriculture. Fisheries 26:6–16.

Sommer, T. R., M. L. Nobriga, W. C. Harrell, W. Batham, and W. J. Kimmerer. 2001b. Floodplain rearing of juvenile chinook salmon: Evidence of enhanced growth and survival. Canadian Journal of Fisheries and Aquatic Sciences 58:325–333.

Sparling, C., M. Lonergan, and B. McConnell. 2017. Harbour seals (*Phoca vitulina*) around an operational tidal turbine in Strangford Narrows: No barrier effect but small changes in transit behaviour. Aquatic Conservation: Marine and Freshwater Ecosystems 28:194–204.

Struthers, D. P., L. F. G. Gutowsky, E. Enders, K. Smokorowski, D. Watkinson, E. Bibeau, and S. J. Cooke. 2017. Evaluating riverine hydrokinetic turbine operations relative to the spatial ecology of wild fishes. Journal of Ecohydraulics 2:53–67.

Taylor, D. L., D. W. Bolgrien, T. R. Angradi, M. S. Pearson, and B. H. Hill. 2013. Habitat and hydrology condition indices for the upper Mississippi, Missouri, and Ohio rivers. Ecological Indicators 29:111–124.

Thresher, R. 2013. Estimated U.S. wave energy resource. Golden, CO: National Renewable Energy Laboratory.

Timilsina, A. B., S. Mulligan, and T. R. Bajracharya. 2018. Water vortex hydropower technology: A state-of-the-art review of developmental trends. Clean Technologies and Environmental Policy 20:1737–1760.

Tuckerman, S., and B. Zawiski. 2007. Case studies of dam removal and TMDLs: Process and results. Journal of Great Lakes Research 33:103–116.

Tupinambás, T. H., R. Cortes, S. J. Hughes, S. G. Varandas, and M. Callisto. 2015. Macroinvertebrate responses to distinct hydrological patterns in a tropical regulated river. Ecohydrology 9:460–471.

US Energy Information Agency. 2016. Renewable energy explained. www.eia.gov/energyexplained/index.cfm?page=renewable_home.

USFWS (US Fish and Wildlife Service). 1997. Gray bat, *Myotis grisescens*. Threatened and endangered species. Fort Snelling, MN: US Fish and Wildlife Service.

van Oort, H., R. A. Gill, and J. M. Cooper. 2011. CLBMON 11B-2 Revelstoke Reach spring songbird effectiveness monitoring.

van Riper, C., K. L. Paxton, C. O'Brien, P. B. Shafroth, and L. J. McGrath. 2008. Rethinking avian response to

Tamarix on the lower Colorado River: A threshold hypothesis. Restoration Ecology 16:155–167.

Ward, J. V., and J. A. Stanford. 1983. The serial discontinuity concept of lotic ecosystems. In Dynamics of Lotic Ecosystems, edited by T. D. Fontaine and S. M. Bartell, 29–42. Ann Arbor, MI: Ann Arbor Science.

Ward, J. V., and J. A. Stanford. 1995. Ecological connectivity in alluvial river ecosystems and its disruption by flow regulation. Regulated Rivers-Research and Management 11:105–119.

Waters, S., and G. Aggidis. 2016. A world first: Swansea Bay tidal lagoon in review. Renewable and Sustainable Energy Reviews 56:916–921.

Wilkes, M. A., M. McKenzie, and J. A. Webb. 2018. Fish passage design for sustainable hydropower in the temperate Southern Hemisphere: An evidence review. Reviews in Fish Biology and Fisheries 28:117–135.

Winemiller, K. O., P. B. McIntyre, L. Castello, E. Fluet-Chouinard, T. Giarrizzo, S. Nam, I. G. Baird, et al. 2016. Balancing hydropower and biodiversity in the Amazon, Congo, and Mekong. Science 351:128–129.

Woodruff, D., V. Cullinan, A. Copping, and K. Marshall. 2013. Effects of electromagnetic fields on fish and invertebrates: Task 2.1.3: Effects on Aquatic Organisms, Fiscal Year 2012 Progress Report. PNNL-22145. Richland, Washington: Pacific Northwest National Laboratory. https://www.pnnl.gov/main/publications /external/technical_reports/PNNL-22154.pdf.

Yates, K. L., D. S. Schoeman, and C. J. Klein. 2015. Ocean zoning for conservation, fisheries and marine renewable energy: Assessing trade-offs and co-location opportunities. Journal of Environmental Management 152:201–209.

Youle, R. 2016. Salmon and sea trout could scupper the Swansea Bay Tidal Lagoon. Wales Online. http://www .walesonline.co.uk/business/business-news/salmon-sea -trout-could-scupper-12304019.

Zhao, Q., S. Liu, L. Deng, S. Dong, Z. Yang, and Q. Liu. 2013. Determining the influencing distance of dam construction and reservoir impoundment on land use: A case study of Manwan Dam, Lancang River. Ecological Engineering 53:235–242.

PART IV **THE FUTURE OF RENEWABLE ENERGY AND WILDLIFE CONSERVATION**

10 — Renewable Energy Policy Directives

Edward B. Arnett

Implications for Wildlife Conservation

Introduction

Global demand for energy has increased more than 50% in the last half century, and production to meet that demand has increased habitat loss and fragmentation, stressed biological diversity, and affected human health and well-being worldwide (Evans and Kiesecker 2014). The interface of energy development—both conventional and renewable—and wildlife conservation has been at the forefront of science, management, and policy decisions for decades, but it is now a mainstream topic in contemporary conservation. Developing renewable energy alternatives has become a global priority, owing to long-term environmental impacts from use of fossil fuels, coupled with a changing climate (Schlesinger and Mitchell 1987, Inkley et al. 2004, National Research Council 2007, Allison et al. 2014). Indeed, numerous consequences for wildlife and their habitats—both predicted and observed—have been well documented (Inkley et al. 2004, Mawdsley et al. 2009), and endeavors like that of the United Nations climate change conference in 2015 and the resulting Paris Agreement (see https://unfccc.int /process-and-meetings/the-paris-agreement/the -paris-agreement) will be critical for mitigating global climate change. The United States, however, recently altered its course on climate policy after withdrawing from the Paris Agreement, repealing the Clean Power Plan set forth by the Obama administration and significantly altering domestic policy to once again favor fossil fuel dominance (see https://www.whitehouse.gov/presidential-actions /presidential-executive-order-promoting-energy -independence-economic-growth/). Still, global renewable energy production has increased dramatically over the past 20 years—even in the United States under new policies—and is projected to continue growing in the coming decades (Ellabban et al. 2014, Kiesecker and Naugle 2017, EIA 2018; this volume, Introduction, Moorman et al.).

Although developing renewable energy sources is important for meeting future energy needs, shifting energy production from fossil fuels to renewable sources that collect more diffuse energy from a broader spatial area will involve environmental trade-offs (Kiesecker et al. 2011). Environmental consequences of renewable energy development are well documented for some wildlife species and renewable sources like wind energy (e.g., Loss et al. 2013, Arnett et al. 2015; this volume, Chapter 5, Dohm and Drake, Chapter 6, Hein and Hale), and yet less is known for other sources like solar energy (e.g., Lovich and Ennen 2011, Walston et al. 2016;

this volume, Chapter 8, Boroski). Effects on wildlife can manifest in the form of direct mortality from strikes with moving wind turbine blades, for example, or indirect effects resulting from habitat loss from construction of the renewable energy facilities and associated infrastructure. Often overlooked when determining effects and possible mitigation requirements, however, are indirect behavioral modifications by wildlife to renewable energy facilities, leading to avoidance of larger spatial areas of otherwise suitable landscapes for many species of wildlife (Arnett et al. 2007a). The scale of proposed landscape change for renewables is indeed unprecedented, and almost all forms of renewable energy development are more land-intensive than are other forms of energy generation (Kiesecker and Naugle 2017).

Renewable energy and wildlife conservation are not necessarily incompatible goals, especially when impacts can be avoided, minimized, or mitigated (e.g., Kiesecker et al. 2010, Cameron et al. 2012, Arnett and May 2016, Allison et al. 2017). However, some authors suggest that it might be preferable to risk individuals and, perhaps, existence of some species in pursuit of rapid renewable energy development (e.g., Allison et al. 2014, Badichek 2015). The latter philosophy assumes that prioritization of species' survival over the conversion from fossil fuels to renewables, and resulting delay or inaction, would increase global climate risk and ultimately threaten more species and habitats in the long term (Allison et al. 2014, Badichek 2015, Kiesecker and Naugle 2017). Regardless of social views and acceptance of such trade-offs, any nation's statutes and administrative policies—or lack thereof—can have dramatic influence on the nature and extent of effects of renewable energy development on wildlife and how these effects may be mitigated through space and time.

In this chapter, I discuss several statutes and administrative policies specific to the United States that affect renewable energy exploration and development and their known or potential influence on wildlife impacts, conservation efforts, and mitigation. My discussions here are by no means exhaustive or global in nature, and they do not include the breadth of policies and statutes governing all sources of renewable energy—wind, solar, waterpower, biomass, and geothermal (DuVivier 2011). Rather, I provide a synthesis of some of the most relevant statutes and policies influencing renewable development and wildlife conservation in the United States while providing reference to international policies, when applicable. Useful reviews of policy and regulations regarding wind energy development also can be found in the Government Accountability Office (2005) and National Research Council (2007) reports, the US Fish and Wildlife Service (USFWS 2012) voluntary land-based wind energy guidelines, and other references cited in this chapter. Moreover, additional information on policy related to bioenergy is provided in Chapter 4 (Rupp and Ribic).

Renewable Energy Standards and Tax Credits

A critically important factor contributing to the rapid growth of renewable energy has been state-driven policies establishing renewable portfolio standards requiring that a predetermined minimum percentage of renewable sources be included in the wholesale electric generation mix (Wiser et al. 2007, DuVivier 2011). By increasing the required amount of renewable energy over time, these portfolio standards set forth a path toward increasing sustainability of renewable sources (Lyon and Yin 2010). The Renewable Fuel Standard is a federal program, established by the US Congress in the Energy Policy Act of 2005 and administered by the US Environmental Protection Agency, requiring a minimum percentage of US transportation fuels sold to be from renewable sources. The Renewable Fuel Standard requires renewable fuel to be blended into transportation fuel in increasing amounts each year, and the Energy Independence and Security Act of 2007 escalates that mix to 36 billion gallons (136 billion liters) by 2022

(US Department of Energy, https://www.afdc.energy.gov/laws/RFS).

The aforementioned renewable portfolio and fuel standards drive policy and timelines that may conflict with wildlife and habitat conservation and the ability to effectively develop land-use plans or collect needed information on wildlife effects to inform decisions and policy. Indeed, the push to meet these state and federal requirements, coupled with often inadequate review or permitting requirements, or lack of input and monitoring on wildlife impacts, may have unintended consequences. For example, corn-based ethanol production—effectively driven by the Renewable Fuel Standard—will require tens of millions of hectares of additional agricultural production, resulting in significant land-use changes and inevitably lost habitat for wildlife (Arnett et al. 2007b, Donner and Kucharik 2008, Fletcher et al. 2011, Holland et al. 2015). Planning for a sustainable energy future also requires comprehensive planning to achieve multiple goals, and "piecemeal" approaches to energy production simply will not work in the long run (Kiesecker and Naugle 2017). If we are to address energy development in a broader land-use planning context, the timelines associated with achieving renewable portfolio and fuel standards and comprehensive planning must be considered in a broader context and integrated with other goals, including wildlife conservation.

The federal renewable electricity production tax credit, a per-kilowatt credit for qualified renewable electricity resources (see https://www.energy.gov/savings/renewable-electricity-production-tax-credit-ptc), is a policy that, like portfolio and fuel standards, seemingly has little to do with wildlife conservation on the surface. The US Congress must reauthorize the tax credit and often does so for only 1 year at a time (Figure 10.1). For example, renewable energy tax credits, including the wind power production tax credit and solar investment tax credit, were recently extended as part of the Consolidated Appropriations Act of 2016 (HR 2029, Sec. 301; https://www.congress.gov/bill/114th-congress/house-bill/2029/text; Mai et al. 2016). The wind tax credit, valued at 2.3¢/kWh for electricity production over the first 10 years of a facility's output, usually allows projects with commercial operation dates after the expiration date to qualify for the tax credit as long as they "commenced construction" prior to the expiration deadline (Mai et al. 2016). This can often create situations where developers are desperately working to get all necessary permits and construction under way before the tax credit expires. The tax credit expiration-extension dynamic—often with unknown

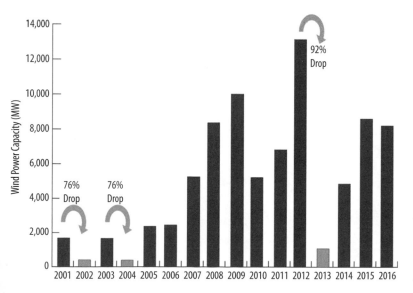

Figure 10.1 Annual installed U.S. wind capacity from 2001–2016, showing years when the production tax credit (PTC) was set to expire (light gray bars) and then extended. This graph demonstrates how development expands up to the year prior to the production tax credit expiration and extension deadline and then declines in that year of expiration. Current legislation to phase down the PTC has ended this boom-bust cycle. *Source: American Wind Energy Association.*

extension length—creates a roller coaster of uncertainty that inadvertently extends into the realm of wildlife science and conservation. Some developers cite the tax credit timeline as a reason why they cannot commit to long-term monitoring or multi-year wildlife research projects. When coupled with timelines to meet renewable energy standards discussed previously, it becomes obvious that these policies and associated timeframes can be problematic and must be integrated into a broader discussion on timing and planning for multiple resources. Although these examples are anecdotal and based solely on my own experiences, I believe it would behoove the US Congress and other governments to consider both short- and long-term consequences of tax credit extensions, renewable portfolio standards, and other energy policies on monitoring and research that support balancing renewable energy development with conservation.

Federal Wildlife Statutes

The US federal government's role in regulating renewable energy development is limited to projects occurring on federal lands or affecting federal trust species, or projects that have some form of federal involvement (e.g., interconnect with a federal transmission line) or require federal permits (Arnett 2012). The Federal Energy Regulatory Commission regulates the interstate transmission of electricity, natural gas, and oil, but it does not approve the physical construction of electric generation, transmission, or distribution facilities, which is currently determined by state and local governments (Government Accountability Office 2005).

The federal regulatory framework for protecting wildlife from most sources of renewable energy development includes, but is not limited to, 3 primary laws—the Migratory Bird Treaty Act, the Bald and Golden Eagle Protection Act, and the Endangered Species Act (Government Accountability Office 2005, National Research Council 2007, USFWS 2012). Importantly, none of these laws require devel-

opers, owners, and operators to follow specific procedures to ensure no harm comes to wildlife during facility construction or operation, but operators may be held liable for any such harm to a protected species that may occur (Lilley and Firestone 2008).

Migratory Bird Treaty Act

The extinction of several migratory bird species, as well as population declines of many others in the nineteenth century, generated public support for enactment of protective legislation for migratory birds early in the twentieth century (Fjetland 2000). The Migratory Bird Treaty Act (MBTA), first enacted in 1918, has been the cornerstone of migratory bird conservation in the United States and implements 4 international treaties with Great Britain, Canada, Mexico, and Japan that provide protection for hundreds of species of migrating birds (USFWS 2012). The MBTA is a strict liability statute that not only prohibits unauthorized and intentional killing of migratory birds, through hunting or poaching, for example, but also has been extended to prohibit the unintentional killing of migratory birds—often referred to as "incidental take"—caused by renewable energy facilities (e.g., wind or solar energy) or even loss of migratory bird habitat (USFWS 2012). The MBTA was instrumental in protecting birds from unrestricted killing, market hunting for meat and plume trade, and wanton waste in the early twentieth century (Fjetland 2000, Lilley and Firestone 2008), and it remains an effective tool in regulating recreational hunting. More recently, the "primary administrative emphasis" of the MBTA has shifted beyond solely managing sport hunting to include regulating a much broader array of activities that could result in the incidental take of migratory birds (Lilley and Firestone 2008), including renewable energy development.

The question of what, if any, legal consequences should apply under the MBTA when a bird protected under the MBTA is killed, intentionally or unintentionally, by anthropogenic activities has been the subject of much legal debate and opinion (e.g., Margolin 1979, Fjetland 2000, Lilley and Firestone 2008, Panarella 2017, Fazio and Strell 2018). For the past century, the MBTA has been consistently interpreted by the Department of Interior to include incidental take (Fazio and Strell 2018), but the court system has struggled to define what limits should be placed on its application to activities that, unlike hunting and poaching, are in no part motivated by intent to kill or injure (e.g., wind turbine, building, or oil pit kills; Martin and Ballard 2013, Panarella 2017).

Although hunting licenses issued to individuals allow the legal "take" of migratory game birds, according to state and federal regulations, no such permit is issued to individuals, developers, corporations, or other entities for incidental take under the MBTA. By criminalizing the take of migratory birds without a permit and simultaneously granting no permits for incidental take, the MBTA creates a conundrum for entities engaged in the many land uses that could result in, albeit unintentionally, migratory bird kills (Lilley and Firestone 2008). In addition to not authorizing incidental take permits, the MBTA does not allow for private citizen suits, and therefore, without the direct involvement of the USFWS, there is no enforcement of the MBTA. Further complicating matters, the opinion of what constitutes a prosecutable activity under the MBTA varies from court to court, leading some to suggest amending the MBTA to allow incidental take permits (Lilley and Firestone 2008), whereas others argue for abolishing the unintentional/incidental take language and legal consequences altogether (Panarella 2017).

The issue of incidental take under the MBTA most recently was addressed by the Obama administration in a Solicitor's Opinion Memorandum (M-37041), issued January 10, 2017 (US Department of Interior 2017a), that concluded, "The MBTA's broad prohibition on taking and killing migratory birds by any means and in any manner includes incidental taking and killing." However, on December 22, 2017, a new Solicitor's Opinion Memorandum (M-37050; US Department of Interior 2017b), suspended the previous memorandum (i.e., M-37041), and concluded that "the MBTA's prohibition on pursuing, hunting, taking, capturing, killing, or attempting to do the same applies only to direct and affirmative purposeful actions that reduce migratory birds, their eggs, or their nests, by killing or capturing, to human control." Thus, incidental take from renewable energy projects, open wastewater and oil pits, or strikes with buildings and communications towers are currently not considered to be violations under the MBTA by the Department of Interior. While the Trump administration's interpretation of the MBTA currently reflects the Department of Interior's policy, the resolution of this question, if not addressed by Congress, will undoubtedly be resolved by the US Supreme Court in the future (Fazio and Strell 2018).

Bald and Golden Eagle Protection Act

The Bald and Golden Eagle Protection Act (BGEPA; 16 U.S.C. § 668 et seq) provides additional legal protection for these 2 species of raptors. This act prohibits the taking by numerous means, sale, and/or transport of live or dead bald eagles (*Haliaeetus leucocephalus*) and golden eagles (*Aquila chrysaetos*), their parts, nests, or eggs, and includes criminal and civil penalties for violating the statute. This act also extends beyond the outright killing of these birds to include disturbance that could lead to, or is likely to cause, injury, decreases in productivity, or nest abandonment (USFWS 2012).

In 2009, the USFWS promulgated a final rule on 2 new permit regulations that authorized under BGEPA, for the first time, the incidental take of eagles and eagle nests in certain situations by companies, government agencies and tribal governments, individuals, and other organizations while conducting lawful activities such as operating utilities and airports (USFWS 2012). The permitting rule also extends to construction and operation of renewable energy facilities, particularly wind projects.

To facilitate issuance of permits under the new regulations and to complement its land-based wind energy guidelines (USFWS 2012), the USFWS developed an Eagle Conservation Plan Guidance document to aid developers of renewable energy and other projects (USFWS 2013). The document provides specific, in-depth guidance for conserving bald and golden eagles during the course of siting, constructing, and operating wind energy facilities. Compliance is voluntary, but the USFWS states that following the guidance aids developers in complying with regulatory requirements to avoid unintentional take of eagles—primarily at wind energy projects (USFWS 2013). The revised ruling and guidance also developed a mitigation strategy for eagle conservation, but it is challenged by a lack of data supporting scientifically rigorous strategies to mitigate unavoidable eagle take (Allison et al. 2017). Allison et al. (2017) summarized options for mitigating take of golden eagles and potentially compensating for unavoidable impacts after avoidance and minimization measures have been taken.

Endangered Species Act

The Endangered Species Act (ESA; 16 U.S.C. § 1531 et seq) is the most comprehensive wildlife conservation and habitat protection legislation in the United States and has been a vital tool for conserving biodiversity (Davison et al. 2005). Enacted by Congress in 1973, the ESA directs the USFWS to identify and protect endangered and threatened species and their critical habitat and to provide a means to conserve their ecosystems (USFWS 2012). To this end, federal agencies are directed to use their authorities to conserve listed species and ensure that their actions are not likely to jeopardize continued existence of these species or destroy or adversely modify their critical habitat (USFWS 2012). Federal agencies are encouraged to do the same with respect to "candidate" species that may be listed in the future.

The ESA was enacted during a period when numerous bedrock environmental laws were passed in

the United States and originally enjoyed widespread support that has since waned, owing to changes in attitudes about economic and social consequences of endangered species protection (Waples et al. 2013). Successive presidential administrations and US Congresses have responded with a series of amendments intended to balance species protection, procedural requirements, and flexibility in implementation, and those amendments shaped the ESA over the past several decades (Waples et al. 2013). Indeed, the rich literature on the ESA covers every conceivable angle and scientific, administrative, and legal opinion on the acts' intent, implementation, effectiveness, and need for improvement or dismantling altogether. Regardless, the act has irrefutably thwarted many species extinctions (< 1% of those listed have been declared extinct; Davison et al. 2005), and has resulted in several success stories for species in the United States, such as the bald eagle, peregrine falcon (*Falco peregrinus*), and Louisiana black bear (*Ursus americanus luteolus*).

As with any development, renewable energy projects are subjected to conditions and requirements of the ESA. In most instances, the secretary of the interior, acting through the USFWS, is the final arbitrator when conflicts between renewable energy projects and biological resources arise (Yung and Sanders 2015). Section 9 of the ESA makes it unlawful to "take" a listed species—defined as to harass, harm, pursue, hunt, shoot, wound, kill, trap, capture, or collect or attempt to engage in any such conduct (USFWS 2012). Regardless of where the activity takes place (i.e., on federal or nonfederal land), any killing or taking of an endangered species would be lawful only if the person responsible was in possession of an incidental-take permit issued under the ESA (Lilley and Firestone 2008). The USFWS may, however, authorize incidental take (i.e., that occurring as a result of an otherwise legal activity) in 2 ways: first, through formal consultation under section 7(a)(2) of the act, whenever a federal agency, federal funding, or a federal permit is involved; and second, through pursuit of an incidental take permit under section 10(a)(1)(B) upon completion of a satisfactory habitat conservation plan for listed species (USFWS 2012; also see https://www.fws.gov /endangered/what-we-do/hcp-overview.html). For example, the Midwest region of the USFWS and a coalition of 8 states, the wind industry, and conservation groups are preparing a multispecies habitat conservation plan that will issue incidental take permits for listed species like the Indiana bat (*Myotis sodalis*) and Kirtland's warbler (*Setophaga kirtlandii*; see https://www.fws.gov/midwest/endangered/esday /wind2010.html).

Federal Administrative Policies and Environmental Review

In addition to obtaining the necessary permits at local (e.g., city or county) and state levels, renewable energy projects usually must undertake either state and/or federal environmental reviews and land-use planning processes. Unlike permitting processes, wherein project proponents are responsible for assembling information and documents and complying with requirements, the public agencies are themselves responsible for environmental review (Morris and Owley 2014). State review and land-use planning processes vary widely, whereas federal approaches for public lands are established in statutes and administrative policy that transcend all federal public land ownership. The Congressional Review Service (2012), Government Accountability Office (2005), Geißler (2013), and Morris and Owley (2014) are examples of useful summaries of environmental review processes specific to the United States.

National Environmental Policy Act

It is the continuing policy of the Federal Government . . . to use all practicable means and measures . . . to create and maintain conditions under which man and nature can exist in productive harmony, and fulfill the social, economic, and other requirements of present and future generations of Americans.
— National Environmental Policy Act of 1969

The National Environmental Policy Act of 1969 (NEPA; 42 U.S.C. § 4321 et seq) articulates a broad national environmental policy that requires federal agencies to analyze the environmental impacts of their actions and of alternatives for all major actions likely to have significant environmental effects (Council on Environmental Quality 2007). The act has had profound influence domestically in the United States, but also globally, as it has been the model for other environmental assessment legislations, including European regulations (Busch and Jörgens 2005, Köppel et al. 2012, Geißler 2013). Several steps in the approval of any renewable energy project involving federal land trigger NEPA review, including Bureau of Land Management approval of rights-of-way and, potentially, issuance of incidental take permits under the ESA (Morris and Owley 2014).

The Council on Environmental Quality has provided guidance to federal agencies and decision makers on the process, implementation, and legal requirements for NEPA (see Council on Environmental Quality 2000, 2005, 2007). The Council on Environmental Quality regulations direct federal agencies to "use the NEPA process to identify and assess the reasonable alternatives to proposed actions that will avoid or minimize adverse effects of these actions" (40 CFR § 1500.2[e]; see 40 CFR § 1500.2[f]). The NEPA process includes forecasting impacts of proposed action(s) and reasonable alternatives and identifying mitigation measures for those impacts prior to making decisions and taking action, and these analyses are documented and made available to the public (Council on Environmental Quality 2007).

Any renewable energy project proposed on federal public lands or connecting to a federal transmission grid triggers the NEPA process, requiring environmental review in the form of an environmental impact statement or an environmental assessment—a less-intensive review (Council on Environmental Quality 2005, Morris and Owley 2014). An environmental impact statement not only evaluates the environmental impacts of a proposed project (and its alternatives), but also outlines possible mitigation measures and assesses cumulative effects (Morris and Owley 2014). In contrast, an environmental assessment is a more concise document prepared under NEPA to provide sufficient evidence and analysis to determine whether a proposed agency action would require preparation of an environmental impact statement or a finding of no significant impact (FONSI).

Salter (2011) and other authors have argued that current law advances wildlife protection and land preservation interests too far at the expense of renewable energy development, owing to the manner in which NEPA commonly is interpreted and applied, favoring the status quo and disfavoring developing land for renewable energy projects. Part of Salter's (2011) argument centers on the significant cost and time it takes to prepare an environmental impact statement, which he feels is inconsistent with policy-making consensus that the United States should be promoting the rapid development of renewable energy. There is no question that the NEPA process takes considerable time and could be streamlined to increase efficiency and reduce costs, but this policy is important for projects on public lands given the extent of known impacts and the need to consider those impacts in project planning, development, and operation.

Land-Use Management Planning

Renewable energy projects proposed on federal public lands are subject to either the Federal Land Policy and Management Act (if on Bureau of Land Management lands; see 43 U.S.C. ch. 35 § 1701 et seq) or the National Forest Management Act (if on US Forest Service lands; see 16 U.S.C. §§1600-1687). Both federal land-use acts require agencies to follow processes in accordance with NEPA, the Council on Environmental Quality's regulations for implementing NEPA (Title 40, Parts 1500–1508 of the Code of Federal Regulations), Department of Interior regula-

tions for implementing NEPA (e.g., 43 CFR Part 46 for US Department of the Interior/Bureau of Land Management), and applicable agency handbook and planning regulations. These processes are extensive, requiring considerable time and resources and often conflict with the timing for renewable project financing, tax credits, and other factors in the development process.

The 94th Congress passed the Federal Land Policy and Management Act in 1976 (43 U.S.C. ch. 35 § 1701 et seq) to establish public-land policy and guidelines for its administration and for the management, protection, development, and enhancement of the public lands managed by the US Department of Interior, namely the Bureau of Land Management (US Department of the Interior/Bureau of Land Management 2016). The Federal Land Policy and Management Act addresses land-use planning, administration of federal land, land acquisition, fees and payments, rights-of-ways, range management, and other management issues. Under the Federal Land Policy and Management Act, the secretary of the interior has authorization to grant rights-of-way on public lands for systems of generation, transmission, and distribution of electric energy (Federal Land Policy and Management Act, Section 501[a][4]). The Bureau of Land Management Land Use Planning Handbook (H-1601-1) requires that land-use planning efforts address existing and potential development areas for renewable energy projects (see H-1601-1, Appendix C, II, Resource Uses, Section E, Lands and Realty; BLM 2005a), consistent with the Energy Policy Act of 2005 and the Bureau of Land Management Energy and Mineral Policy implemented in 2008.

To address the expansion of renewable energy and desires to develop on federal lands, the Bureau of Land Management prepared programmatic environmental impact statements for wind and solar energy in the western United States (BLM 2005b; http://windeis.anl.gov/; BLM 2012; http://solareis.anl.gov/). In 2005, the Bureau of Land Management issued a Record of Decision to implement a comprehensive wind energy development program in 11 western states, excluding Alaska, and to amend 52 Bureau of Land Management land-use plans to adopt the new program (BLM 2005b). The decision established policies and best management practices for the administration of wind energy development activities and established minimum requirements for mitigation measures (BLM 2005b). In 2012, the Bureau of Land Management issued a Record of Decision to implement utility-scale solar (>20 MW capacity) energy development on Bureau of Land Management–administered lands in 6 states in the US Southwest (Arizona, California, Colorado, Nevada, New Mexico, and Utah; BLM 2012). This Record of Decision described updated and revised Bureau of Land Management policies and procedures related to solar energy development on public lands, providing internal administrative guidance to the Bureau of Land Management regarding the processing of rights-of-way applications for utility-scale solar energy projects (BLM 2012). The Bureau of Land Management Programmatic Solar Environmental Impact Statement established "solar energy zones" designed to accelerate development while minimizing environmental impacts and avoiding sensitive natural areas. This approach and upfront planning encouraged cross-agency collaboration and has reduced project permitting time by more than half (Krueger et al. 2017).

Importantly, neither the solar nor wind programmatic environmental impact statement and Record of Decision authorize any development projects or eliminate the need for site-specific environmental reviews for any future project. The Bureau of Land Management makes separate decisions as to whether or not to authorize individual renewable energy projects in conformance with existing land-use plans, as amended by these Records of Decision (BLM 2005b, 2012). Once a renewable energy project is proposed, the Bureau of Land Management completes a site-specific environmental review of all projects and rights-of-way applications in accordance with NEPA prior to issuing an authorization.

On US Forest Service public lands—either national forests or grasslands—renewable energy projects are subjected to the National Forest Management Act of 1976 and an amendment to the Forest and Rangeland Renewable Resources Planning Act of 1974 (https://www.fs.fed.us/emc/nfma/includes/NFMA1976.pdf). The US Forest Service describes its renewable energy program and associated regulations under "special uses," and the direction for the agency is presented in its Strategic Energy Framework (USFS 2011; https://www.fs.fed.us/specialuses/special_energy.shtml). The strategic framework and agency's handbooks and manual set forth specific requirements for project components similar to the Bureau of Land Management and in accordance with NEPA and the National Forest Management Act.

Mitigation

In the context of this chapter, mitigation means reducing the negative impacts of a proposed renewable project. The mitigation hierarchy, which seeks to minimize impacts through the application of avoidance, minimization, and compensation, is an essential tool for advancing fish, wildlife, and natural resources conservation and balancing it with renewable energy development (Council on Environmental Quality 2000, Kiesecker et al. 2010). Comprehensive landscape planning that minimizes risks for developers and natural resource stakeholders should clearly define the following areas: (1) those to be avoided because of irreplaceable natural resource values; (2) those where development can occur with minimal controversy; and (3) those where compensatory mitigation should be carried out to best replace or enhance lost ecological function. Such planning efforts also reduce conflicts and communicate what is expected early in the process, thus reducing time and costs while better assuring effectiveness. Avoidance, the first step in the hierarchy, is the best form of mitigation because a resource not yet affected yields the least cost and amount of work required to offset impacts.

The Council on Environmental Quality regulations direct federal agencies to "use the NEPA process to identify and assess the reasonable alternatives to proposed actions that will avoid or minimize adverse effects of these actions" (see 40 CFR § 1500.2[e] and 40 CFR § 1500.2[f]). Under NEPA, mitigation (1) may be included in the proposed project design to minimize impacts (referred to as "design features"); (2) must be included in the range of alternatives and analyzed in the "Environmental Consequences" section of an environmental impact statement; and (3) may be analyzed in an environmental assessment and support what is known as a "mitigated finding of no significant impact" (FONSI) (40 CFR §§ 1502.14[f] and 1502.16[h]; see 40 CFR § 1508.25[b][3]). Frequently, a project as originally proposed will appear to have significant environmental effects, but during the assessment process, the agency will develop mitigation measures that render those effects insignificant in the final analysis, resulting in a mitigated FONSI. In such a case, a critical issue is whether the agency must prepare an environmental impact statement or whether it can justify a FONSI, and thus a timelier and less expensive environmental assessment, by adopting mitigation measures. However, Council on Environmental Quality regulations only require that agencies discuss mitigation measures in a full environmental impact statement, but do not discuss the issue of whether mitigation measures can justify a FONSI (Herson 1986). Indeed, the courts have interpreted NEPA to be a procedural statute, requiring completion of an environmental review process, but without any substantive requirements regarding actions, alternatives, or mitigation (Morris and Owley 2014). In the absence of clear statutory or regulatory requirements, most federal appellate courts have allowed agencies to justify FONSIs with mitigation measures that reduce impacts to less than significant levels, thus providing agencies with time and cost incentives to mitigate a project's significant environmental impacts (Herson 1986).

The Trump administration, notably the Department of Interior, has questioned the legal authority of the Bureau of Land Management to, for example, require mitigation. While it is true that the Federal Land Policy and Management Act does not mandate the Bureau of Land Management to require application of the full mitigation hierarchy, the agency certainly has ample discretion to use the full mitigation hierarchy as directed by the Council on Environmental Quality, Federal Land Policy and Management Act, and NEPA. Recently, the secretary of the interior issued Secretarial Order 3349 that directed the Department of Interior to review its mitigation policies, and on December 22, 2017, the Department of Interior issued Secretarial Order No. 3360 (https:// www.eenews.net/assets/2018/01/05/document_gw _04.pdf) that rescinded 2 Bureau of Land Management mitigation policies: (1) Bureau of Land Management, Manual Section 1794—Mitigation (December 22, 2016); and (2) Bureau of Land Management, Mitigation Handbook H-1794-1 (December 22, 2016). The order also directed the Bureau of Land Management to revise and reissue Instruction Memorandum No. 2008-204, "Offsite Mitigation" (https://www.blm.gov/policy/im-2008-204), within 30 days. Interestingly, the 2008 memorandum notes that "the Bureau of Land Management's authority to address the mitigation of impacts on public lands associated with a use authorization issued by the Bureau of Land Management derives from the Federal Land Policy and Management Act." The revised Instruction Memorandum ordered by the secretary of the interior has yet to be issued at the time of this writing.

The future of mitigation requirements on Department of Interior lands remains unclear, creating confusion and uncertainty. Application of the full mitigation hierarchy, however, is seemingly an important tool for Bureau of Land Management in 3 critical ways: (1) assisting Bureau of Land Management in achieving land health standards, which are mandated under the Federal Land Policy and Management Act; (2) allowing Bureau of Land Management to avoid "undue and unnecessary damage"; and (3) reducing proposed impacts so as to achieve a defensible mitigated FONSI described above, resulting in much quicker permitting. For the Bureau of Land Management to achieve permitting efficiencies and better achieve its goals and mandates under the Federal Land Policy and Management Act, a well-implemented mitigation hierarchy is advisable.

State and Local Jurisdictions

Because development of renewable energy facilities has occurred primarily on nonfederal land, regulating such facilities is largely the responsibility of state and local governments (Government Accountability Office 2005). Each state may enforce its own laws for siting, permitting, and wildlife protection, in addition to any applicable federal statutes, making it an impossibility to cover all states, and multiple counties and local permitting authorities within each state, in this chapter. States also may establish cooperative efforts to address renewable development and wildlife impacts. For example, the Pennsylvania Game Commission (2007) developed a cooperative agreement with wind energy companies that allows for prosecutorial discretion by the Game Commission for the take of state-listed species when cooperating companies comply fully with conditions of the agreement.

The primary permitting jurisdiction for renewable energy projects in many instances is a local planning commission, zoning board, city council, or county board of supervisors or commissioners; typically, these local jurisdictional entities regulate wind projects, for instance, under zoning ordinances and building codes (Government Accountability Office 2005), often without the basic knowledge needed to make informed environmental decisions. In some states, one or more state agencies play a role in regulating renewable energy development, such as natural resource and environmental protection agencies, public utility commissions, or siting boards. Further-

more, some states have environmental laws that impose requirements on many types of construction and development, including wind power, that state and local agencies must follow (Government Accountability Office 2005). All states have their own threatened and endangered species lists, as well as laws prohibiting the killing of nongame animals. However, it does not appear that states readily enforce these laws for the unintentional killing of bats or birds, for example, perhaps because of funding and personnel constraints, other enforcement priorities, or other factors.

It is apparent that most local jurisdictional entities lack experience in wildlife science and, unless they coordinate with their respective state wildlife or natural resource agency, concerns about wildlife issues may fall by the wayside and never enter the discussion during decision making. Although most developers proactively coordinate with state wildlife agencies to address concerns, a gap remains between the decision-making authority and those with expertise and jurisdiction over wildlife issues. Potential for miscommunication or no communication at all is apparent, and while policies and regulations vary among states, it seems prudent to suggest that stronger coordination and perhaps even policy and regulation may be warranted in many states, especially where wildlife agencies have limited authority in such decision making. The intent, of course, should not be to increase process or extend timelines, but rather to ensure environmental concerns are fully articulated and addressed throughout the permitting process. However, neither agencies nor local jurisdictional entities may have the financial wherewithal or resources necessary to act as an effective regulatory body. Merely having environmental regulations in place to carry out their mission might sometimes make for untenable situations whereby renewable energy developers are forced to comply with the law without the agency achieving the conservation goals they desire or the ability to effectively manage permitting programs, which, in turn, can have negative impacts on business.

Renewable Development Guidelines

Several countries have developed voluntary guidelines for siting, developing, and operating renewable energy facilities, especially for wind energy (e.g., Canada, Ontario Ministry of Natural Resources 2011; Ireland, Bat Conservation Ireland 2012; the European Union, Rodrigues et al. 2015). Methods and metrics are also well-established for studying impacts of wind energy development, for example, on wildlife (e.g., Kunz et al. 2007, Strickland et al. 2011), that inform land-use planning, management, operations, mitigation, and policy decisions. The USFWS issued voluntary guidelines in 2003 that were later updated through recommendations from a federal advisory committee commissioned by the secretary of the interior (USFWS 2012). These guidelines are intended to (1) promote compliance with relevant wildlife laws and regulations; (2) encourage scientifically rigorous survey, monitoring, assessment, and research designs proportionate to the risk to species of concern; (3) produce potentially comparable data across the nation; (4) mitigate for potential adverse effects on species of concern and their habitats; and (5) improve the ability to predict and resolve impacts at local, regional, and national scales (USFWS 2012).

As previously discussed, the Pennsylvania Game Commission (2007) developed an exemplary cooperative agreement with wind energy companies that includes a set of methods and metrics required by the agency. Entering into the agreement is voluntary, but cooperating companies must adhere to the guidance, methods, and analyses set forth by the Game Commission. By doing so, participating companies are given prosecutorial discretion for the incidental take of state-listed species by the state agency.

Financing renewable projects that pose environmental threats is risky for institutions and private investors hoping to make returns on those investments in a timely manner. Some financial institutions have developed their own principles and guidelines for development of renewable energy sources.

The World Bank issued its own guidance in 2011 (Ledec et al. 2011) that identified best practices for managing key environmental and social issues associated with wind power development and provided recommendations for addressing project planning, construction, and operation of wind facilities. Educating and engaging financial institutions regarding environmental risks associated with poorly designed and sited projects is critical to avoiding those risks and properly mitigating unforeseen impacts once projects are developed.

Research and guidance on siting and developing solar resources remains in its infancy, and few studies have addressed impacts empirically. However, a few authors have summarized either known or postulated impacts from solar development and offer suggestions for siting and mitigating impacts (e.g., Lovich and Ennen 2011, Cameron et al. 2012, Hernandez et al. 2014, Walston et al. 2016; this volume, Chapter 8, Boroski). From a habitat fragmentation and degradation perspective, the principles and guidance for siting, avoidance, and mitigation of such impacts provided by Kiesecker et al. (2010, 2011), the USFWS (2012), and other sources provided in this chapter for wind are prudent for solar and geothermal projects. Cameron et al. (2012, 2017) and Hernandez et al. (2015) discussed approaches for developers of solar projects that could reduce impacts on areas of higher conservation value. These landscape-scale approaches to planning have been effective in balancing the trade-offs between renewable energy development and conservation of biodiversity, at least in California, but can be effective anywhere (Cameron et al. 2017).

Regulations and Lack Thereof

Arnett (2012) discussed how lack of regulations can sometimes create unique challenges for responsible and scientifically credible development of wind energy projects—an issue that still occurs for most if not all renewable energy technologies today. Hypothetically, a landscape that has reasonable commercial wind or solar energy potential, for instance, might have several different developers "prospecting" in the area to secure landowner leases and begin the permitting process. Several important wildlife concerns may be brought to the attention of each developer independently through consultation with local, state, and federal wildlife agencies or academic and nongovernment experts. After reviewing the situation with the relevant agencies and their company officials, each developer might decide that this landscape is too environmentally sensitive for development and all might abandon their respective projects. This hypothetical landscape has now been unofficially deemed as a high-risk site for development, but the question remains: "Is it now safe from future development of a wind or solar facility?" The answer is "no," because there is no regulatory statute in any state that I am aware of that protects an area because a conscientious, proactive company (or companies) walked away from a project because of environmental concerns. Another developer can proceed with prospecting and, within the existing laws, permit, build, and operate a facility right where others did not, and, in many instances, with no consequence or requirements to assess or monitor impacts (depending on the state). Arnett (2012) also noted that any other industry or land-use developer could pursue development of this hypothetical environmentally sensitive landscape unless the land is put under, for example, a conservation easement, as the landowner has the right to realize the highest and best economic use of their lands within the limits of the law. This scenario, although having little to do with the biology or science of the matter, could have extensive implications for renewable energy (and other land-use development) and wildlife impacts. Situations like these are not well documented, and most never reach public dissemination, but they undoubtedly occur and threaten wildlife and all stakeholders, including the renewable industry. Creating policy that is consistent within and across industries would "level the playing field" for all developers and their requirements for environmentally sensitive ar-

eas and minimize or perhaps eliminate bad behavior, rather than reward it.

Arnett (2012) also discussed a scenario in which regulations create unique challenges centered on ethics. Let's say, for example, there are 2 adjacent sites being developed by different companies that occur in an area occupied by a federally listed threatened or endangered species. One company is proactive and decides to conduct extensive surveys for the listed species on their proposed site, and they do so even though it is not legally required, but they believe it is the right thing to do. Meanwhile, a neighboring developer chooses not to conduct such surveys and proceeds with development. Upon finding the listed species occupying its site, the proactive company reports to the USFWS, as it should, but soon finds that the surveys and reporting now seriously jeopardize their project. This company may need to complete an expensive and time-consuming habitat conservation plan and may never get to complete this project. While this is an appropriate course of action, the less conscientious company continues to develop unhindered without expenditure on surveys or consultation with appropriate agencies. Their only risk would be to actually "take" a listed species and, though unlikely, be caught doing so. With no consistent requirements in place regarding surveys for listed species, the proactive company has been penalized for good behavior, and the less conscientious company is rewarded for bad behavior. Hence, there is a need for further policy discussions and decisions to ensure a consistent environmental process among all renewable energy companies. Creating incentives for responsible development and rewarding companies that adhere to guidelines, mitigate impacts, implement voluntary efforts, and support research will be a critically important component for resolving potential threats to wildlife conservation from renewable energy in the future.

Another key issue is consistent application of regulations. In my discussions with wildlife professionals and industry developers, this appears to be an overarching problem with many federal agencies across the county. A state agency colleague conveyed to me that 4 different USFWS offices had different interpretation and application of the ESA, given the same issue brought forth to each of them. This creates untenable situations, considerable uncertainty, consternation, and lack of trust among stakeholders that seems completely unnecessary. Clear guidance and training would help improve consistency among professionals interpreting and implementing statutes, regulations, and administrative policies.

Closing Thoughts

Elliott (2013) pointed out that frequent changes of control of government by our political parties and the shifting policies that result are an impediment to having a successful national renewable energy policy. I would argue that the same is true for implementing consistent policy for balancing wildlife conservation with renewable energy development. Indeed, recent changes to the MBTA, to mitigation requirements, and to other related policies by the Department of Interior during the Trump administration may seem beneficial to proponents in the short term, but future administrations may swing the proverbial pendulum back the other direction, per Elliot's (2013) observation. Lawmakers need to consider long-term certainty and stability—financial, ecological, and social—as they decide to amend, abolish, or develop statutes and regulatory and administrative policies for renewable energy, if we are to achieve balance with wildlife conservation.

Arnett et al. (2007a) speculated that given projected increases in multiple sources of energy development, including biomass, wind, and oil and gas development, future conflicts surrounding land-use change, mitigation measures, and conservation strategies should be anticipated. They noted that habitat mitigation options, for example, when developing wind in open prairie, may be compromised by development of other energy sources and their needs to mitigate impacts (i.e., there are only so many acres in any given landscape and not all can accommodate development and needed mitigation of impacts). Arnett et al. (2007) recommended that regional assessments of existing and multiple forecasts of possible land uses be conducted and that planning of regional conservation strategies among industries, agencies, and private landowners be implemented to reduce conflicts and increase options for mitigation and conservation. Evans and Kiesecker (2014) corroborated these recommendations when they suggested that cumulative impacts from any 1 or multiple sources of energy development pose the greatest challenge for achieving landscape-level conservation goals. They also pointed out that, unfortunately, assessments of environmental impacts usually are made at the project scale rather than the landscape scale. Landscape-scale modeling of different scenarios can allow regulators to examine the potential consequences of development objectives quickly and inexpensively (Evans and Kiesecker 2014). This will be an important next step for improving land-use planning and integrating with policy for renewable energy development and wildlife conservation in the future.

International treaties and existing or potentially needed regulations should be addressed, given that fatalities at wind and solar facilities, in particular, involve migratory species of wildlife. Voigt et al. (2012) reported that wind turbines kill bats not only of sedentary local populations but also of distant populations, potentially having a negative impact beyond political borders. They suggested that international regulations for implementing mitigation measures to prevent large-scale detrimental effects on endangered bat populations are warranted. The same is true globally for migratory wildlife, and countries sharing populations of migratory wildlife should heed this advice from Voigt et al. (2012).

Arnett (2012) summarized a number of issues surrounding the interface of science, management, and policy that continue to plague renewable energy and wildlife conservation. A number of policy (statutory and regulatory), information, and communication challenges face those attempting to protect wildlife and mitigate impacts while responsibly developing renewable energy. Unless there is a government-based nexus, federal and/or state statutes and administrative policies dictating development standards, monitoring and research, siting, and mitigation efforts by energy developers and operating companies will be voluntary, likely without regard for cumulative effects (Arnett 2012). Radzi and Droege (2014) identified a growing challenge with lack of coordination between central and local policies and between numerous entities with vested interest in renewable energy development. The likelihood of miscommunication or no communication—both within and among state and federal agencies, counties, and other local governing and permitting authorities—must be improved. Stronger coordination is essential, and perhaps policy revisions and regulation will be necessary, especially where wildlife agencies have little or no authority in decision making or where there is no legal protection, for bats, for example, and other wildlife (Arnett 2012). Decision making must be grounded in the best available science. Also, consistent policy, accountability, effective siting and mitigation strategies, and a "level-playing field" for the industry (i.e., consistent requirements and incentives for all companies within and among energy sectors) are fundamental if we are to successfully develop renewable energy that conserves wildlife and its habitats.

ACKNOWLEDGMENTS
I wish to thank C. Moorman, S. Grodsky, and S. Rupp for leading efforts to prepare this book and chapters within. Their comments helped improve this manuscript. My thoughts on energy policies and wildlife conservation have been shaped by discussions with numerous colleagues, notably J. Anderson, S. Enfield, J. Lindsay, B. Thresher, T. Kunz, M. Tuttle, M. Bean, J. Lyons, C. Plumer, S. Kline, J. Wilkensen, T. Hasson, and L. Barsen.

LITERATURE CITED

Allison, T. D., T. L. Root, and P. C. Frumhoff. 2014. Thinking globally and siting locally: Renewable energy and biodiversity in a rapidly warming world. Climate Change 126:1–6.

Allison, T. D., J. F. Cochrane, E. Lonsdorf, and C. Sanders-Reed. 2017. A review of options for mitigating take of golden eagles at wind facilities. Journal of Raptor Research 51:319–333.

Arnett, E. B. 2012. Impacts of wind energy development on wildlife: Challenges and opportunities for integrating science, management, and policy. In Wildlife Science: Connecting Research with Management, edited by J. P. Sands, S. J. DeMaso, L. A. Brennan, and M. J. Schnupp, 213–237. New York: Taylor and Francis.

Arnett, E. B., and R. F. May. 2016. Mitigating wind energy impacts on wildlife: Approaches for multiple taxa. Human-Wildlife Interactions 10:28–41.

Arnett, E. B., D. B. Inkley, R. P. Larkin, S. Manes, A. M. Manville, J. R. Mason, M. L. Morrison, M. D. Strickland, and R. Thresher. 2007a. Impacts of wind energy facilities on wildlife and wildlife habitat. Wildlife Society Technical Review 07-2. Bethesda, MD: Wildlife Society.

Arnett, E. B., M. D. Strickland, and M. L. Morrison. 2007b. Renewable energy resources and wildlife: Impacts and opportunities. Transactions of the 72nd North American Wildlife and Natural Resources Conference 72, 65–95. ISSN: 0078-1355. Washington, DC: Wildlife Management Institute.

Arnett, E. B., E. F. Baerwald, F. Mathews, L. Rodrigues, A. Rodriguez-Duran, J. Rydell, R. Villegas-Patraca, and C. C. Voigt. 2015. Impacts of wind energy development on bats: A global perspective. In Bats in the Anthropocene: Conservation of Bats in a Changing World, edited by C. C. Voigt, and T. Kingston, 295–323. New York: Springer Science Press.

Badichek, G. 2015. Resolving conflicts between endangered species conservation and renewable energy siting: Wiggle room for renewables? Consilience: The Journal of Sustainable Development 14:1–24.

Bat Conservation Ireland. 2012. Wind turbine/wind development bat survey guidelines, version 2.8. Bat Conservation Ireland, https://www.batconservation ireland.org/wp-content/uploads/2013/09/BCIreland -Wind-Farm-Turbine-Survey-Guidelines-Version -2-8.pdf.

BLM (Bureau of Land Management). 2005a. Land use planning handbook, H-1601-1. Washington, DC: US Department of the Interior, Bureau of Land Management. https://www.ntc.blm.gov/krc/uploads/360/4 _BLM%20Planning%20Handbook%20H-1601-1.pdf.

BLM (Bureau of Land Management). 2005b. Final Programmatic Environmental Impact Statement on wind energy development on BLM administered land in the western United States. Washington, DC: US Department of the Interior, Bureau of Land Management. http://windeis.anl.gov/documents/fpeis/index.cfm.

BLM (Bureau of Land Management). 2012. Approved Resource Management Plan Amendments/Record of Decision (ROD) for solar energy development in six southwestern states. Washington, DC: US Department of the Interior, Bureau of Land Management. http://solareis.anl.gov/documents/docs/Solar_PEIS_ROD.pdf.

Busch, P. O., and H. Jörgens. 2005. International patterns of environmental policy change and convergence. European Environment 15:80–101.

Cameron, D. R., B. S. Cohen, and S. A. Morrison. 2012. An approach to enhance the conservation-compatibility of solar energy development. PLOS One 6:e38437. doi: 10.1371/journal.pone.0038437.

Cameron, D. R., L. Crane, S. S. Parker, and J. M. Randall. 2017. Solar energy development and regional conservation planning. In Energy Sprawl Solutions, edited by J. M. Kiesecker and D. E. Naugle, 67–75. Washington, DC: Island Press.

Congressional Review Service. 2012. The role of the environmental review process in federally funded highway projects: Background and issues for Congress. CRS Report to Congress 7-5700. Washington, DC: Congressional Review Service. https://environment.transportation.org/pdf/proj_delivery_stream/crs_report_envrev.pdf.

Council on Environmental Quality. 2000. Protection of the environment (under the National Environmental Policy Act). Report No 40 CFR 1500-1517. Washington, DC: Council on Environmental Quality.

Council on Environmental Quality. 2005. Regulations for implementing the procedural provisions of the National Environmental Policy Act. Reprint No 40 CFR 1500–1508. Washington, DC: Council on Environmental Quality. https://www.energy.gov/sites/prod/files/NEPA-40CFR1500_1508.pdf.

Council on Environmental Quality. 2007. Aligning National Environmental Policy Act processes with environmental management systems. Washington, DC: Council on Environmental Quality. https://www.energy.gov/sites/prod/files/CEQ_NEPA_EMS_Guide-04-2007_1.pdf.

Davison, R. P., W. P. Burger, H. Campa III, P. J. Conry, K. D. Elowe, G. Frazer, D. C. Mason, D. E. Moore III, and R. D. Nelson. 2005. Practical solutions to improve the effectiveness of the Endangered Species Act for wildlife conservation. Wildlife Society Technical Review 05–1. Bethesda, MD: Wildlife Society.

Donner, S. D., and C. J. Kucharik. 2008. Corn-based ethanol production compromises goal of reducing nitrogen export by the Mississippi River. Proceedings of the National Academies of Science USA 105:4513–4518.

DuVivier, K. K. 2011. The renewable energy reader. Durham, NC: Carolina Academic Press.

EIA (Energy Information Administration). 2018. Annual energy outlook 2018. Washington: DC: US Energy Information Administration. https://www.eia.gov/outlooks/aeo/.

Ellabban, O., H. Abu-Rub, and F. Blaabjerg. 2014. Renewable energy resources: Current status, future prospects and their enabling technologies. Renewable and Sustainable Energy Reviews 39:748–764.

Elliott, E. D. 2013. Why the United States does not have a renewable energy policy. Environmental Law Reporter News and Analysis 43:1095–10101.

Evans, J. S., and J. M. Kiesecker. 2014. Shale gas, wind and water: Assessing the potential cumulative impacts of energy development on ecosystem services within the Marcellus Play. PLOS One 9:e89210. doi:10.1371/journal.pone.0089210.

Fazio, C., and E. Strell. 2018. Abrupt policy change on century-old Migratory Bird Treaty Act. New York Law Journal, February 21. Law.Com https://www.law.com/newyorklawjournal/2018/02/21/abrupt-policy-change-on-century-old-migratory-bird-treaty-act/?slreturn=20180519140713.

Fjetland, C. A. 2000. Possibilities for expansion of the Migratory Bird Treaty Act for the protection of migratory birds. Natural Resources Journal 40:47–68.

Fletcher, R. J., B. A. Robertson, J. Evans, P. J. Doran, J. R. R. Alavalapati, and D. W. Schemske. 2011. Biodiversity conservation in the era of biofuels: Risks and opportunities. Frontiers in Ecology and the Environment 9:161–168.

Geißler, G. 2013. Strategic environmental assessments for renewable energy development: Comparing the United States and Germany. Journal of Environmental Assessment Policy and Management 15. doi:10.1142/S1464333213400036.

Government Accountability Office. 2005. Wind power: Impacts on wildlife and government responsibilities for regulating development and protecting wildlife. Report to Congressional Requesters, GAO-05-906. Washington, DC: US Government Accountability Office.

Hernandez, R. R., S. B. Easter, M. L. Murphy-Mariscal, F. T. Maestre, M. Tavassoli, E. B. Allen, C. W. Barrows, J. Belnap, R. Ochoa-Hueso, S. Ravi, and M. F. Allen. 2014. Environmental impacts of utility-scale solar energy.

Renewable and Sustainable Energy Reviews 29:766–779.

Hernandez, R. R., M. K. Hoffacker, M. L. Murphy-Mariscal, G. C. Wu, and M. F. Allen. 2015. Solar energy development impacts on land cover change and protected areas. Proceedings of the National Academy of Sciences 112: 13579–13584.

Herson, A. I. 1986. Project mitigation revisited: Most courts approve Findings of No Significant Impact justified by mitigation. Ecology Law Quarterly 13:51–72.

Holland, S. P., J. E. Hughes, C. R. Knittel, and N. C. Parker. 2015. Unintended consequences of carbon policies: Transportation fuels, land-use, emissions, and innovation. Energy Journal 36:35–74.

Inkley, D. B., M. G. Anderson, A. R. Blaustein, V. R. Burkett, B. Felzer, B. Griffith, J. Price, and T. L. Root. 2004. Global climate change and wildlife in North America. Wildlife Society Technical Review 04-1. Bethesda, Maryland: Wildlife Society.

Kiesecker, J. M., and D. E. Naugle, eds. 2017. Energy sprawl solutions. Washington, DC: Island Press.

Kiesecker, J. M., H. Copeland, A. Pocewicz, and B. McKenney. 2010. Development by design: Blending landscape level planning with the mitigation hierarchy. Frontiers in Ecology and the Environment 8:261–266.

Kiesecker, J. M., J. S. Evans, J. Fargione, K. Doherty, K. R. Foresman, T. H. Kunz, D. Naugle, N. P. Nibbelink, and N. D. Niemuth. 2011. Win-win for wind and wildlife: A vision to facilitate sustainable development. PLOS One 4:e17566. doi:10.1371/journal.pone.0017566.

Köppel, J, G. Geißler, J. Helfrich, and J. Reisert. 2012. A snapshot of Germany's EIA approach in light of the United States archetype. Journal of Environmental Assessment Policy and Management 14:1250022-1–1250022-21.

Kruger, L. B. McKenney, G. Watkins, and A. Amin. 2017. Policies, practices, and pathways for sustainable energy. In Energy Sprawl Solutions, edited by J. M. Kiesecker and D. E. Naugle, 131–141. Washington, DC: Island Press.

Kunz, T. H., E. B. Arnett, B. M. Cooper, W. P. Erickson, R. P. Larkin, T. Mabee, M. L. Morrison, M. D. Strickland, and J. M. Szewczak. 2007. Methods and metrics for studying impacts of wind energy development on nocturnal birds and bats. Journal of Wildlife Management 71:2449–2486.

Ledec, G. C., K. W. Rapp, and R. G. Aiello. 2011. Greening the wind: Environmental and social considerations for wind power development in Latin America and beyond. Washington, DC: World Bank. http://documents .worldbank.org/curated/en/239851468089382658/pdf/6 62330PUB0EPI00e0wind09780821389263.pdf.

Lilley, M. B., and J. Firestone. 2008. Wind power, wildlife, and the Migratory Bird Treaty Act: A way forward. Environmental Law 38:1167–1214.

Loss, S. R., T. Will, and P. P. Marra. 2013. Estimates of bird collision mortality at wind facilities in the contiguous United States. Biological Conservation 168:201–209.

Lovich, J. E., and J. R. Ennen. 2011. Wildlife conservation and solar energy development in the desert southwest, United States. BioScience 61:982–992.

Lyon, T. P., and H. Yin. 2010. Why do states adopt renewable portfolio standards? An empirical investigation. Energy Journal 31:133–157.

Mai, T., W. Cole, E. Lantz, C. Marcy, and B. Sigrin. 2016. Impacts of federal tax credit extensions on renewable deployment and power sector emissions. Technical Report NREL/TP-6A20-65571. Golden, CO: National Renewable Energy Laboratory. https://www.nrel.gov /docs/fy16osti/65571.pdf.

Margolin, S. 1979. Liability under the Migratory Bird Treaty Act. Ecology Law Quarterly 7:989–1010.

Martin, R. J., and R. Ballard. 2013. Reconciling the Migratory Bird Treaty Act with expanding wind energy to keep the big wheels turning and endangered birds flying. Animal Law 20:145–166.

Mawdsley, J. R., R. O'Malley, and D. S. Ojima. 2009. A review of climate-change adaptation strategies for wildlife management and biodiversity conservation. Conservation Biology 23:1080–1089.

Morris, A. W., and J. Owley. 2014. Mitigating the impacts of the renewable energy gold rush. Minnesota Journal of Law, Science and Technology 15:293–388. https:// scholarship.law.umn.edu/mjlst/vol15/iss1/18/?utm _source=scholarship.law.umn.edu%2Fmjlst%2Fvol15%2 Fiss1%2F18&utm_medium=PDF&utm_campaign =PDFCoverPages.

National Research Council. 2007. Ecological impacts of wind-energy projects. Washington, DC: National Academies Press.

Ontario Ministry of Natural Resources. 2011. Bats and bat habitats: Guidelines for wind power projects. Ontario: Ministry of Natural Resources and Forestry. https://www .ontario.ca/document/bats-and-bat-habitats-guidelines -wind-power-projects.

Panarella, S. J. 2017. A bird in the hand: Shotguns, deadly oil pits, cute kittens, and the Migratory Bird Treaty Act. Virginia Environmental Law Journal 35:153–212.

Pennsylvania Game Commission (PGC). 2007. Pennsylvania Game Commission wind energy voluntary cooperation agreement. http://www.pgc.pa.gov/Information Resources/AgencyBusinessCenter/WindEnergy/Pages /WindEnergyVoluntaryCooperativeAgreement.aspx. Accessed 18 June 2018.

Radzi, A., and P. Droege. 2014. Latest perspectives on global renewable energy policies. Current Sustainable Renewable Energy Report 1:85–93.

Rodrigues, L., L. Bach, M. J. Dubourg-Savage, B. Karapandza, D. Kovac, T. Kervyn, J. Dekker, et al. 2015. Guidelines for consideration of bats in wind farm projects, revision 2014. EUROBATS Publication Series no. 6 (English version). Bonn, Germany: UNEP/ EUROBATS Secretariat. http://www.eurobats.org/sites /default/files/documents/publications/publication_series /pubseries_no6_english.pdf.

Salter, T. 2011. NEPA and renewable energy: Realizing the most environmental benefit in the quickest time. University of California Environmental Law and Policy Journal 34:173–187.

Schlesinger, M. E., and J. F. B. Mitchell. 1987. Climate model simulations of the equilibrium climate response to increased carbon dioxide. Reviews of Geophysics 25:760–798.

Strickland, M. D., E. B. Arnett, W. P. Erickson, D. H. Johnson, G. D. Johnson, M. L. Morrison, J. A. Shaffer, W. Warren-Hicks. 2011. Comprehensive guide to studying wind energy/wildlife interactions. Washington, DC: National Wind Coordinating Collaborative. http://www .batcon.org/pdfs/wind/National%20Wind%20Coordinating%20Collaborative%202011_Comprehensive%20 Guide%20to%20Studying%20Wind%20Energy%20 and%20Wildlife%20Interactions.pdf.

US Department of Interior. 2017a. Incidental take prohibited under the Migratory Bird Treaty Act. Memorandum M-37041 to the director of the US Fish and Wildlife Service, Office of the Solicitor, US Department of Interior, Washington, DC. https://www .eenews.net/assets/2017/02/21/document_ew_01.pdf.

US Department of Interior. 2017b. The Migratory Bird Treaty Act does not prohibit incidental take. Memorandum M-37050 to the Secretary, Office of the Solicitor, US Department of Interior, Washington, DC. https:// www.doi.gov/sites/doi.gov/files/uploads/m-37050.pdf.

US Department of the Interior/Bureau of Land Management, ed. 2016. The Federal Land Policy and Management Act of 1976, as amended. Washington, DC: US Department of the Interior, Bureau of Land Management, Office of Public Affairs. file:///C:/Users/Ed/ Downloads/FLPMA2016.pdf.

USFS (US Forest Service). 2011. Strategic energy framework. Washington, DC: US Forest Service. https://www .fs.fed.us/specialuses/documents/Signed_Strategic Energy_Framework_01_14_11.pdf.

USFWS (U S Fish and Wildlife Service). 2012. Land-based wind energy guidelines. Arlington, VA: US Fish and Wildlife Service. https://www.fws.gov/ecological -services/es-library/pdfs/WEG_final.pdf.

USFWS (US Fish and Wildlife Service). 2013. Eagle conservation plan guidance: Module 1, Land-based wind energy, version 2. Arlington, VA: US Fish and Wildlife Service. https://www.fws.gov/migratorybirds/pdf /management/eagleconservationplanguidance.pdf.

Voigt, C. C., A. G. Popa-Lisseanu, I. Niermann, and S. Kramer-Schadt. 2012. The catchment area of wind farms for European bats: A plea for international regulations. Biological Conservation 153:80–86.

Walston, L. J. Jr., K. E. Rollins, K. E. LaGory, K. P. Smith, and S. A. Meyers. 2016. A preliminary assessment of avian mortality at utility-scale solar energy facilities in the United States. Renewable Energy 92:405–414.

Waples, R. S., M. Nammack, J. F. Cochrane, and J. A. Hutchings. 2013. A tale of two acts: Endangered species listing practices in Canada and the United States. BioScience 63:723–734.

Wiser, R., C. Namovicz, M. Gielecki, and R. Smith. 2007. The experience with renewable portfolio standards in the United States. Electricity Journal 20:8–20.

Yung, J. E. C., and M. J. Sanders. 2015. Reconciling the Endangered Species Act and large-scale renewable energy projects. Natural Resources and Environment 29:1–5.

11 — Renewable Energy Ecology
The Next Frontier in Wildlife Science

Steven M. Grodsky,
Sarah R. Fritts, and
Rebecca R. Hernandez

Introduction

The progression of renewable energy enterprises and their physical manifestations (hereafter collectively referred to as "renewable energy development") is a critical wildlife conservation issue. This book provides a foundation on which wildlife professionals and researchers can build their understanding of renewable energy and wildlife conservation in theory and practice. This concluding chapter synthesizes key principles and some trends in renewable energy and wildlife conservation and serves as a springboard for research and educational opportunities in the emerging field of renewable energy ecology. Indeed, the interface between renewable energy development and wildlife conservation has created a new frontier in wildlife science, complete with the environmental challenges associated with other contemporary global changes like climate change.

Significant economic and industrial inertia of renewable energy development in North America, coupled with growing social and political support, warrant the timely publication of this book and prudent consideration of interplays between renewable energy and wildlife conservation. Environmental concerns regarding climate change continue to provide further impetus to develop renewable energy with fewer greenhouse gas emissions. Renewable energy development also provides an obvious and realizable answer to the lingering quandary of finite fossil fuel resources ultimately running out. Meanwhile, implementation of renewable energy technologies throughout the landscapes of North America is outpacing our understanding of its effects on wildlife. Current knowledge already suggests that renewable energy development may act in conjunction with other global changes, including urbanization and climate change, to restrict land resources available to wildlife species. Additionally, some wildlife species may be killed or displaced, for instance, by site-specific infrastructure and activities for renewable energy production. In light of this current situation, every wildlife professional and student should possess a firm understanding of the basic principles of renewable energy and wildlife conservation and a willingness to consider the future of wildlife conservation with respect to renewable energy development.

Principles of Renewable Energy and Wildlife Conservation

The chapters in this book consolidate important themes centered on renewable energy and wildlife conservation that transcend effects of individual energy technologies on specific wildlife taxa. From these themes, we have synthesized the basic principles of renewable energy and wildlife conservation:

1. Wildlife are likely to be directly (e.g., mortality) and/or indirectly (e.g., habitat loss) affected by renewable energy development, regardless of the type.
2. Wildlife response to renewable energy development varies both spatially and temporally.
3. Some wildlife species may be negatively affected by renewable energy development, whereas other wildlife species may benefit from it.

Direct and Indirect Effects

The technology of today's renewable energy infrastructure does not yet preclude wildlife mortality. Such mortality may invoke emotional responses from concerned citizens and scientists alike, proliferated by images of decapitated eagles at wind facilities and visions of songbirds bursting into flames at solar facilities. In an age of increased environmental awareness and scarcity of undeveloped lands, wildlife mortality at renewable energy facilities may be scrutinized far more than wildlife fatalities caused by comparatively older, conventional energy technologies (e.g., oil and coal). The Deepwater Horizon oil spill in the Gulf of Mexico serves as a recent reminder that conventional energy technologies kill wildlife, often in large numbers. Yet, when considering the scope of avian mortality from renewable energy production, fatality comparisons often are made to other anthropogenic factors such as feral house cats (*Felis catus*), roads, glass windows, and communication towers. Such comparisons of wildlife mortality can be a red herring for prioritizing conservation goals and values related to renewable energy development. Specifically, lumping wildlife species into a common taxonomic representation like "birds" and generalizing wildlife mortality over varied landscapes often fail to capture site- and species-specific phenomena related to wildlife mortality at renewable energy facilities.

Potential population-level effects of wildlife mortality at renewable energy facilities likely vary by taxa, and more research linking disparate mortality estimates at individual facilities to population-level impact assessments certainly is needed. For highly abundant and fecund wildlife, like many migratory passerines, mortality associated with renewable energy infrastructure and activities likely causes minor population-level effects. Conversely, mortality associated with renewable energy facilities may significantly decrease populations of already rare birds of prey with slow reproductive rates. Of course, understanding the effects of wildlife mortality from renewable energy production on animal populations is predicated by previous knowledge of species population levels. For wildlife like bats, this presents a formidable problem. Bats are nocturnal and can be small and migratory. As such, little is known about bat population numbers in general, making quantification of population-level, additive mortality via wind turbines, for instance, difficult to achieve. Similar issues arise with invertebrates for many of the same reasons (Grodsky et al. 2015). Yet another challenge presented by wildlife mortality at renewable facilities is amalgamating mortality estimations from different studies that used different methods to generate regional estimates of wildlife mortality from renewable energy production.

Indirect effects of renewable energy development on wildlife are more numerous, linked to a greater diversity of wildlife, and more difficult to quantify than direct effects. Although indirect effects may lead to wildlife mortality, they typically are associated with nonlethal effects such as displacement. For example, disturbances caused by the construction and operation of a wind facility may cause raptor species to leave the area (Garvin et al. 2011). Habitat alteration is another indirect effect of renewable energy development. In industrial forests of the southeastern United States, woody biomass harvesting for forest bioenergy reduces coarse woody debris, which, in turn, provides essential food and cover resources for some invertebrates (Grodsky et al. 2018a). In aquatic systems, river flow altered by dams for hydropower generation often has negative effects on migratory fish and aquatic invertebrate habitat. Indirect effects of renewable energy development on spread of disturbance-dependent, invasive species are largely unknown. Anthropogenic disturbance associated with renewable energy development (e.g., disturbed soils) can facilitate the spread of invasive species, particularly plants and invertebrates, which colonize newly disturbed areas (e.g., red imported fire ants, *Solenopsis invicta*; Grodsky et al. 2018b). The ecological ramifications of these species invasions facilitated by renewable energy development may be exacerbated in areas with endemic flora and fauna that are maladapted to novel disturbances and unable to compete with rapidly colonizing invasive species.

Temporal and Spatial Variation

Effects of renewable energy development on wildlife may vary temporally, and questions regarding adaptive capacity of species to respond to the long-term ecological effects of renewable energy development remain largely unanswered. Land cover—defined here as the biophysical capacity of a defined geographic area—may change over time owing to alterations to the environment and human land-use history. Abrupt anthropogenic changes along this timeline, such as conversion of natural areas to renewable energy production, are likely to have noticeable effects on wildlife communities. On the other hand, areas suitable for coproduction of renewable energy, like industrial forests (e.g., use of low-value wood products for bioenergy) and agricultural lands (e.g., siting wind turbines in active crop fields), were converted from natural lands for timber and crop production, respectively, well prior to renewable energy development. Wildlife response to renewable energy development in these pre-converted areas must be teased apart from their response to effects of concurrent and temporally variable land uses, in

this case timber production and agriculture. Each milestone along the timeline of renewable energy development—siting, construction, operation and maintenance, and decommissioning—may have different effects on wildlife. For example, disturbance from construction and decommissioning of some renewable energy facilities may be more severe than that which occurs during operation and maintenance, leading to variable wildlife response throughout the life cycle of a renewable energy facility. For many wildlife species, their susceptibility to effects from renewable energy development varies seasonally in relation to life history events (e.g., breeding, migration).

We currently know little about wildlife adaptation to renewable energy production over time, mostly because of a lack of long-term studies. Local wildlife populations that are negatively affected by renewable energy development may recover over time. Conversely, some wildlife species may not possess the adaptive capacity to recover from disturbance associated with renewable energy development, and hence these species may be locally extirpated around energy development projects. Ongoing research on the adaptive capacity of species to climate change provides a conceptual framework to guide similar studies involving species adaption to renewable energy development.

Concomitant with time, the effects of renewable energy development on wildlife vary over space, and the cumulative space occupied by renewable energy facilities is increasingly vast. Siting of renewable energy facilities on the landscape may be a major driver of wildlife response to renewable energy development. For example, renewable energy facilities sited along migration routes or within or near protected natural areas likely cause greater negative effects on wildlife than those sited on marginalized lands or within urban environments. Spatial proximity of individual renewable energy facilities also influences effects on wildlife. As the density of facilities increases across the landscape, habitat fragmentation and impedance to animal migrations may restrict animal movement and increase wildlife mortality. For example, migratory birds and bats may encounter a series of individual wind energy facilities sited along their migration routes, which, in turn, may lead to high cumulative mortality risk at the landscape level. Local-scale responses of wildlife to renewable energy development may differ from those that occur at the landscape level and vice versa. Similarly, the spatial scale at which wildlife respond to renewable energy development may vary among taxa, depending on life history, behavior, and distribution. Large, mobile species and migratory species (e.g., bats, birds) may respond to renewable energy development across large areas, whereas smaller and less motile species like invertebrates and some herpetofauna may exhibit more local-scale responses (Fritts et al. 2015).

"Winners" and "Losers"

Similar to disturbances associated with agricultural expansion and urbanization, disturbances associated with renewable energy development may benefit generalist species but may be detrimental to specialist species. Basic principles of disturbance ecology suggest that generalist species with plastic habitat requirements are better adapted to persist through and thrive on disturbance than are specialist species that occupy narrower niches and are tightly linked to specific habitat requirements. For example, solar energy development in desert ecosystems may be a boon for generalist species like coyotes (*Canis latrans*) and ravens (*Corvus corax*), whereas desert specialists such as the desert tortoise (*Gopherus agassizii*) and Gila monster (*Heloderma suspectum*) may be negatively affected by disturbance caused by facility construction and by the increased abundance of their aforementioned predators (Moore-O'Leary et al. 2017). Given their restricted ranges and susceptibility to disturbance, specialist species often are designated as species of conservation concern. As such, disturbance related to renewable energy development may affect threatened or endangered species, which, in turn, brings to light pertinent wildlife management and policy considerations. Classification of "generalist" and "specialist" species may be a helpful exercise for categorizing interactions between wildlife and renewable energy development. However, other classifications based on morphology, taxonomy, and especially habitat selection may be important for identifying the "winning" and "losing" species of a particular energy development endeavor; yet, use and application of these prospective types of classifications have not been fully explored.

Trends in Renewable Energy and Wildlife Conservation

Beyond the basic principles outlined above, several noteworthy trends revealed by the chapters of this book may shed light on future research and management directions of renewable energy and wildlife conservation. Although each renewable energy technology has its own subset of potential effects on wildlife, management and mitigation implications, geographic extent, and policies, generalizations from the current state of knowledge of effects of renewable energy development on wildlife may provide a more holistic understanding of cumulative, current, and future wildlife impacts. Similarly, reviewing some trends in renewable energy and wildlife conservation can help identify research needs moving forward.

Geographic Spread of Renewable Energy

The geographic distribution of renewable energy development may directly influence the degree to which wildlife populations are affected, yet the cumulative spread of renewables across land and water is naturally predictable in some cases and unintuitive in many others. The predictability of renewable energy development is linked to renewable resource availability on the landscape. For example, hydropower is predictably developed along rivers, whereas forest bioenergy production naturally occurs in areas with abundant forest resources. Bioenergy production from agricultural crops, for instance, commonly occurs in areas that support soil resources compatible to agricultural production, such as the US Great Plains area. For renewable energy technologies like wind energy and photovoltaic solar energy, development may occur in a variety of ecosystems and thereby affect various species throughout North America or affect the same species differently in various regions of North America.

The unintuitive nature of renewable energy development is derived from the fact that more than just renewable resource availability affects the geographic distribution of renewable energy. For example, the political and economic environment of individual states in the United States, as represented by state-level renewable portfolio standards, may influence the type and intensity of local or regional renewable energy development. As an example of such unintuitive distribution of energy sectors, North Carolina, which is predominantly made up of forested or urban land cover, ranks among the top US states for solar energy development, with production greater than several western states that would appear to have greater opportunity for industrial-scale solar energy. Renewable energy development in the built environment or on marginalized lands (e.g., Hoffacker et al. 2017) may be an intuitive solution to mitigate adverse effects of renewable energy on wildlife by replacing renewable energy development in more sensitive environments. Yet, industrial-scale renewable energy production has yet to follow the trail of many other industry players (e.g., manufacturing, technology) and center development in urban landscapes.

Whether the geography of renewable energy development is intuitive or not, renewables certainly are spreading throughout most landscapes of North America at a rapid rate. Resultant land-use and land-cover changes from renewable energy development may significantly reduce wildlife habitat and biodiversity because of habitat conversion and competition for finite land resources to support conservation.

Understanding the effects of current and projected geographic distribution of renewable energy development on wildlife may become progressively important as the number of facilities continues to increase across the landscapes of North America.

Fueled by Technology

Renewable energy and an understanding of its effects on wildlife are inherently linked to technology. Indeed, technological advancement is the primary effector responsible for the rapid expansion of renewable energy development. The size and capacity of renewable energy infrastructure may have direct effects on wildlife, and these factors continuously change as technology develops. For example, wind turbines are becoming larger and taller, and these turbines may affect birds and bats differently than their historically smaller predecessors. Concentrating solar energy using mirrors may require vast acreages for suitable energy production, albeit with potentially low-intensity site preparation (e.g., mowing), whereas photovoltaic solar technologies may require slightly less land area but high-intensity site preparation like bulldozing (Murphy-Mariscal et al. 2018). Of course, technological advancements in genetic engineering may create more viable or totally new bioenergy crops, each having variable potential effects on wildlife and wildlife-habitat relationships.

In addition to technology associated with operational renewable energy, wildlife studies and mitigation at renewable energy facilities require advanced technology and innovation. Whereas most wildlife studies are enhanced by technology, to some degree, renewable energy development presents specific challenges and unique situations that call for technological innovation; oftentimes, these innovations must be installed or implemented among or within existing renewable energy infrastructure. For example, acoustic deterrents may be installed on top of wind turbines to mitigate bat mortality at wind energy facilities. More recently, bat researchers initiated

Figure 11.1   Texture coating being applied to the monopole of a wind turbine at the Wolf Ridge wind energy facility, Texas, US; this textured coating may help mitigate bat mortality at wind energy facilities by deterring bats from interacting with smooth tower surfaces. *Photo credit: Amanda Hale.*

an experiment to test the viability of repainting surfaces of wind turbine monopoles with a textured paint to reduce bat mortality at wind facilities (Figure 11.1). Importantly, successful mitigation of negative wildlife impacts and effective wildlife conservation at renewable energy facilities is predicated by a sound working knowledge of how different renewable energy technologies are deployed and the socioeconomic drivers of deployment regionally and throughout North America.

Tip of the Iceberg

Renewable energy technologies have received variable attention from wildlife researchers to date, and these research efforts have not necessarily aligned with the pace and intensity of development of renewable energy sectors. In general, wind energy and wildlife conservation has been the most researched topic, with studies published on birds and wind energy beginning in the 1990s. However, we still do not know the reason for high bat mortality at wind facilities, for example, and potential effects of offshore wind energy development on wildlife in North America are largely unknown. And as evidenced by the chapters of this book on wind energy and wildlife, wind energy–wildlife research has predominantly focused on birds (Chapter 5, Dohm and Drake) and bats (Chapter 6, Hein and Hale), whereas effects on other wildlife remain mostly unknown (but see Chapter 7, Korfanta and Zero). Although hydropower still is the most used renewable energy resource in North America, there exists a mismatch between the longevity of hydropower production and a paucity of published wildlife-hydropower studies (but see Chapter 9, Jager and Wickman). Although the concept of burning wood for fuel is that it's "so easy a caveman can do it," studies on effects of forest bioenergy on wildlife have only recently started to appear in the literature at the time of this writing (but see Chapter 1, Greene, Martin, and Wigley, and Chapter 2, Homyack and Verschuyl). While we can borrow, to a certain degree, from the well-established literature on effects of agricultural production on wildlife, research on explicit effects of actual bioenergy production on wildlife is lacking (but see Chapter 3, Otto, and Chapter 4, Rupp and Ribic). With advances in technology, solar energy is quickly becoming a predominant renewable energy technology; apart from scattered studies throughout the last few decades, research on effects of solar energy development (both concentrating solar and photovoltaic) is woefully lacking (but see Chapter 8, Boroski).

In short, there is much applied research needed explicitly centered on understanding effects of individual renewable energy technologies and cumulative renewable energy development on wildlife globally. In the meantime, policy decisions regarding renewable energy and wildlife conservation may be informed by the growing number of peer-reviewed publications on renewable energy and wildlife conservation (Chapter 10, Arnett). Additionally, policy may be influenced by proactive expert opinion and involvement, as empirical evidence continues to amass.

Opportunities in Renewable Energy Ecology

Through time, wildlife professionals have witnessed (and hopefully experienced) a variety of advancements and refinements in wildlife science necessitated by global change. Once Aldo Leopold shivered with delight at the sound of a Canada goose (*Branta canadensis*) passing overhead, and urban ecology was an abstraction. Now, Canada geese are permanent fixtures of mini-malls across America; they'd rather inhabit a drainage pond next to Home Depot than migrate. Technological developments have facilitated integration of fields like landscape ecology and molecular ecology into wildlife studies, strengthening our ability to more holistically understand wildlife ecology and management. Indeed, failure to adapt as a wildlife professional may ultimately lead to the same fate as that of some other species—extinction!

Renewable energy ecology is an emerging field of study integral to the evolution of modern wildlife science. Nearly every wildlife professional and every wildlife study uses energy, specifically electricity, and generation of electricity has ecological ramifications. As global changes continue to affect wildlife at accelerated rates, failure to acknowledge and study effects of renewable energy development on wildlife would be a major conservation oversight. Fortunately, renewable energy ecology presents a myriad of opportunities for wildlife professionals to creatively apply their expertise and inform sustainable renewable energy development because it is an inherently interdisciplinary and collaborative field of study (e.g., Moore-O'Leary et al. 2017).

Renewable energy ecology goes beyond documentation of wildlife response to renewable energy—it is solutions-oriented to match the urgency of the exigent sustainability issues of today. We can explore exciting opportunities to expand studies beyond renewable energy effects on a single wildlife taxon or groups of wildlife taxa through interdisciplinary studies of ecosystem response, including ecological and trophic interactions among soils, plants, and animals. Techno-ecological synergies (i.e., mutually beneficial relationships across technological and ecological systems) exist that may facilitate wildlife conservation through protection of habitat for species and maintenance of genetic diversity (e.g., ground-mounted solar energy generation coupled with ecological restoration on degraded lands). Globally relevant issues like land sparing are especially applicable to renewable energy ecology, as competition for land resources for energy, food production, housing of people, and conservation stiffens. Additionally, we would do current and future wildlife professionals a disservice if we did not provide contemporary educational resources on ecological effects of renewable energy development because renewable energy is already woven into the fabric of modern-day life throughout most of human-inhabited North America, and its proliferation across both human-dominated and undeveloped landscapes is likely to increase. In the sections below, we elaborate on some opportunities in renewable energy ecology that can collectively lead to more informed wildlife conservation efforts in the face of renewable energy development.

Interdisciplinary and Applied Research

The interface between renewable energy development and wildlife conservation offers rich opportunities for interdisciplinary research. Understanding renewable energy technologies and policies that guide their sustainable development is paramount to designing and executing wildlife studies that generate results with real-world applications. For example, empirical work conducted by Walker et al. (2007) showed that current lease stipulations that prohibit development of coal-bed natural gas near greater sage-grouse (*Centrocercus urophasianus*) habitat imposed by regulatory agencies in Wyoming and Montana did not meet minimum spatial restrictions to preclude negative impacts on the species. Similar studies could be conducted to inform sustainable renewable energy development. Further, a great diversity of wildlife taxa may be affected by renewable energy development, leaving room for every "ologist" to contribute to and collaborate within renewable energy ecology. For example, invertebrates are excellent ecological indicators that may be used in renewable energy and wildlife studies to elucidate trends that may be otherwise unrevealed through the study of vertebrate species alone. Recently, offshore wind energy development has opened doors for marine biologists to conduct research on renewable energy and marine wildlife (including fish) conservation. As renewable energy development continues to expand into novel landscapes, ecologists with specific taxonomic foci won't need to look far for opportunities to contribute to renewable energy ecology.

Renewable energy ecology is, by nature, a field foundationally based in applied research facilitated by interdisciplinary and interinstitutional collaboration. For example, well-curated and well-maintained collaborations among wildlife researchers, renewable energy industry partners, and agency land managers offer great opportunities to apply scientific results to on-the-ground wildlife management at renewable energy sites. Technological innovations implemented by the renewable energy industry may have explicit and identifiable effects on wildlife, both in terms of potential negative effects and mitigation of those effects. As such, clear communication between wildlife researchers and industry technologists could increase the likelihood of minimizing negative effects of renewable energy on wildlife moving forward. Effects of specific renewable technologies on particular wildlife taxa may vary regionally;

thus, interregional collaborations among academia, agencies, and industry could be developed to better understand landscape-level effects of wide-ranging renewable energy expansion.

Ecological and Trophic Interactions

Renewable energy development ultimately affects entire ecosystems, thus effective conservation in and near renewable energy facilities requires comprehensive, ecosystem-based management approaches informed by species-species and species-process interactions (Grodsky et al. 2017). To date, research on renewable energy and wildlife conservation has focused primarily on direct (e.g., mortality) or indirect (e.g., displacement) effects of renewable energy systems on a single taxon. Although these research efforts may effectively guide management for certain taxa and may cumulatively inform conservation efforts for wildlife communities, they do not necessarily grow our understanding of the ecological mechanisms responsible for ecosystem-level responses to renewable energy development. In contrast, an ecosystem-based approach may elucidate effects of renewable energy development on biota by focusing on "bottom-up" or trophic interactions among soils, plants, and animals and energy production, which may, in turn, subsequently reveal mechanisms behind ecological responses to renewable energy-associated disturbance and environmental change (Figure 11.2).

Land Sparing for Wildlife-Friendly Renewable Energy Development

Development of novel, low-carbon sources of fuel and electricity for human use has engendered a shift in global resource use, in both type and magnitude, from that associated with fossil fuel production. However, one common resource required for the production of all energy types is physical space— either on terrestrial or aquatic surfaces (hereafter called "land," for simplicity). Land-use intensity (i.e.,

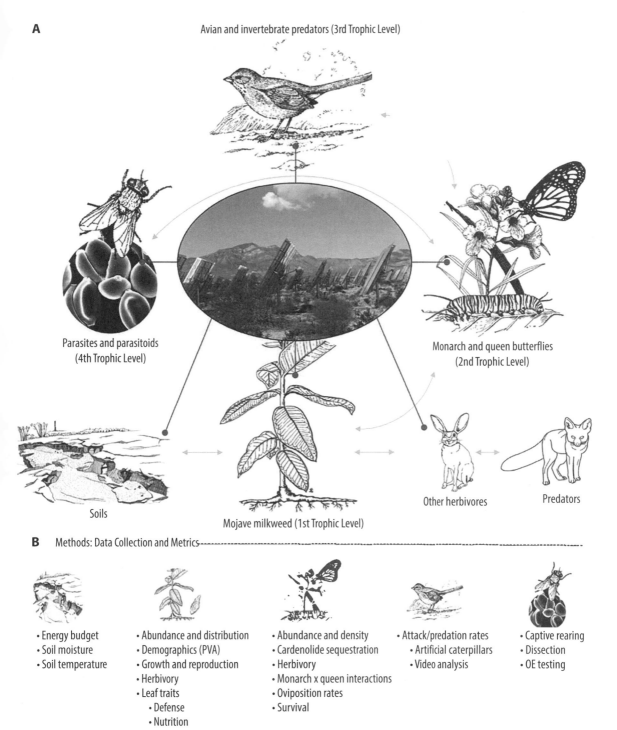

A Avian and invertebrate predators (3rd Trophic Level)

Parasites and parasitoids
(4th Trophic Level)

Monarch and queen butterflies
(2nd Trophic Level)

Soils

Other herbivores

Predators

Mojave milkweed (1st Trophic Level)

B Methods: Data Collection and Metrics

- Energy budget
- Soil moisture
- Soil temperature

- Abundance and distribution
- Demographics (PVA)
- Growth and reproduction
- Herbivory
- Leaf traits
  - Defense
  - Nutrition

- Abundance and density
- Cardenolide sequestration
- Herbivory
- Monarch x queen interactions
- Oviposition rates
- Survival

- Attack/predation rates
  - Artificial caterpillars
  - Video analysis

- Captive rearing
- Dissection
- OE testing

Figure 11.2 (*A*) Conceptual model of the Milkweed-Danaus Trophic System, including potential interactions among a concentrating solar power plant, soils, plants, and animals in the US Mojave Desert; (*B*) List of interdisciplinary measurements to examine trophic interactions in the Milkweed-Danaus Trophic System. *Figure by Steven M. Grodsky.*

the amount of land needed to generate a finite quantity of energy) varies across renewable energy types (Lovering et al. forthcoming). For example, dedicated biomass for renewable energy generation requires more land per unit of electricity generation (60,000 ha/TWh/y) than all other renewable (and carbon-intensive) sources of electricity for human use, some by several orders of magnitude. Hydroelectric and wind power (including spacing between turbine footprints) are approximately 6 times greater (11,000 and 12,000 ha/TWh/y, respectively) in land-use intensity than ground-mounted solar energy. Residue biomass, geothermal, nuclear, and integrated solar energy have the lowest (in decreasing order) land-use intensity across renewables.

Indeed, land is a commonly overlooked albeit integral resource for production of renewable energy, owing to the diffuse nature of the sun's radiation for solar energy, the dispersed nature of kinetic energy from moving water for tidal and hydropower, the geographically constrained nature of wind and geothermal resources, and the proliferation of soil used as a growing medium for bioenergy crops. Operation and transmission require human accessibility and passageways for electricity flows, respectively; thus, construction of roads and transmission corridors associated with renewable energy facilities requires additional land. In an era of looming land scarcity, space allocations for infrastructure and activities associated with renewable energy may be an unintended driver of habitat destruction and fragmentation (Lambin and Meyfroidt 2011).

Previous studies showed that preventing habitat loss is more important to curtailing species population extinction events than preventing fragmentation, the latter of which, by definition, can actually decrease when habitat patches are "removed" (destroyed) from the landscape (i.e., fewer patches overall; Fahrig 1997). Although habitat area is easily calculated with geospatial data used by wildlife ecologists, at least for some species, it can be a poor predictor of species richness. Instead, habitat accessibility, including both habitat area and configuration of

nearby obstructions often associated with renewable energy development (e.g., roads, transmission lines), may be a better predictor of species richness than habitat area (Eigenbrod et al. 2008). Lastly, recent studies have shown that intact protected areas for wildlife may defend against habitat loss from renewable energy development; however, land-use and land-cover changes within 5, 25, and 75 km beyond the perimeter of protected areas can still elicit negative ecological impacts (Hernandez et al. 2015).

Considering the habitat sensitivities of some wildlife, efforts to use space within areas that have already been developed by humans (i.e., biophysical capacity already lost) is a key approach toward sustainable, wildlife-friendly renewable energy development. For example, a developer who decides, on behalf of her company, to site a photovoltaic solar energy installation over a company-owned commercial building and associated parking lots is sparing land under the counterfactual scenario that land supporting an equivalent energy capacity would be developed in an undeveloped or agricultural landscape used by wildlife. Under appropriate assumptions of this counterfactual scenario, the amount of land spared from land conversion for renewable energy development elsewhere can be enumerated and serve as a valuable sustainability metric (i.e., land sparing or habitat sparing) for all stakeholders of renewable energy.

Techno-Ecological Synergies of Renewable Energy

Land sparing is one of several potential synergistic outcomes when renewable energy industry and ecological goals align. Techno-ecological synergy is a systems-based approach to sustainable development. When applied to renewable energy technologies, techno-ecological synergies produce synergistic outcomes favorable for both technology (e.g., photovoltaic module efficiency) and support of ecological systems, including habitat for wildlife species (Bakshi et al. 2015). Synergies (and their ecological-centric

Figure 11.3   Pollinator habitat planted and managed by Prairie Restorations, Inc., at a Connexus Energy photovoltaic solar power facility in Ramsey, Minnesota, US. *Photo credit: Rob Davis.*

outcomes germane to wildlife) may include the following (from Hernandez et al. forthcoming):

- Using severely degraded and contaminated land for renewable energy development (potential benefits: climate regulation; land sparing).
- Coupling renewable energy development with ecological restoration (potential benefits: carbon sequestration and storage; climate regulation; erosion prevention and/or maintenance of soil fertility; habitat for species; maintenance of genetic diversity, pollination, and water quality; Figure 11.3).
- Co-locating renewable energy systems (potential benefit: land sparing).
- Utilizing surfaces within the built environment (e.g., parking lots) for solar energy or other renewable energy systems (potential benefit: land sparing).
- Using residential and commercial rooftop surfaces for solar energy production (potential benefit: land sparing).

Education

Renewable energy and wildlife conservation issues provide contemporary case studies of present and future wildlife science applications that are accessible to current wildlife students. Indeed, modern wildlife science education has the opportunity to broaden its scope through the study of renewable energy and wildlife conservation. For example, most college curricula for wildlife ecology majors involve learning historical wildlife laws and acts, but modern applications of policy are not always included. Thus, there is a need for focused discussions on new policy issues related to renewable energy development relevant to today's conservation concerns. Case studies centered on wildlife conservation and renewable energy may illustrate many modern applications of policies, including the Farm Bill, Migratory Bird Treaty Act, Endangered Species Act, Bald and Golden Eagle Protection Act, and the National Environmental Policy Act (Chapter 10, Arnett). For example, classroom discussions regarding recent changes to the interpretation of "take" at renewable energy facilities under the Migratory Bird Treaty Act could help students realize the role of politics and regulation in wildlife conservation, motivating students to get involved in more holistic conversations about sustainable renewable energy development.

Wildlife conservation in the context of renewable energy development includes a wide range of human behaviors affecting energy policy that illustrate current applications of human dimensions of wildlife. Environmental concerns regarding the sustainability of renewable energy are generated at the nexus of economics, politics, technology, and biology. Wildlife undergraduate (and often graduate) students may consider wildlife conservation in the absence of societal and/or economic factors. Renewable energy ecology brings to light the complex interplay between conservation and practical effectors like industry and economy. The complexity of most issues facing wildlife today, including anthropogenic disturbance from renewable energy development, requires adeptness in human dimensions and educated communication with various stakeholders. The increase in human dimension specialist positions in university wildlife programs and state agencies across the United States is evidence that incorporating human-dimensions training into curricula is vital to successful wildlife management education. We encourage continued and advanced training from a diversity of programs to apply wildlife conservation in the context of renewable energy development.

We challenge professors in wildlife programs to stay current and implement new courses that incorporate renewable energy ecology. Students will then be better prepared for a wildlife career for having taken courses that focused on current events in wildlife conservation, including sustainable renewable energy development, thereby developing novel problem-solving skills and knowledge applicable to modern-day wildlife conservation and management. For example, incorporating renewable energy ecology in wildlife education may involve corresponding with federal and/or state agencies to prioritize challenges, such as bat and bird mortality at wind facilities and aquifer depletion through biocrop irrigation. After identifying these challenges, students can undertake projects that address modern issues in renewable energy ecology. Another example of a classroom activity to integrate renewable energy and wildlife conservation is asking students to write a 1- or 3-minute elevator speech that informs the public of the magnitude and scope of specific issues like

land sparing for biodiversity or the food versus fuel debate and its impacts on wildlife. Ultimately, present and future wildlife students should be prepared to effectively communicate with stakeholders, persuade people to care about wildlife conservation, and find real-world solutions to today's wildlife issues, including renewable energy and wildlife conservation.

In turn, we challenge wildlife students to actively educate themselves about the sustainability of renewable energy development in relation to wildlife conservation. Students may enter the wildlife profession with the idea that wildlife sustainability is centered on preservation. In the philosophy of conservation, students should learn to balance environmental considerations with those of real-world societal needs like energy. Further, they can learn to collaboratively resolve wildlife issues encountered during renewable energy development. By using renewable energy and wildlife conservation as a case study, students may enhance their communication and problem-solving skills and conflict-resolution abilities. Students—take charge and be active in learning contemporary wildlife ecology and management skills, such as those accessed by studying renewable energy ecology!

Academia is not solely responsible for providing educational opportunities to future wildlife biologists; industry also can play an active role. Increased institutional engagement between academia and the renewable energy industry could facilitate educational opportunities that better prepare students for real-world wildlife issues. These educational opportunities could help students develop abilities that broaden their employability, like problem-solving skills specific to environmental issues faced by the renewable energy industry. Representatives from the renewable energy industry may have a vested interest in training and developing future renewable energy ecologists and wildlife managers. Incorporating a diversity of guest speakers and lecturers from renewable energy business and industry may be a good first step for integration of the renewable energy industry into academia. Meanwhile, the renewable energy industry can support site visits or provide paid internships for students to show them real-world applications of the concepts discussed in the classroom. To promote applied learning opportunities, wildlife programs could facilitate internships, undergraduate projects, and theses with renewable energy industry partners for course credit.

Professional Development

Professional development opportunities specifically related to renewable energy and wildlife conservation are lacking, with a few exceptions. For example, the Association of Fish and Wildlife Agencies, Bat Conservation International, American Wind Wildlife Institute, and the National Renewable Energy Laboratory sponsor an in-person professional development workshop on wind energy and wildlife conservation. The Wildlife Society has published publicly available technical reviews on effects of bioenergy production on wildlife (Rupp et al. 2012) and impacts of wind energy facilities on wildlife (Arnett et al. 2007). Wind energy–wildlife interactions also have been addressed by the National Wind Coordinating Collaborative (Strickland et al. 2011) and the National Wind and Wildlife Research Plan (American Wind Wildlife Institute 2017). Professional development in the form of on-the-job training may occur, but such training is often biased regionally or by the organizations providing it. We strongly recommend that the Wildlife Society and other similar groups invest in professional development opportunities centered on renewable energy and wildlife conservation. The Wildlife Society supports a Renewable Energy Working Group; we encourage readers to join and energize this important working group.

Conclusion

This book serves as a collection of the current state of knowledge for renewable energy and wildlife conservation and provides a platform for future wildlife research in the burgeoning field of renewable energy ecology. Although wildlife professionals have made progress toward better understanding and mitigating effects of renewable energy development on wildlife, we have only uncovered the tip of the iceberg! We are presented with great opportunities to further develop our knowledge of renewable energy ecology during a period of unprecedented renewable energy development in North America and beyond. We hope this book serves not only as a reference, but also as a creative springboard for building renewable energy ecology to conserve wildlife into the future.

LITERATURE CITED

American Wind Wildlife Institute. 2017. National Wind Wildlife Research Plan. Washington, DC. www.awwi.org.

Arnett, E. B., D. B. Inkley, D. H. Johnson, R. P. Larkin, S. Manes, A. M. Manville, J. R. Mason, M. L. Morrison, M. D. Strickland, and R. Thresher. 2007. Impacts of wind energy facilities on wildlife and wildlife habitat. Wildlife Society Technical Review 07-2. Bethesda, MD: Wildlife Society.

Bakshi, B. R., G. Ziv, and M. D. Lepech. 2015. Techno-ecological synergy: A framework for sustainable engineering. Environmental Science and Technology 49:1752–1760.

Eigenbrod, F., S. J. Hecnar, and L. Fahrig. 2008. Accessible habitat: An improved measure of the effects of habitat loss and roads on wildlife populations. Landscape Ecology 23:159–168.

Fahrig, L. 1997. Relative effects of habitat loss and fragmentation on population extinction. Journal of Wildlife Management 61:603–610.

Fritts, S. R., S. M. Grodsky, D. W. Hazel, J. A. Homyack, S. B. Castleberry, and C. E. Moorman. 2015. Quantifying multi-scale habitat use of woody biomass by southern toads. Forest Ecology and Management 346:81–88.

Garvin, J. C., C. S. Jennelle, D. Drake, and S. M. Grodsky. 2011. Response of raptors to a windfarm. Journal of Applied Ecology 48:199–209.

Grodsky, S. M., R. B. Iglay, C. E. Sorenson, and C. E. Moorman. 2015. Should invertebrates receive greater inclusion in wildlife research journals? Journal of Wildlife Management 79:529–536.

Grodsky, S. M., K. A. Moore-O'Leary, and R. R. Hernandez. 2017. From butterflies to bighorns: Multi-dimensional species-species and species-process interactions may inform sustainable solar energy development in desert ecosystems. Proceedings of the 31st Annual Desert Symposium, April 14–15, edited by R. L. Reynolds. Zzyzx: California State University Desert Studies Center.

Grodsky, S. M., C. E. Moorman, S. R. Fritts, J. W. Campbell, M. A. Bertone, C. E. Sorenson, S. B. Castleberry, and T. B. Wigley. 2018a. Invertebrate community response to coarse woody debris removal for bioenergy production from intensively managed forests. Ecological Applications 28:135–148.

Grodsky, S. M., J. W. Campbell, S. R. Fritts, T. B. Wigley, and C. E. Moorman. 2018b. Variable responses of non-native and native ants to coarse woody debris removal following forest bioenergy harvests. Forest Ecology and Management 427:414–422.

Hernandez R. R., M. K. Hoffacker, M. Murphy-Mariscal, G. Wu, and M. F. Allen. 2015. Solar energy development impacts on terrestrial ecosystems. Proceedings of the National Academy of Sciences 112:13579–13584.

Hernandez R. R., A. Armstrong, J. Burney, K. Moore-O'Leary, G. Ryan, S. M. Grodsky, L. Saul-Gershenz, et al. Forthcoming. Techno-ecological synergies of solar energy produce beneficial outcomes across industrial-ecological boundaries to mitigate global environmental change. Nature Sustainability.

Hoffacker M. K., M. F. Allen, and R. R. Hernandez. 2017. Land sparing opportunities for solar energy development in agricultural landscapes: A case study of the Great Central Valley, CA, USA. Environmental Science and Technology 51:14472–14482.

Lambin, E. F., and P. Meyfroidt. 2011. Global land use change, economic globalization, and the looming land scarcity. Proceedings of the National Academy of Sciences 108:3465–3472.

Lovering J., M. Swain, L. Blomqvist, R. R. Hernandez, and T. Nordhaus. Forthcoming. Land-use intensity of electricity of production and tomorrow's energy landscape.

Moore-O'Leary, K., R. R. Hernandez, D. S. Johnston, S. R. Abella, K. E. Tanner, A. C. Swanson, J. Kreitler, and J. E. Lovich. 2017. Sustainability of utility-scale solar energy: Critical ecological concepts. Frontiers in Ecology and the Environment 15:385–394. doi: 10.1002/fee.1517.

Murphy-Mariscal, M., S. M. Grodsky, and R. R. Hernandez. 2018. Solar energy development and the biosphere. In A Comprehensive Guide to Solar Energy Systems, edited by T. Letcher and V. Fthenakis, 387–401. London: Academic Press / Elsevier.

Rupp, S. P., L. Bies, A. Glaser, C. Kowaleski, T. McCoy, T. Rentz, S. Riffell, J. Sibbing, J. Verschuyl, and T. Wigley. 2012. Effects of bioenergy production on wildlife and wildlife habitat. Wildlife Society Technical Review 12-03. Bethesda, MD: Wildlife Society.

Strickland, M. D., E. B. Arnett, W. P. Erickson, D. H. Johnson, G. D. Johnson, M. L. Morrison, J. A. Shaffer, and W. Warren-Hicks. 2011. Comprehensive guide to studying wind energy/wildlife interactions. Prepared for the National Wind Coordinating Collaborative, Washington, DC.

Walker, B. L., D. E. Naugle, and K. E. Doherty. 2007. Greater sage grouse population response to energy development and habitat loss. Journal of Wildlife Management 71:2644–2654.

Index